MECHANISM AND REGULATION OF DNA REPLICATION

Edited by
Alan R. Kolber
Virus Research Institute
German Cancer Research Center
Heidelberg, Germany

and
Masamichi Kohiyama
Institute of Molecular Biology
University of Paris
Paris, France

PLENUM PRESS • NEW YORK AND LONDON

Library of Congress Cataloging in Publication Data

Main entry under title:

Mechanism and regulation of DNA replication.

"Proceedings of two recent NATO Advanced Study Institutes on the mechanism and regulation of DNA replication."
Includes bibliographies.
1. Deoxyribonucleic acid synthesis—Congresses. 2. Chromosome replication—Congresses. I. Kolber, Alan R., ed. II. Kohiyama, Masamichi, ed. III. Nato Advanced Study Institute. [DNLM: 1. DNA replication—Congresses. QU58 M485]
QP624.M42   574.8'732   74-14571
ISBN 0-306-30818-5

QP
624
.M 42

Proceedings of two recent NATO Advanced Study Institutes on the mechanism and regulation of DNA replication

© 1974 Plenum Press, New York
A Division of Plenum Publishing Corporation
227 West 17th Street, New York, N.Y. 10011

United Kingdom edition published by Plenum Press, London
A Division of Plenum Publishing Company, Ltd.
4a Lower John Street, London W1R 3PD, England

Printed in the United States of America

PREFACE AND ACKNOWLEDGMENTS

The content of this volume represents the proceedings of two recent NATO Advanced Study Institutes on the mechanism and regulation of DNA replication. We discussed, at these meetings, the similarities and differences between DNA replication in prokaryotic and eukaryotic cells, and attempted to discover whether or not the models for DNA replication, largely derived from the bacterial and phage systems could be applied to the higher organisms. Of interest also were the mammalian DNA viruses; could the integration-excision phenomenon of the lysogenic bacteriophages serve as a model for transformation of animal cells by DNA oncogenic viruses? Other sections of this volume are devoted to the relationship between DNA replication and cell division, and to the role of the cell membrane in DNA replication, the latter topic being a rather controversial subject which has not been satisfactorily resolved. The editors have enjoyed the tasks of organizing these Advanced Study Institutes and compiling the proceedings into this volume, and we hope the reader will find herein some stimulation for further research on these compelling subjects.

The editors are indebted to the North Atlantic Treaty Organization for their very generous financial support and to the managements of the Aspen Square, Aspen, Colorado, and the Cristallo Palace Hotel, Cortina d'Ampezzo, for their fine service to the participants. We especially wish to thank the following persons for their excellent editorial assistance: Jan McInroy and Cynthia Snow of the Professional Manuscript Service, Kathryn Hunter, Janet Kolber, Louis Marshall, Rhonda Mitchell, and Elizabeth Milewski. We acknowledge the kind assistance of Frauline Webler and Herr Scheid of the Deutsches Krebsforschungzentrum, Heidelberg, for figure reproductions, and the administration of this center for the provision of editorial equipment. Lastly we express the hope that the long suffering of our publisher's representatives, Seymour Weingarten and Phyllis Straw, was rewarded.

Chapel Hill, North Carolina   A. K.

Paris University VII   M. K.

CONTENTS

PART 1 - CHROMOSOME REPLICATION IN PROCARYOTES

Enzymatic Aspects of Chromosome Replication in E. coli

DNA Replication in Bacteriophage and Episomes

Chromosome Structure and Mode of Replication

MECHANISM
AND REGULATION
OF DNA REPLICATION

ESCHERICHIA COLI DNA POLYMERASES II AND III

Thomas Kornberg, Malcolm L. Gefter, and
Ian J. Molineux
Department of Biological Sciences, Columbia University
New York, New York
Department of Biology, Massachusetts Institute of
Technology, Cambridge, Massachusetts

When cells of an E. coli mutant (pol Al⁻)[1] which is deficient
in DNA polymerase I (pol I) are lysed in a French pressure cell,
they yield a cell-free extract with a small but demonstrable amount
of DNA polymerizing activity.[2] Following the removal of nucleic
acid from such an extract, phosphocellulose chromatography can
resolve two separate DNA polymerases.[3] Both of these DNA polymerases, called DNA polymerase (pol II) and DNA polymerase III (pol
III), have been shown not only to be physically, catalytically, and
genetically distinct from each other, but that they are similarly
distinct from Pol I.[4-6] It will be the subject of this talk to
review the work which led to these conclusions.

DNA POLYMERASE II

The resolution of polymerases II and III by phosphocellulose
chromatography is shown in Figure 1. The early eluting peak is
pol III; the activity emerging at higher salt concentration, Pol II.
Nuclease activity (dotted line) eluting from the column is active
with double stranded DNA only, and has been identified as Exonuclease III.

DNA polymerase II has been purified and characterized by several laboratories (2-7). The procedure we employ is described in
Table I. The enzyme is purified more than 2,000 fold relative to
the S100 crude extract. Analysis on polyacrylamide gels under native conditions reveals a single band of protein. We have shown
that polymerizing activity comigrates with this band. The enzyme
is free of detectable contaminating nuclease or phosphatase activity. An approximate molecular weight of 90,000 has been obtained

1

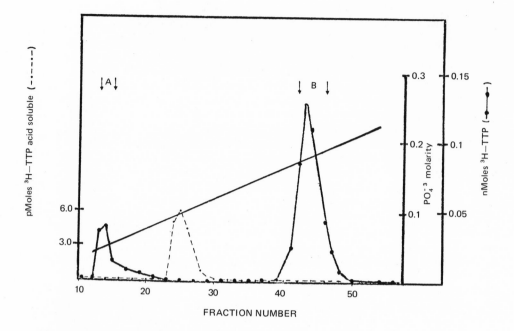

Fig. 1. Phosphocellulose chromatography of fraction III. Each fraction was assayed for DNA polymerase activity (————) and Exonuclease III activity (- - - - -). Fractions bounded by arrows were pooled and are referred to as Pol III and Pol II enzymes.

from density gradient sedimentation, using Pol I as a marker. From the extent of purification and the approximate molecular weight, we calculate that there are less than 100 molecules of Pol II per bacterial cell.[3]

Pol II requires the presence of both a primer and a template to synthesize DNA.[4,5] It requires deoxynucleoside-5'-triphosphates and cannot utilize other phosphorylated nucleosides. Addition of ATP is without effect. The enzyme is active only in the presence of Mg++ ion and is abolished by the presence of the sulfhydryl reagent, N-ethylmaleimide, by 0.2 M KCl, or by DNAase. Pol II activity is insensitive to anit-DNA polymerase I antiserum.[4]

Studies of the template requirements of Pol II show that the enzyme cannot utilize fully helical DNA, nicked DNA, or fully denatured DNA efficiently as template. "Gapped" DNA - DNA pretreated briefly with exonuclease III - is the most active template studied to date. Studies employing the synthetic polymer poly (dA), poly

(dC) and the copolymer d(TG)$_{400}$ show these polymers to be inactive
as templates unless the appropriate complementary oligonucleotide
is provided. Similarly, the single-stranded circular DNA from the
phages ϕX174 and M13 will not support synthesis by Pol II. However,
the addition of a short primer oligonucleotide, either RNA or DNA,
will render these molecules active as template. We conclude from
these studies that DNA polymerase is unable to initiate synthesis
de novo, but requires the presence of a primer.[4]

The direction of synthesis and the role of the primer in
polymer-directed synthesis by Pol II was assessed by the use of the
polymer poly(dA) and the acid soluble oligomer pT_{10}.[4] Using poly
(dA), [^{3}H]-dTTP, and pT_{10} labelled at the 5' end with ^{32}P incubation
with Pol II renders both ^{3}H and ^{32}P acid insoluble, indicating that
the acid insoluble product is covalently linked to the primer. The
product of the reaction was also subjected to filtration at 65°C
on G-50 Sephadex. Figure 2A shows that ^{32}P has increased in size
relative to the elution volume of pT_{10} determined prior to application of the sample, and that it runs coincident with some of the ^{3}H.
Treatment of the reaction product with alkaline phosphatase prior
to application of the column revelas that all of the ^{32}P remains
susceptible to the phosphomonoesterase (Figure 2B). We conclude
therefore that all of the TMP has been added onto the 3' end of
the primer and the direction of DNA polymerase II catalyzed synthesis is exclusively 5' to 3'.

Pol II can catalyze the exonuclease degradation of single-
stranded DNA.[4,6] The nuclease activity, like that of Pol I and the
T4-induced DNA polymerase, removes only mis-matched bases - nucleotides which cannot base-pair properly - from the termini of helical,
double-stranded DNA.[8] Because both the nuclease and polymerase
activity co-chromatograph on phosphocellulose and also show identical heat denaturation kinetics, we conclude that both the polymerase
and nuclease activities are part of the same molecule.

To determine the role of Pol II in DNA metabolism, pol II was
purified from E. coli mutants harboring DNA related lesions.[9] In
collaboration with Dr. Y. Hirota, mutants thermosensitive for DNA
replication in vivo, as well as mutants deficient in recombination
were examined. All mutants examined were found to yield normal
Pol II. The failure to associate Pol II activity with any dna
or rec locus does not prove that this enzyme is not involved in
either DNA replication or recombination. Isolation of a mutant defective in Pol II is required to resolve this question. A search
for such a mutant was undertaken, and one mutant defective in Pol II
was isolated. The mutant has less than 0.1% of the amount of Pol II
activity present in wild type cells. That in addition this polymerase activity is thermolabile indicates that the lesion is in the
structural gene for Pol II.[10] Dr. Hirota has described the phenotype of this mutant.

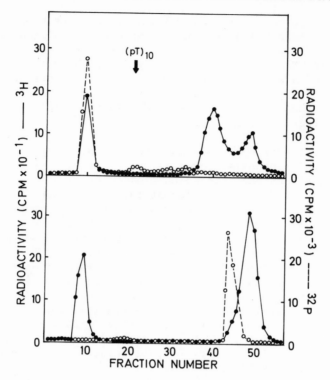

Fig. 2. Sephadex filtration of poly (dA) directed product. The details of the reaction are given in reference 4.11. After incubation, the reaction mixture was heated to 100°C for 3 minutes. One-half was then incubated with alkaline phosphatase and the other half remained unheated. The top panel (2A) represents the results obtained with untreated samples and the bottom panel (2B) the phosphatase-treated sample. A portion of each fraction was analyzed for ^{3}H (———) and ^{32}P (- - -). The arrow marks the elution position of [^{32}P]pT$_{10}$ determined on each column prior to application of the experimental samples.

DNA POLYMERASE III

Pol III, the polymerizing activity which elutes earlier from phosphocellulose, has also been purified free of contaminating nuclease and phosphatase activities, and its basic catalytic functions characterized.[11] The purification procedure is detailed in Table 2. The enzyme is purified more than 10,000 fold relative to the crude extract; polyacrylamide gel analysis indicates a maximum purity of 10%. Gel filtration and sedimentation studies indicate an approximate molecular weight of 155,000. With the degree of purification, the relative yield, a knowledge of the molecular weight allows a rough calculation of the number of molecules of

Pol III per E. coli cell, as well as the rate of chain elongation
per Pol III molecule. We calculate that approximately 10 moles of
Pol III are present per bacterial cell. Under optimal conditions,
the rate of nucleotides addition is about 10 to 20% of the in vivo
rate of replication.

The basic catalytic properties of DNA polymerase III are shown
in Table III. The enzyme requires the presence of a DNA template,
of all four deoxynucleosides triphosphates, and of magnesium ion.
The enzyme activity is sensitive to the presence of N-ethylmaleimide
and to low concentrations of salt, but is not inhibited by antiserum directed against DNA polymerase I. The titration of ethanol
and salt for the three E. coli DNA polymerases is shown in Figure 3.
DNA polymerase I (open triangles) is relatively unaffected by the
presence of ethanol or salt. Pol II (solid line) shows a two-fold
stimulation of rate at low concentrations of salt and inhibition
at higher concentrations. Pol III (heavy line) shows a unique
response both to ethanol and ionic strength. While the addition of
salt to the reaction mixture is inhibiting, the rate of Pol III
catalyzed synthesis is stimulated two-fold in the presence of 10%
ethanol. Pol III can catalyze the pyrophosphorolysis of DNA as well
as a pyrophosphate exchange reaction with deoxynucleonide triphosphates in the presence of DNA. Both reactions proceed at less
than 1% the rate of polymerization.

The effect of temperature on DNA polymerase III catalyzed
synthesis is shown in Figure 4. The insert shows that the initial
rate of synthesis by DNA polymerase III increases with temperature
over the range we have tested. However, if the extent of the
reaction is followed, we find that DNA polymerase III is labile in
the reaction mixture at 37°. Within 10 minutes, the enzyme no
longer synthesizes DNA. Addition of BSA, glycerol or other possible
stabilizing agents is not successful in rendering DNA polymerase III
stable for long incubation. However, at lower temperature (23°C),
catalysis by DNA polymerase III is linear for more than two hours.
In contrast, both DNA polymerases I and II are active at 37° for
many hours.

Primer requirement and direction of synthesis similar to those
conducted with DNA polymerase II were conducted for Pol III. We
conclude from these experiments that, like DNA polymerases I and
II and the T4-induced DNA polymerase, Pol III requires the presence
of a primer to initiate synthesis on single-stranded templates, and
polymerization proceeds in the 5' to 3' direction by covalent
extension of the primer. Like Pol II, Pol III is maximally active
with a "gapped" DNA template, fully helical, nicked, or denatured
DNA being relatively inactive.

Highly purified preparations of Pol III catalyze the exonucleolytic degradation of single-stranded DNA in the 5' direction. The

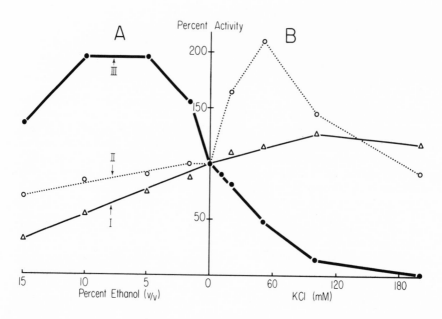

Fig. 3.   A. effect of ethanol on reaction rate.   Ethanol was added
to standard reaction mixture, unsupplemented with ethanol and containing 0.4 unit of Pol I ($\triangle$-$\triangle$), 0.04 unit of Pol II (0 – – – 0),
or 0.1 unit of Pol III (0———0).   Insulations were performed as
described in reference 11.   B. effect of KCl on reaction rate.
KCl was added to standard reaction mixtures, unsupplemented with
ethanol, containing Pol I, Pol II, and Pol III as above.

nucleolytic activity, inactive with double stranded DNA, is capable
of digesting a mis-base paired 5' nucleotide from an otherwise
duplex structure before commencing polymerization.   This indicates
that like Pol I, Pol II, and T4 DNA polymerase, Pol III has the
capacity to edit its synthetic product.   That the nuclease and
polymerase activities co-sediment in sucrose density gradient centrifugation and are both temperature-sensitive when isolated from
mutants harboring ts lesions in the Pol III gene, indicates that
both the nuclease and polymerase activities are part of the same
molecule.

     DNA Polymerase III was isolated from mutants thermosensitive
for DNA synthesis in an attempt to correlate the genetic lesions
with altered DNA polymerase activity in vitro.[9]   Pol III activity
appears to be normal in all strains carrying mutations at the dna
A, B, C, D, and G loci.   Of four independently isolated dna E
mutants, all had altered Pol III activity in vitro.   These results
suggest that the structural gene for Pol III is located at the
dna E locus.   Furthermore, since dna E mutants fail to replicate

TABLE 1: PURIFICATION OF DNA POLYMERASE II

	FRACTION	VOLUME ml	PROTEIN mg/ml	SPECIFIC ACTIVITY units/mg	TOTAL UNITS
I	S100	250.0	14.7	0.051	187
II	Phosphocellulose 1	500.0	6.5	0.062	200
III	DEAE cellulose	550.0	3.3	0.151	275
IV	Phosphocellulose 2	2.8	1.2	58.000	195
V	Sephadex G200*	46.0	0.019	133.000	115

*Fifty percent of the enzyme preparation was filtered through Sephadex G200.

The values in the Table are corrected to 100%.

Methods for preparation are described in reference 3.

TABLE 2: PROPERTIES OF DNA POLYMERASE II

Additions	pmoles Incorporated
* Complete System	25.5
" – DNA	<0.5
" – dATP, dGTP, dCTP	3.5
" – Mg^{++} + E.D.T.A. (3mM)	<0.5
" – 2-mercaptoethanol + N.E.M. (10mM)	4.0
" – DNA + "Activated" Calf Thymus DNA	3.1
" – DNA + Calf Thymus DNA	5.3
" – DNA + Denatured DNA	1.1
" – DNA + Native T7 DNA	<0.5
" + DNAase (75 µg/ml)	<0.5
" + KCL (0.2 M)	<0.5
" + Antiserum (20 µl) 5 min	17.5
" + Antiserum (20 µl) 0 min	18.0

*The components of the reaction mixture and the details of the assay are described in reference 3. Reaction mixtures containing antiserum had reduced values of incorporation due to counting error. Antiserum added after the incubation period (5 min) and just prior to TCA precipitation, served as a control. NEM is N-ethylmaleimide.

TABLE 3:   PURIFICATION OF DNA POLYMERASE III

	Fraction	Units	Protein mg/ml	Specific Activity units/mg protein	Yield %
I	S100*	5,440	11.0	1.16	100
II	DEAE cellulose I	6,048	3.4	4.2	110
III	Ammonium Sulfate	3,080	10.0	14.4	57
IV	DEAE-cellulose II	1,320	0.05	120.	24
V	Phosphocellulose	220	<0.01	12,000.	4

*Methods for preparation are described in reference 11.

TABLE 4: PROPERTIES OF DNA POLYMERASE III

Reactants	Incorporation (p moles)
Complete System*	56
+ ethanol (10% v/v)	112
- DNA	<2
- dATP, dGTP, dCTP	5.6
- Mg++, + EDTA (3mM)	<2
- 2-Mercaptoethanol, + N-ethylmaleimide (10 mm)	<2
+ Kce (0.15 m)	<2
+ DNAase (15 µg/ml)	<2

*The reaction mixture and the details of the assay are described in reference 11.

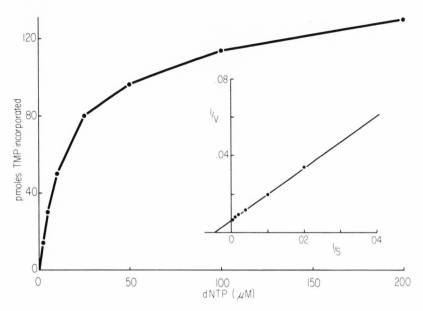

Fig. 4. Kinetic analysis of enzyme activity. Reaction mixture as
described in reference 11 was incubated and aliquots were withdrawn
at the indicated times, precipitated with 5% TCA and radioactivity
determined. Insert Dependence of the initial rate on temperature.
Pol III was incubated for 5 minutes at the indicated temperature.

their DNA at 42°C in vivo and contain a thermolabile Pol III in
vitro, we conclude that this enzyme if required for DNA replication
in E. coli.

REFERENCES

De Lucia, P. and Cairns, J. Nature. 224: 1164 (1969).
Kornberg, T., and Gefter, M. L. Biochem. Biophys. Res. Comm. 40:
    1348 (1970).
Kornberg, T. and Gefter, M. L. Proc. Nat. Acad. Sci. 68: 761
    (1971).
Gefter, M. L., Molineux, I. J., Kornberg, T., and Khorana, H. G.
    J. Biol. Che. 247: 3321 (1972).
Moses, R. E. and Richardson, C. C. Biochem. Biophys. Res. Comm.
    41: 1565 (1970)
Wickner, L. B., Ginsberg, B., Berkower, I., and Hurwitz, J. J. Biol.
    Che. 247: 489 (1972).
Knippers, R. and Stratling, W. Nature. 226: 713 (1970).
Brutlag, D. and Kornberg, A. J. Biol. Chem. 247: 232 (1972).

Gefter, M. L., Hirota, Y., Kornberg, T., Wechsler, J. A., and
    Barnoux, C. Proc. Nat. Acad. Sci. 68: 3150 (1971).
Hirota, Y., Gefter, M. L., and Mindich, L. Proc. Nat. Acad. Sci.
    69: 3238 (1972).
Kornberg, T. and Gefter, M. L. J. Biol. Chem. 247: 5369 (1972).

INITIATION OF DNA SYNTHESIS

Arthur Kornberg

Department of Biochemistry, Stanford University Medical

School, Stanford, California

Basic features of polymerase action (1)

All the polymerases studied have two basic features: (i) a
phosphodiester bridge is formed between the 5'-phosphate group of
a deoxyribonucleotide and the 3'-hydroxyl group at the growing end
of a DNA chain called _primer_ DNA; the chain growth is 5'——> 3';
and (ii) each deoxynucleotide added to the chain is selected by
base-pair matching to a DNA chain called _template_ DNA.

Hypotheses for DNA chain initiation (2)

What DNA polymerases cannot do is initiate chains. With single
stranded circles of M13 and φX174 phage DNA as templates for _E. coli_
DNA polymerase, it appeared that chain initiation did take place
(3). Subsequently however, the presence and participation of small,
linear DNA fragments as primers to initiate synthesis was recognized (4).

How then is new DNA synthesis initiated? One possibility is
that all DNA synthesis takes place by covalent extension of pre-
existing DNA chains. Thus, oligonucleotide fragments of a DNA
chain or ends produced by endonucleolytic scissions of a chain might
serve as primers. Experience with available enzymes favors this
scheme. An alternative to this possibility is that a new enzyme
initiates chains by itself, or in conjunction with one of the known
DNA polymerases. Studies with intact cells favor this suggestion,
but no enzyme has yet been found to do this job.

Another alternative occurred to us. Since RNA polymerase
starts new RNA chains, and DNA polymerase is known to covalently
extend a ribonucleotide terminus during DNA synthesis, a brief
transcriptional operation by RNA polymerase might provide an RNA

primer for DNA synthesis. This priming piece of RNA could later
be recognized and excised by nuclease action. Thus, a de novo
initiation event catalyzed by RNA polymerase would be an essential
step for the start of DNA synthesis. The synthesis of the double-
stranded replicative forms of M13 DNA provided an excellent system
in which to test this possibility. The first step in M13 DNA synthesis involves the conversion of the phage single-stranded DNA
(SS) to a double-stranded replicative form (RF). This step relies
on host-cell enzymes; it does not require the expression of any
known viral gene or any new protein synthesis. The next stage, in
which the replicative forms multiply (RF ——⟶ RF), requires, among
other things, the product of a viral gene.

A role for RNA polymerase in initiation of M13 DNA synthesis, in vivo (2)

The conversion of single-stranded DNA of bacteriophage M13
to the double-stranded replicative form in Escherichia coli is
blocked by rifampicin, an antibiotic that specifically inhibits
the host-cell RNA polymerase. Chloramphenicol, an inhibitor of
protein synthesis, does not block this conversion. The next stage
in phage DNA replication, multiplication of the double-stranded
forms, is also inhibited by rifampicin; chloramphenicol, although
inhibitory, has a much smaller effect. An E. coli mutant whose
RNA polymerase is resistant to rifampicin action does not show inhibition of M13 DNA replication by rifampicin. These findings
indicate that a specific rifampicin-RNA polymerase interaction is
responsible for blocking new DNA synthesis. It thus seems plausible
that RNA polymerase has some direct role in the initiation of DNA
replication, perhaps by forming a primer RNA that serves for covalent attachment of the deoxyribonucleotide that starts the new
DNA chain.

RNA synthesis initiates in vitro conversion of M13 DNA to its replicative form (5)

The observations that RNA polymerase action is required in
the conversion of M13 single strands to RF were strengthened and
extended by studies with a soluble enzyme preparation from uninfected cells (Fig. 1).

Soluble enzymes catalyze the formation of full-length linear
strands on an M13 circular template. The production of this
replicative form (RF II) can be separated into two stages: an
initial RNA synthesis followed by DNA synthesis. The first requires the four ribonucleoside triphosphates and is inhibited by
rifampicin. A macromolecular product separated after this stage
supports the synthesis of the complementary DNA strand in the
absence of ribonucleoside triphosphates and in the presence of
rifampicin.

We presume that a short segment of RNA synthesized in the

Uninfected E. coli

   1. Freeze in liquid N_2, thaw

   2. Lysozyme

   3. Lyse by brief warming

   4. High-speed centrifugation

Soluble extract

FIGURE 1

Preparation of Soluble Extract from Uninfected E. coli

first stage of RF formation serves as a primer for the extensive
DNA synthesis in the second. However, our attempts to determine
the size and composition of this initiating RNA segment have been
frustrated by the synthesis of nonspecific RNA by these relatively
crude enzyme fractions. Unlike the DNA synthesis, this RNA synthesis
(70 pmol/mg protein) is independent of the presence of M13 DNA; the
RNA resists pancreatic RNase (25% persists) and contaminates the
RF II product in density gradient separation. Further purification
of the enzymes involved is required to clarify this problem.

Even though the RNA priming segment has not been identified,
its existence is supported by evidence in addition to that already
cited or reported here. The isolated RF II contains a $5' \longrightarrow 3'$
phosphodiester link of a deoxyribonucleotide to a ribonucleotide.
Alkaline hydrolysis of the RF II synthesized in the presence of
four (α-^{32}P) deoxyribonucleoside triphosphates, yielded ^{32}P-labeled
ribonucleotide (1.47 mol/mol RF II) of which 84% was a mixture of

2'- and 3'rAMP:

$$
\begin{array}{cccc}
X & A & Y & Z \\
-OH & -OH & -H & -H
\end{array}
$$

- - - - P   P*   P*   P*- - - -

alkali cleavages

Another line of evidence suggesting the role of an RNA primer is the presence at the 5' end of the synthesized DNA of a ribonucleotide structure. These studies will be presented below.

Synthesis of φX174 replicative form requires RNA synthesis resistant to rifampicin (6)

Replication of φX174, a virus of base composition similar to that of M13, is not inhibited by rifampicin in vivo or in vitro. Does this absence of rifampicin inhibition imply a different mechanism of DNA-strand initiation or is there an RNA synthetic system that, unlike RNA polymerase, is resistant to rifampicin?

We have found that conversion of single-stranded DNA of phage φX174 to the double-stranded replicative form in Escherichia coli uses enzymes essential for initiation and replication of the host chromosome. These enzymes can now be purified by the assay that this phage system provides. The φX174 conversion is distinct from that of M13. The reaction requires different host enzymes and is resistant to rifampicin and streptolydigin, inhibitors of RNA polymerase. However, RNA synthesis is essential for φX174 DNA synthesis: the reaction is inhibited by low concentration of actinomycin D, all four ribonucleoside triphosphates are required, and an average of one phosphodiester bond links DNA to RNA in the isolated double-stranded circles. Thus, we presume that, as in the case of M13, synthesis of a short RNA chain primes the synthesis of a replicative form by DNA polymerase. Initiation of DNA synthesis by RNA priming thus appears to be a mechanism of wide significance.

Incorporation of the RNA primer into the phage RF (7)

In the conversion of M13 and φX174 single-stranded DNAs to the duplex RF, the product has a discontinuity in the synthetic strand. We observed that sealing of this RF II product to the covalently closed, alkali-stable duplex through the joint action of E. coli DNA polymerase I and E. coli ligase (Fig. 2). Failure of T4 DNA polymerase to substitute for the E. coli enzyme implied the presence of an RNA priming segment at the 5' end which the unique 5'——> 3' exonuclease function of the E. coli enzyme can excise. This suggestion was confirmed by conversion of the RF II product to an alkali-labile RF I through the joint action of T4 DNA polymerase and T4 ligase, an enzyme which can join RNA ends, and by location of the alkali-labile linkage of the RF I in the complementary (synthetic) strand. These results extend earlier evidence for RNA priming of DNA replication. They also call attention to a

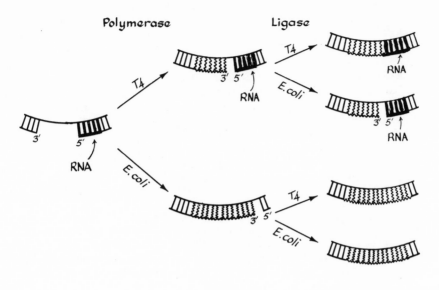

FIGURE 2

Scheme illustrating joint action of DNA polymerases and
ligases in sealing synthetic φX and M13 RF II to RF I

possible physiological role of <u>E. coli</u> DNA polymerase I in the
excision of such RNA primers during DNA replication.

   With the recognition that RNA fragments prime DNA synthesis
it is important to consider mechanisms for excising such fragments.
A prominent candidate for such a role is, of course, DNA polymerase
I with its built-in 5'——> 3' exonuclease activity. The enzyme
has been demonstrated in model experiments to be effective in removing the RNA priming fragments synthesized by RNA polymerase on
M13 viral DNA. Furthermore, when the cell extract was prepared
from wild type <u>E. coli</u> rather than from the <u>pol A</u> mutant, 50% of
the product was in the form of an alkali-stable RF I. This result
suggests that the lack of polymerase I in the extract from mutant
cells is responsible for the accumulation of RF II with an RNA
fragment at the 5' end. Finally, <u>in vivo</u> results also indicate
an essential function for polymerase I that may exploit its capacity
for coordinated 5'—— 3' excision and gap filling. This function
could account for the excision of the RNA fragments now known to
initiate the nascent 10 S DNA pieces in discontinuous replication
and for filling the gaps that this excision creates. <u>Pol A</u> mutants,

deficient in polymerase I, accumulate abnormal amounts of 10 S DNA and gaps. The fact that such mutants previously regarded as lacking polymerase I remain viable may be attributable to residual levels of the enzyme activity.

There are other possibilities for removal of part or all of the priming RNA. The triphosphate might be degraded by the 5'-triphosphatase that acts on the origins of RNA transcription. The RNA itself might be removed by nucleases such as RNase III and RNase H which degrade the RNA of RNA-DNA duplex hybrids. A precise characterization of the full size and composition of the priming fragment as well as the mechanisms for excising it must await further purification of the enzymes responsible for the single strand conversion to RF.

Comparison of the M13 and φX174 replication system

As shown in Fig. 3 both the M13 and φX174 appear to depend on RNA priming for initiation of DNA synthesis. However, the two systems are distinct from one another as judged by response to inhibitors and requirements for different proteins.

		φX	M13
RNA Priming			
1.	rNTP required before DNA chain growth	+	+
2.	Isolation of a "primed SS"	+	+
3.	One RNA-DNA covalent link per RF	+	+
Responsible Enzymes			
1.	Sensitive to rifampicin, Ab to RNAP, streptolydigin	−	+
2.	Sensitive to AF/ABDP in extracts of rif^R cells	+	−
3.	Requires dnaA, B, C-D and G gene products	+	−
4.	Requires dnaE gene product	+	+

FIGURE 3

φX and M13 SS——— RF in Soluble Extracts

Purification of Replication Enzymes Catalyzing φX SS⟶ RF

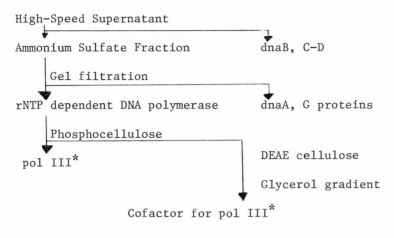

FIGURE 4

Purification of the enzyme components of the M13 and φX174 replicative assemblies

Conversion of M13 SS⟶ RF involves: RNA polymerase, a form of DNA polymerase III, called pol III* (dna E gene product), and spermidine or a DNA-unwinding protein. The same form of pol III* is also required for φX SS⟶ RF. In addition, other enzymes utilized in replication of the coli chromosome are required for φX SS⟶ RF in vitro. These include the products of dna genes A, B, C-D and G. The requirement for these enzymes for φX and M13 SS⟶ RF has provided assays for their purification. We have developed one scheme for such a purification (Fig. 4). The availability of a reaction with defined primer-template and product which will require these purified enzymes should permit analysis of the role of each in catalysis.

REFERENCES

1. Kornberg, A. Active Center of DNA Polymerase, Science 163: 1410 (1969).

2. Brutlag, D., Schekman, R. and Kornberg, A. A Possible Role for RNA Polymerase in the Initiation of M13 DNA Synthesis, Proc. Nat. Acad. Sci. USA 68: 2826 (1971).

3. Mitra, S., Reichard, P., Inman, R. B., Bertsch, L. L. and Kornberg, A. Enzymatic Synthesis of Deoxyribonucleic Acid. XXII. Replication of a Circular Single-stranded DNA Template by DNA Polymerase of E. coli. J. Mol. Biol. 24: 429 (1967).

4. Goulian, M.   Initiation of DNA Chains.   Cold Spring Harbor
    Symposium 33: 11 (1968).

5. Wickner, W., Brutlag, D., Schekman, R. and Kornberg, A.   RNA
    Synthesis Initiates In Vitro Conversion of M13 DNA to its
    Replicative Form.   Proc. Nat. Acad. Sci. USA 69: 965 (1972).

6. Schekman, R., Wickner, W., Westergaard, O., Brutlag, D.,
    Geider, K., Bertsch, L. L. and Kornberg, A.   Initiation of
    DNA Synthesis: Synthesis of φX174 Replicative Form Requires
    RNA Synthesis Resistant to Rifampicin.   Proc. Nat. Acad. Sci.
    USA 69: 2691 (1972).

7. Westergaard, O., Brutlag, D. and Kornberg, A.   Initiation of
    Deoxyribonucleic Acid Synthesis.   IV.   Incorporation of the
    Ribonucleic Acid Primer into the Phage Replicative Form.
    J. Biol. Chem. 248: 1361 (1973).

IN VITRO REPLICATION OF DNA

F. Bonhoeffer

Max Planck Institut für Virusforschung

Tübingen, Germany

In the lecture on in vitro replication of DNA work with the cellophan disc system has been discussed. Most of this work has been published elsewhere (1-9), therefore only a brief summary of the major results will be given here.

The system consists of a crude bacterial lysate containing all macromolecular components of the cell at very high concentrations, which are probably similar to the in vitro concentrations. DNA synthesis in this system is different from repair synthesis and has almost all properties of in vivo replication which have been tested. One of the most critical tests concerns the temperature sensitivity of lysates prepared from mutants which are temperature sensitive with respect to DNA replication. The isolation and characterisation of such mutants has been described (10). Another observation arguing strongly for replication and against repair synthesis is the finding of DNA intermediated like Okazaki pieces (11), which during the process of replication become joined to each other when polynucleotide ligase action is allowed (presence of NAD). Whereas the synthesis of DNA in one strand occurs always via synthesis of Okazaki pieces, the other strand is synthesized in pieces whose size depends very much on experimental conditions. This observation is explained by a model for discontinuous replication, which assumes that the propagation of a growing DNA strand can compete against the initiation of new DNA strands (3).

The cellophan disc system prepared from temperature sensitive replication mutants can be complemented by wild type extracts to yield temperature resistant synthesis. With the use of such complementation tests as assay the dnaE and dnaG proteins which are able to complement dnaE and dnaG mutant lysates have been isolated.

The purification steps and the characterisation of their biochemical and physical properties have been described (2,7,8 and Nusslein provate communication). Whereas the function of the dnaE protein is known to be a polymerising activity the function of the dnaG protein is not yet clear. It seems to be involved in initiation of DNA intermediates (5).

Finally some in vitro experiments on initiation of λ DNA replication have been described (9). An active of RNA polymerase is required during initiation and the initiation process is insensitive against RNase. Reinitiation of λ DNA replication be observed in vitro.

REFERENCES

1. Schaller, H., Otto, B. Nublein, V., Juf, J. Herrmann, R. and Bonhoeffer, F. J. Mol. Biol. 63: 183, 1972.

2. Nublein, V., Otto, B., Bonhoeffer, F. and Schaller, H. Nature (New Biol.) 234: 285, 1971.

3. Olivera, B. M. and Bonhoeffer, F. Nature (New Biol.) 240: 233, 1972.

4. Herrmann, R., Huf, J. and Bonhoeffer, F. Nature (New Biol.) 240: 235, 1972.

5. Lark, K. G. Nature (New Biol.) 240: 237, 1972.

6. Olivera, B. M. and Bonhoeffer, F. PNAS 69: 25, 1972.

7. Otto, B., Bonhoeffer, F. and Schaller, H. Eur. J. Biochem. 34: 440, 1973.

8. Klein, A., Nublein, V., Otto, B. and Powling, A. In DNA Synthesis in vitro, R. D. Wells and R. B. Inman (ed), University Park Press, Baltimore, 1973.

9. Klein, A. and Powling, A. Nature (New Biol.) 239: 71, 1972.

10. Wechsler, J. A., Nublein, V., Otto, B., Klein, A., Bonhoeffer, F. Herrmann, R., Gloger, L. and Schaller, H. J. Bact. 113: 1381, 1973.

11. Okazaki, R., Okazaki, T., Sakabe, K., Sugimoto, K., Kainuma, R., Subino, A. and Iwatzuki, N. Cold Spring Harbor Symp. Quant. Biol. 33: 129, 1968.

THE ROLE OF ATP IN CHROMOSOME REPLICATION STUDIED IN TOLUENIZED ESCHERICHIA COLI

Patrick Forterre and Masamichi Kohiyama

Laboratoire des Biomembranes, Institut de Biologie

Moleculaire, Paris, France

Introduction: Moses et al. (1970) have found an absolute ATP requirement, besides the four dNTP, for DNA replication in a toluene treated polA$_1^-$ mutant. None of the known E. coli DNA polymerases are affected by ATP (Kornberg et al. 1970). Two major hypotheses can be considered to explain the role of ATP: ATP is a cofactor for some enzyme necessary to replication and/or it is a precursor for RNA synthesis involved in the priming of Okazaki pieces (Sugino et al. 1972).

In this report, we wish to present experimental results which describe some characteristics of this ATP effect and to discuss both hypotheses.

A - ATP Specificity

In agreement with Pisetsky et al. (1972), we have found that ATP stimulates DNA synthesis in toluene treated polA$^-$ mutants more efficiently than any other NTP throughout a wide concentration range. The mixture of the four NTP at equal concentration ratios stimulates DNA synthesis less efficiently than ATP alone at any given concentration (Fig. 1). This specificity is less pronounced when the incubation time is less than 10 minutes at 30° C (Fig. 2). The synthesis which occurs in the presence of GTP, UTP and CTP is called "non-specific DNA synthesis."

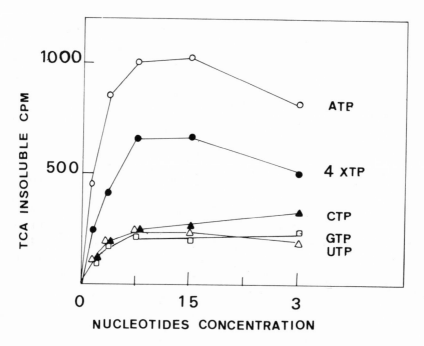

Figure 1. Influence of NTP Concentration on DNA Synthesis E. coli.
PA 3364 (polA⁻, thy⁻, endo I⁻), harvested at logarithmic phase,
were treated with toluene as previously described (Kohiyama et al.
1971). ^{3}H–TTP incorporation was tested in 0.15 ml of reaction
mixture containing 0.05 M Tris buffer pH 7.4, 0.005 M MgCl$_2$, 0.14 M
KCl, 0.001 M β-mercaptoethanol, 3.5 10^{-5} M dNTP, 1.5 µCi/ml of
^{3}H TTP and in this experiment, increasing concentration of NTP or
of the equal ratio mixture of the four NTP. The reaction was
started by addition of 25 µl toluenized bacteria (3.10^8 cells)
and was incubated for 30 minutes at 29° C. After addition of 0.5
ml 5% ice cold TCA, the samples were filtered through GFB Whatman
filter, rinsed with ice cold 5% TCA and ice cold EtOH dried
and placed in a vial containing 5 ml toluene POPOP. Radioactivity
was measured with a Nuclear Chicago liquid scintillation counter.

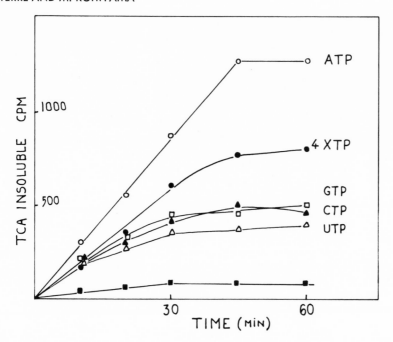

Figure 2: Kinetics of DNA Synthesis in Toluene Treated PA 3364.
The same procedure as cited in Figure 1 was followed. NTP concentration was 10^{-3} M.— 0 — ATP; — ● — 4XTP (ATP + GTP +
CTP + UTP);   □   GTP; ——— ▲ ———CTP; ——— △ — UTP; — ■ —
no XTP.

          The following observations support the assumption that the
DNA synthesis occurring in the presence of GTP, CTP and UTP for
short periods represents chromosome replication; first in a
toluenized dna G ts mutant (NY 73) (Gross, 1971) non-specific
DNA synthesis does not occur at non-permissive temperature nor
does ATP dependent DNA synthesis (Fig. 3) and secondly, nalidixic
acid, an inhibitor of in vivo DNA replication, also inhibits DNA
synthesis in vitro in the presence of CTP, GTP, UTP or ATP.

          The following hypothesis may be advanced to explain non-
specific DNA synthesis: first, the NTP protects the dNTP and/or
preexisting ATP against phosphatase degradation; second, they

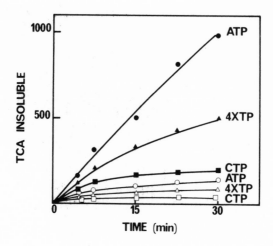

Figure 3: Kinetics of DNA Synthesis in Toluene Treated NY 73 (polA⁻ dnaG⁻) at Two Temperatures. NTP concentration was 10^{-3} M. Samples were divided in two groups. One was incubated at 29° C (closed symbols) and the other at 39° C (open symbols).

act as RNA precursors necessary for DNA synthesis (Sugino et al. 1972) and finally, they act as phosphate donor for ATP formation.

We think the latter hypothesis is the correct interpretation. In fact, we could eliminate the phosphatase hypothesis because GDP, UDP, CDP were unable to replace NTP. Also if the RNA precursor hypothesis is correct, we should observe more DNA synthesis in the presence of NTP and rifampicin since rifampicin does not block the RNA synthesis required for replication (Sugino et al. 1972) making available more RNA precursors for DNA synthesis. However the addition of rifampicin (6.6 mg/ml) does not affect the non-specific DNA synthesis.

We have been able to show that endogeneous ATP and/or ATP formed from NTP are responsible for non-specific DNA synthesis because, by adding various quantities of hexokinase and glucose we observed reduction of DNA synthesis in the presence of GTP, CTP, UTP or ATP (Table I).

TABLE I

Effect of Glucose and Hexokinase on DNA Synthesis

Glucose	Hexokinase	Nucleotides					
		0	ATP	GTP	CTP	UTP	4NTP
0	0	100%	100	100	100	100	100
5%	0.25 /ml	127	13.5	70	76	72	36
	5 /ml	98	7	40	34	45	18
0*	0*	100*	13*	44*	36*	39*	23*

NTP concentration is 10^{-3} M. Bacteria were incubated 30 minutes at 29° C with various concentrations of hexokinase and glucose.
*Activity minus indicated nucleotide.

Among the nucleotides tested, ADP replaces ATP with half efficiency. This phenomenon is due to the production of ATP by myokinase: AMP (promoting reverse reaction of myokinase) can inhibit DNA synthesis dependent on ADP (Fig. 4) and a temperature

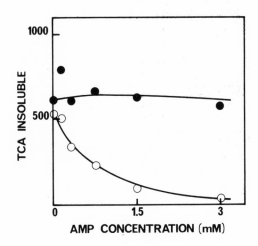

Figure 4: Effect of AMP Concentration on DNA Synthesis in Presence of ATP or ADP. ATP or ADP concentration was 10^{-3} M. The reaction was stopped after 30 minutes of incubation at 29° C.

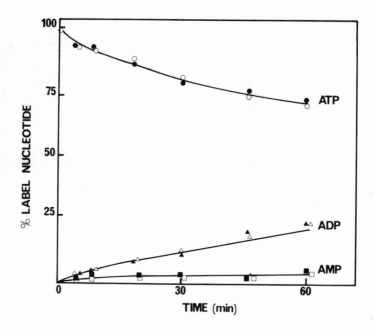

<u>Figure 5</u>: ATP Degradation in Toluene-Treated Cells. 0.1 ml of
toluene-treated bacteria are incubated with 0.6 ml of the reaction mixture with or without dXTP and 0.2 µc/ml of ^{14}C ATP at
29° C. At the time indicated 50 γ are removed, mixed with 50 γ
of 50% formic acid containing cold ATP, ADP, AMP and adenosine
2.10^{-2} M. The extract was deposited on Whatman paper n° 3 and
chromatographed with a 35% ethanolamine, 65% butyric acid solvent.
Nucleotide positions were determined by UV light, samples were
cut out and placed in a vial containing 10 ml of BRAY cocktail.
Radioactivity is determined and percentage of radioactivity in
each nucleotide calculated. Open symbols: +4 dNTP; closed symbols:
−4dNTP.

sensitive myokinase mutant <u>CR341 T28</u> (Cousin <u>et al</u>. 1969) can
replicate the chromosome in the presence of ADP at permissive
temperature but not at non-permissive temperature.

B – Metabolism of ATP

The possibility has been mentioned that ATP is transformed
into some metabolite which in turn is used for DNA synthesis (Smith
1970). We explored this possibility in two ways: first ATP degradation in toluenized bacteria was studied and second the effect of
metabolic inhibitors of ATP and ADP was examined.

Degradation of ATP in Toluenized Bacteria

In order to study ATP degradation in toluenized bacteria,
nucleotides were extracted by treating the mixture of toluenized
bacteria and ^{14}C ATP with formic acid and chromatographing, in
order to separate nucleoside mono-, di- and triphosphate. Thus,
we can see the kinetics of ATP degradation into the primary product, ADP (Fig. 5). This degradation is not stimulated by the
presence of the four dNTP, suggesting a certain autonomy between
ATPase action and DNA synthesis. Therefore, it is impossible
to measure the stoichiometry of DNA snythesis and ATP degradation.
However, because ATP is present in the reaction mixture even after
30 minutes, we can exclude a complete transformation of ATP into
some metabolite.

Effect of ATP and ADP Analogues

When added at the beginning of reaction, the ATP analogue
AOPOPCP (β-γ-methylene ATP) and the ADP analogue AOPCP (α-β-
methylene ADP) can inhibit DNA synthesis to various degrees
depending upon the ratio ATP/analogue (Fig. 6-A). When plotted
according to the Lineweaver-Burke formula, a group of straight
lines representing competitive inhibition are obtained (Fig. 6-B).
Ki for AOPCP and AOPOPCP is 2.5 (±0.5) x 10^{-4} M, equivalent to
the Km for ATP. Since no phosphate transfer from ^{32}P γ ATP to
AOPCP was found, the AOPCP inhibition does not result from the
formation of an ATP analogue.

Because of high concentration of analogue, complete inhibition of DNA synthesis is observed; we can determine whether an
ATP metabolite which is necessary for DNA replication accumulates.
As shown in Fig. 7, AOPOPCP inhibits DNA synthesis immediately
after addition; thus the possibility that an ATP derivative
accumulates is excluded.

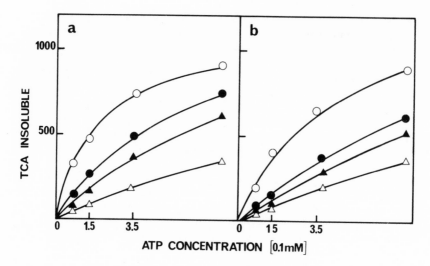

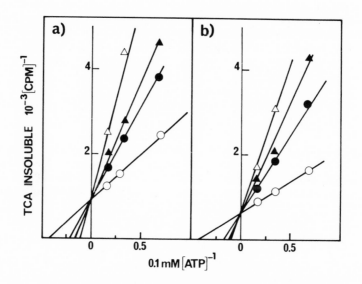

Figure 6: Competitive Inhibition with AOPOPCP and AOPCP.
    (A) The bacteria were incubated 30 minutes at 29° C with
increasing concentrations of ATP in presence of various amounts
of a) AOPCP; b) AOPOPCP. Analogue concentration is: zero   0 ;
3.5 x 10⁻⁴ M — ● — ; 7.5 x 10⁻⁴ M — ◐ — ; 1.5 x 10⁻³ M — ◓ — .
    (B) Lineweaver-Burke curves are plotted from the above
results. a) AOPCP; b) AOPOPCP. Michaelis constant and inhibition
constant are calculated from these data.

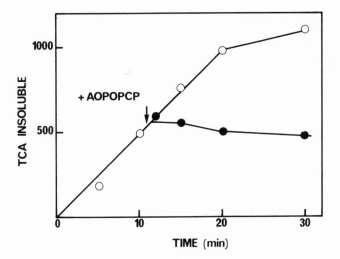

Figure 7:   Continuous requirement for ATP.   Toluenized bacteria
were incubated at 29° C for various time and divided in two groups.
In one of them 1.5 x 10⁻3 M of AOPOPCP is added after ten minutes
of reaction. ──0── - AOPOPCP; ── ● ── + AOPOPCP.

C – DNA Replication and RNA Synthesis
        Schekmann et al. (1972) have succeeded in replicating
single-strand DNA from phage φ χ 174, using a crude extract obtained from a polA⁻ mutant.   Their system is inhibited by actinomycin D, stimulated by spermidine and requires the four NTPs.
These observations are consistent with the idea of an RNA synthesis necessary for initiation of DNA replication.

In toluenized bacteria, ATP alone can support DNA synthesis. Likewise, spermidine (10^{-4} to 5.10^{-3} M) does not affect either DNA synthesis nor RNA synthesis in toluene treated bacteria. Actinomycin D at 2 γ/ml reduces RNA synthesis drastically while it affects DNA synthesis with less efficiency (Fig. 8). These results are in agreement with those found in vivo in a mutant bacteria permeable to actinomycin D (Sekiguchi et al. 1967).

Inhibition of DNA synthesis at high concentration of actinomycin D should be a pleiotropic effect, resulting from complex formation between the antibiotic and DNA. Artificially induced repair type DNA synthesis performed by polymerase I (Moses et al. 1970) is inhibited by increasing amount of actinomycin D with the same kinetics as ATP dependent DNA synthesis. These results indicate that DNA synthesis is unaffected by known RNA synthesis.

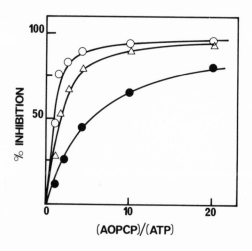

Figure 8: Effect of Actinomycin D. Toluene treatment and ^{3}H TMP incorporation are given in the legend to Fig. 1. The same method is used to measure RNA synthesis except that in the reaction mixture, dNTP and ^{3}H TTP were replaced by the four NTP 6×10^{-3} M and 1.5 µCi/ml ^{3}H GTP. Polymerase activity is measured in a G 34 strain (polA$^+$ dnaG$^-$). The reaction mixture for repair synthesis was the same as previously described except minus β-mercaptoethanol and plus N-ethyl maleimide 10^{-2} M and DNAse 10 γ/ml. Toluenized bacteria were incubated with increasing concentration of actinomycin D for 30 minutes at 29° C in the case of RNA synthesis and DNA replication and repair synthesis was measured at 39° C. — 0— RNA synthesis; —●— DNA synthesis; —△— repair synthesis.

The AOPCP inhibition pattern for DNA and RNA synthesis is the inverse of the antibiotic pattern: DNA synthesis is more sensitive than RNA synthesis to the AOPCP (Fig. 9).   Poly A synthesis is not inhibited by AOPCP.

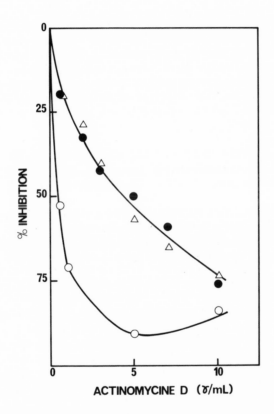

Figure 9:   Comparison of Inhibition of DNA Synthesis and RNA Synthesis by ADP Analogue in Toluene-Treated Bacteria.   DNA synthesis and RNA synthesis are measured from the same toluene treated cells.   Various amounts of AOPCP were added and the percentage inhibition was plotted against the ratio between AOPCP and ATP concentration: — 0— DNA synthesis in presence of the four NTP 10^{-3} M; — △— DNA synthesis in presence of ATP 10^{-3} M; — ●— RNA synthesis.

Discussion

The report of Sugino et al. (1972) raised the possibility that ATP is used to synthesize the RNA precursors which are incorporated in the short stretch of RNA covalently linked to the nascent DNA fragments. The fact that the mixture of four NTP is unable to fully replace ATP indicates that this cannot fully explain the ATP requirement.

The relative resistance of DNA synthesis to actinomycin D contradicts an involvement of RNA synthesis in DNA replication. However, it may be possible that initiation sites have few G-C bases.

One might also imagine that ATP is used for poly A synthesis necessary for production of Okasaki pieces. If this is the case, we may modify ATP specificity by substituting thymidine moiety with 5-bromouridine. However bacteria grown in presence of 5-bromouridine also require ATP with the same specificity for DNA synthesis, and moreover, AOPCP inhibits DNA synthesis without affecting poly A synthesis.

If ATP is effectively used for the production of RNA precursors, there will be a competition between total RNA synthesis and DNA replication for use of ATP. We could not find any modification of optimal ATP concentration for DNA synthesis in the presence of rifampicin. All these results indicate that ATP is most likely a specific cofactor for some unknown enzyme. In fact, a preliminary experiment on ATP degradation in a toluenized mutant lacking ATPase (Daniel et al., in preparation) showed that ATP to ADP degradation is stimulated in the presence of dNTP; this means that ATP is not incorporated into RNA but that ATP acts as a cofactor for DNA replication. Likewise, this facilitates understanding of the mechanism of inhibition by AOPCP of DNA synthesis.

The ATP enzyme should work with polymerase III, the presence of which is absolutely necessary for DNA replication (Gefter et al. 1972). Recently Milewski et al. (1973) isolated a protein complex containing polymerase III and which, when ATP is supplied, can synthesize DNA using a single strand DNA as template. This complex shows the same specificity for ATP as tolunized bacteria.

BIBLIOGRAPHY

Cousin, O. et Buttin, G. (1969) Annales de l'Institut Pasteur, Tome 117.
Gefter, M. L., Hirota, Y., Kornberg, T., Wechsler, J. A. and Barnoux, C. (1971) Proc. Nat. Acad. Sci. (USA) 68, 3150-53.

Gross, J. (1972) Current Topics in Microbiology and Immunology, 57.

Kohiyama, M. and Kolber, A. (1970) Nature, $\underline{228}$, 1157.

Kornberg, T. and Gefter, M. L. (1970) Biochem. Biophys. Res. Commun. $\underline{40}$, 1348-55.

Moses, R. E. and Richardson, C. C. (1970) Proc. Nat. Acad. Sci. (USA) $\underline{67}$, 674-81.

Pizetsky, D., Berkower, I., Wickner, R. and Hurwitz, J. (1972) J. Mol. Biol. $\underline{71}$, 557-572.

Schekman, R., Wickner, W., Westergaard, O., Brutlag, D., Geider, K., Bertsch, L. and Kornberg, A. (1972) Proc. Nat. Acad. Sci. (USA) $\underline{69}$, 2691-95.

Sekiguchi, M. and Lida, S. (1967) Proc. Nat. Acad. Sci. (USA) $\underline{58}$, 2315-20.

Smith (1971) Cold Spring Harbour DNA Meeting.

Sugino, A., Hirose, S. and Okazaki, R. (1972) Proc. Nat. Acad. Sci. (USA) $\underline{69}$, 1863-67.

Sugino, A., Okazaki, R. (1973) Proc. Nat. Acad. Sci. (USA) $\underline{70}$, 88-92.

Milewski, E. and Kohiyama, M. (1973) J. Mol. Biol. $\underline{78}$, 229-235.

MEMBRANE PROTEIN COMPONENTS AND DNA SYNTHESIS IN <u>ESCHERICHIA COLI</u>

Antonio G. Siccardi

Instituto di Genetica

Pavia, Italy

Andrée Lazdunski

C.N.R.S. Laboratoire de Chimie Bacterienne

Marseille, France

Yukinori Hirota

Institut Pasteur

Paris, France

Bennet M. Shapiro

Department of Biochemistry

Seattle, Washington, U.S.A.

INTRODUCTION

In order to correlate chromosomal replication with other
aspects of bacterial cell division, Jacob, Brenner and Cuzin (1963)
proposed a model (the replicon hypothesis) which suggested that
these processes could be temporally coordinated if the DNA were
spatially oriented <u>via</u> a liaison with the cell surface. According
to this model, DNA synthesis would occur with the chromosome attached

37

to the membrane, and subsequent membrane synthesis would separate the genetic replicas; the occurrence of separation between the DNA copies then leads to two daughter cells of identical genetic constitution.   In its original formulation the model relies upon localized membrane growth to ensure chromosome partition, but experimental evidences have shown that most of the new membrane material is immediately randomized over the whole membrane surface, suggesting that growth follows a non-conservative, dispersive pattern, which is also in accordance with the current ideas about the "fluid" state of the membrane material (Mindich and Dales, 1972). This is not enough to disprove the replicon hypothesis since these observations necessarily concern membrane synthesis "in toto" and not the fate of any minor fraction that still might behave as predicted by the model.

Over the past several years a host of evidence has accumulated which suggests that the bacterial chromosome is attached to the cytoplasmic membrane and that DNA replication itself might be associated with it, since newly synthetized DNA is found in membrane fractions in several systems (Tremblay, Daniels and Schaechter, 1969; Snyder and Young, 1969; Sueoka and Quinn, 1968).   The biochemical role of the membrane in chromosomal replication and segregation is still obscure and a more detailed comprehension of the elements and mechanisms involved will probably have to await a more precise biochemical knowledge of the membrane material itself, of its components and of their supra-molecular arrangement.

Kaback (and Stadtman, 1966; 1971; 1972) has devised a technique to purify bacterial membrane vescicles devoid of cell walls, ribosomes and soluble cytoplasmic material and still able to carry out active transport, oxidative phosphorilation and other membrane functions.   When disaggregated in 1% SDS (at 40°) and analyzed by SDS polyacrylamide electrophoresis the membrane vescicles reveal between 30 and 40 "bands" of protein material (Shapiro et. al., 1970). Soluble proteins migrate in SDS gels on the basis of their molecular weight and by comparison with standards the MW of a protein can be evaluated from its migration in the system.   There is no direct evidence that this holds also for membrane proteins, but for practical purposes we called MP60, MP40 etc. membrane protein material migrating in the system as would soluble proteins of MW 60,000, 40,000 etc.   Disaggregated in 1% SDS at 100°C the membranes show a different electrophoresis pattern with fewer bands in the high molecular weight area and more bands of low molecular weight; since the conditions do not cause proteolysis of the most common soluble proteins, it has been suggested that the different pattern might be due to more complete disaggregation of protein complexes rather than to polypeptide chain breaks (Siccardi et al., 1971).

We undertook the task of investigating whether membranes ex-

tracted from cultures deficient in DNA synthesis (because of mutation or of various treatments) were abnormal with respect to their
protein composition as analyzed by SDS electrophoresis. An analogous approach was taken by Inouye and collaborators (Inouye and
Guthrie, 1969; Inouye and Pardee, 1970).

The cultures of E. coli to be compared were labelled with
radioactive leucine (one with 3^H-leu, the other with 14^C-leu);
growth was ended at 4°C in an excess of unlabelled leucine, the
cultures were then pooled, the membranes were extracted by the
Kaback procedure, disaggregated in SDS and run in SDS-gel electrophoresis; the gels were fractionated and the radioactivity of both
isotopes evaluated by scintillation counting. The data were nor-
malized as fraction of the total radioactivity in order to make
the double labelling directly comparable (Shapiro et al., 1970;
Siccardi et al., 1971; Siccardi, Lazdunski and Shapiro, 1972). A
typical electrophoretic run is shown in Fig. 1.

MEMBRANE ALTERATIONS IN DnaA MUTANTS

DnaA mutants (2 non identical mutations, T46 and T83, that map
near ilv) are unable to initiate new rounds of chromosomal replication at 40°, yet grow normally at 30° (Hirota, Mordoh and Jacob,
1970).

Membranes isolated from mutant cultures grown at 40° (and disaggregated at 40°) show a relative deficiency of a protein fraction
MP60 and a relative increase of MP30 when compared with the membranes
of wild type cultures grown in the same conditions or of mutant
cultures grown at 30°C (Shapiro et al., 1970). If the disaggregation is carried out at 100° the protein composition is still abnormal, but the relative deficiency now involves a fraction that
migrates as MP35, while the MP30 excess persists (Siccardi et al.,
1971).

Another abnormality, initially overlooked by us because of its
minor entity, is a relative deficiency of MP40, seen with both kinds
of disaggregation, and probably corresponding to the "Y" band
described by Inouye and Guthrie (1969).

Nishimura, Caro, Berg and Hirota (1971) and Lindahl, Hirota and
Jacob (1971) have shown that the thermosensitive "initiator"
system of dnaA mutants can be replaced by other initiator systems
provided by integrated episomes or prophages ("integrative suppression" of the dnaA phenotype). (P_2) lysogenic derivatives of dnaA
mutants are suppressed by the insertion of the phage and grow at
40°C, but their membranes are still abnormal in protein composition
(deficiency of MP60, excess of MP30) and only the MP40 defect

disappears (Lazdunski and Shapiro, 1973).

The membrane alterations of the dnaA mutants appear thus to be a direct consequence of the mutation per se and not a secondary defect consequent to the lack of DNA synthesis. They have been observed also in several different derivatives of E. coli K12 in which the mutations were introduced by transduction.

The two different membrane alterations (deficiency of MP60 or of MP35) observed with the two kinds of disaggregation (at 40° or at 100°) are in fact representative of the same deficiency of membrane material. If a membrane preparation is disaggregated at 40°, the MP60 of an unfixed gel can be sliced, eluted, disaggregated at 100°C and rerun by SDS electrophoresis: the relative deficiency of protein material is in this way shifted from MP60 to MP35 (Siccardi et al., 1972).

MEMBRANE ALTERATIONS IN DnaB MUTANTS

DnaB mutants (three non identical mutations, T313, T42 and T266 that map near malB) stop DNA synthesis altogether as soon as they are shifted to 41°C (immediate-stop, or "DNA elongation" mutants) (Hirota, Ryter and Jacob, 1968).

Again, the membranes of cultures grown at non permissive temperature and disaggregated at 40°C show an abnormal protein composition when compared with otherwise isogenic wild type strains grown in the same conditions: the MP60, MP40 and MP30 fractions are altered in a manner qualitatively and quantitatively indistinguishable from dnaA situation.

However, when harsher conditions of disaggregation are used, by heating at 100°C, the dnaA and dnaB membrane preparations may be differentiated, since with dnaB the differences disappear and the protein pattern is indistinguishable from the wild type one, with the exception of the MP40 deficiency that persists (Siccardi et al., 1971).

DnaB mutants are a heterogeneous class with respect to their membrane protein composition. One of the mutants, dnaB-T42 shows some composition alteration even at permissive temperature and the MP30 alteration is quite variable quantitatively in the various mutants. Inouye and Guthrie (1969) have described the altered membrane composition of a DNA-stop thermosensitive mutant, whose mutation was later recognized to map in the dnaB locus. The mutant does not show any abnormality in the MP60 fraction and is altered in the MP40 ("Y" band) and in the MP30 (Inouye and Guthrie, 1969; Lazdunski and Shapiro, 1973).

Some of the dnaB mutants (BT313 and Inouye's MX74T2ts27) can
be phenotypically corrected if grown at high temperature in the
presence of high ionic strength and/or osmotic pressure (Richard
and Hirota, 1969). Membranes from cultures grown at 41°C in 2%
NaCl are corrected in the MP40 defect for both mutants: the membranes of MX74T2ts27 are normal, while the ones of BT313 are still
altered in the MP60 and MP30 fractions.

EFFECT ON MEMBRANE PROTEIN COMPOSITION OF VARIOUS
TREATMENTS AFFECTING DNA SYNTHESIS

Starvation of aminoacid auxotrophs of E. coli is a non-lethal
treatment which stops protein synthesis immediately and allows a
limited amount of DNA synthesis to continue. Maaloe and Hanawalt
(1961) and Lark, Repko and Hoffman (1963) have concluded that protein synthesis is required for DNA synthesis initiation and not for
chain elongation and that the treatment leads to the arrest of
chromosomal replication at a discrete, heritable chromosomal locus
(the "terminus").

Aminoacid restoration to the starved culture causes immediate
resumption of protein synthesis and a 30 min lag before DNA synthesis
resumes the rate characteristic of logarithmically growing cultures.

Analysis of membranes labelled prior to the starvation and for
various times after the restoration of aminoacids has shown that the
membrane composition is specifically altered only during the first
30' after aminoacid restoration being highly deficient of MP60 and
showing several other minor abnormalities (among which a slight
increase in MP40).

"Chase" experiments have shown that the MP60 deficiency is
probably due to lack of synthesis and not merely to lack of integration in the membrane of a synthetized precursor (Siccardi et al.,
1971).

Most of the results up to this point, could be optimistically
gathered together in a model that implies that 2 different components, MP30 and MP35 must aggregate in a MP60 complex. 1) If both
are missing the resulting membrane is deprived only of MP60; 2) if
MP35 is missing, but MP30 is present, there will be a deficiency of
MP60 and an excess of non-aggregated MP30; 3) if both are present
but do not aggregate, there will be a deficiency in MP60 and an
excess in the MP35-MP30 area.

The model is probably too optimistic and studies with various
inhibitors of DNA synthesis have failed either to support it or to
disprove it.

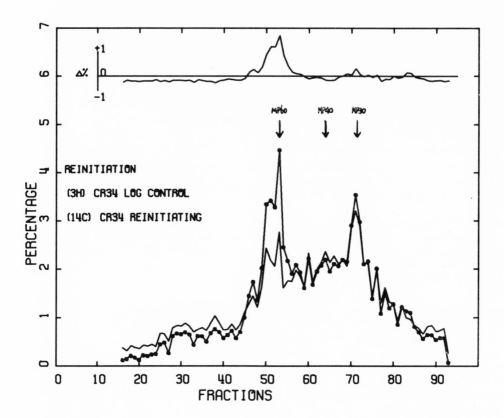

Fig. 1. The effect of labelling a reinitiating culture after readdition of aminoacids. Strain CR34 (<u>E. coli</u> K12F⁻ <u>thr</u> <u>leu</u> <u>lac</u>$_y$ λ^S <u>thy</u>) growing exponentially in minimal medium at 37°C was starved for aminoacids for 150 min. Upon subsequent readdition of the required aminoacids, the culture was labelled with ^{14}C leucine from 0 to 30 min. after starvation. The control culture was an exponentially growing culture of CR34 labelled with ^{3}H leucine for 30 min. The radioactive labelling was of the order of 1 x 10^7 cpm/mg protein of either isotope. The membrane fractions were disaggregated in 1% SDS at 40°C and run in SDS electrophoresis. The 12 cm gel was fractionated by Autogel divider (Savant Co.), the radioactivity was evaluated by liquid scintillation counting, the data were analyzed by a computer and the graph was generated by a Calcomp plotter.

The inhibitors tested are nalidixic acid, mitomycin C, hydroxyurea, fluorodeoxyuridine and thymine starvation of thymine auxotrophs.

None of these treatments (all of which are lethal and cause
several secondary phenomena, like DNA degradation and inhibition
of other macromolecular syntheses) mimics exactly the specific
alterations observed with dnaA, dnaB or reinitiation after amino-
acid starvation. All show extensive alterations of the membrane
protein composition and although an MP60 deficiency is common to
all (except to mitomycin C treatment) and MP40 deficiencies are
frequent, several other regions of the electrophoretogram are also
disturbed in a rather inconsistent manner (Siccardi et al., 1972).

EFFECT OF INHIBITION OF CELL DIVISION ON MEMBRANE
PROTEIN COMPOSITION

We have found no significant change in membrane composition
when cell division is inhibited by either penicillin or diazouracil.
Nor have we found a difference when a temperature sensitive fila-
ment forming mutant is examined (Hirota et al., 1968).

SEPARATION OF INTERNAL AND EXTERNAL MEMBRANES OF
THE E. COLI ENVELOPE

E. coli membrane vescicles contain 2 types of membranes
corresponding to the two separate lipoprotein layers of the E. coli
envelope. Lazdunski and Shapiro (1973) have applied the techniques
developed by Osborn et al. (1972) to separate the two layers and
have purified three fractions that, on the basis of their chemical
composition and enzyme activities, correspond to the two pure membranes and to a mixture of the two.

The MP40 difference belongs only to the cytoplasmic membrane
while the MP60 and MP30 differences are common to both cytoplasmic
and outer membranes. This finding, while supporting the role of the
MP40 component, sheds some doubt on the direct involvement of the
MP60 and MP30 components in DNA replication.

CONCLUSIONS

Alterations in membrane protein composition specifically related
to the DNA replication cycle have been identified in E. coli. One
of them, the deficiency of an MP40 component ("Y" band of Inouye and
collaborators)(Inouye and Guthrie, 1969; Inouye and Pardee, 1970;
Lazdunski and Shapiro, 1973) appears every time DNA synthesis is

TABLE 1.   Summary of the changes seen in membrane protein
components (disaggregation in 1% SDS at 40°C)

		MP60	MP40	MP30
dnaA mutants:	at 30°C	NC(2/2)	NC(2/2)	NC(2/2)
	at 41°C	-(2/2)	-(2/2)	+(2/2)
	(P2) lysogen at 41°C	-	NC	+
dnaB mutants:	at 30°C	NC(3/4) -(1/4)[a]	NC(4/4) NC(1/4)[a]	NC(3/4) +(1/4)[a]
	at 41°C	-(3/4) NC(1/4)[b]	-(4/4)	-(4/4)
	at 41°C with 2% NaCl	-(1/2) NC(1/2)[b]	NC(2/2)	+(2/2)
Reinitiation after aminoacid starvation	0'-30'	-	NC	NC
	30'-60'	NC	NC	NC

- indicates a deficiency, + indicates an excess, NC indicates no
change in comparison with the wild type in normal conditions.

The fractions indicate the number of different mutants at this
locus which showed the indicated difference over the total number of mutants examined.

a refers to the mutant T42.
b refers to the mutant MX74T2ts27.

blocked and it is corrected if DNA synthesis is permitted. Another more complex alteration of several membrane fractions (MP60–MP30 differences) can arise from different situations all related to DNA synthesis and can probably derive both from lack of synthesis of protein components or from lack of aggregation in higher multimolecular complexes.

Since the protein components involved in the MP60–MP30 alteration are common to cytoplasmic and outer membrane layers, it might be possible that they are structural membrane proteins not directly involved in DNA synthesis. The fact that "integrative suppression" in one case and "salt suppression" in another case do not correct the membrane composition while restoring normal DNA synthesis also speaks for a role related to, but not absolutely necessary for DNA synthesis.

It is important to emphasize the limitations of the experimental techniques employed so far. For one thing, the electrophoresis peaks probably represent contributions of many protein components which migrate together under the conditions employed and it has not been established if membrane proteins migrate in the system on the basis of their molecular weight as soluble proteins do; secondly, the differences are observed in regions of the electrophoresis pattern that represent a large fraction (10-30%) of the total membrane protein material. Hence we cannot be precise in molecular weight evaluations of the differences observed and it is possible that any changes involving the same area of the pattern do not reflect changes in identical molecular species. Thus the differences seen may be generated by coordinate changes in several different molecular species which migrate together. It is evident that a more meaningful interpretation of the results supplied by these techniques awaits further purification of membrane components and other independent insights on their function.

REFERENCES

Hirota, Y., A. Ryter and F. Jacob. 1968. Cold Spring Harbor Symp. Quant. Biol. 32:677.

Hirota, Y., J. Mordoh and F. Jacob. 1970. J. Mol. Biol. 53:369.

Inouye, M. and J. P. Guthrie. 1969. Proc. Nat. Acad. Sci., U.S.A. 64:957.

Inouye, M. and A. B. Pardee. 1970. J. Biol. Chem. 245:5813.

Jacob, F., S. Brenner and F. Cuzin. 1963. Cold Spring Harbor Symp. Quant. Biol. 28:329.

Kaback, H. R. and E. R. Stadtman. 1966. Proc. Nat. Acad. Sci., U.S.A. 55:920-927.

Kaback, H. R. 1971. Methods in Enzymology. Vol. XXII, ed. W. B. Jacoby. Academic Press, New York. p. 99.

Kaback, H. R. 1972. Biochim. Biophys. Acta. 265:367-416.

Lark, K. G., T. Repko and E. J. Hoffman. 1963. Biochim. Biophys. Acta. 76:9.

Lark, K. G. 1969. Ann. Rev. Biochem. 38:569.

Lazdunski, A. and B. M. Shapiro. 1973. Biochim. Biophys. Acta Biomembranes, in the press.

Lindahl, G., Y. Hirota and F. Jacob. 1971. Proc. Nat. Acad. Sci., U.S.A. 68:2407-2411.

Maaloe, O. and P. C. Hanawalt. 1961. J. Mol. Biol. 3:144.

Mindich, L. and S. Dales. 1972. J. Cell Biol. 55:32-41.

Nishimura, Y., L. Caro, C. M. Berg and Y. Hirota. 1971. J. Mol. Biol. 55:441.

Osborn, M. J., J. E. Gander, E. Parisi and J. Carson. 1972. J. Biol. Chem. 247:3962.

Richard, M. and Y. Hirota. 1969. C. R. Acad. Sci. Ser. A, 268:1335.

Shapiro, B. M., A. G. Siccardi, Y. Hirota and F. Jacob. 1970. J. Mol. Biol. 52:75.

Siccardi, A. G., B. M. Shapiro, Y. Hirota and F. Jacob. 1971. J. Mol. Biol. 56:475.

Siccardi, A. G., A. Lazdunski and B. M. Shapiro. 1972. Biochemistry. 11:1573.

Snyder, R. W. and F. E. Young. 1969. Biochem. Biophys. Res. Commun. 35:354.

Sueoka, N. and W. G. Quinn. 1968. Cold Spring Harbor Symp. Quant. Biol. 33:695.

Tremblay, G. Y., M. J. Daniels and M. Schaechter. 1969. J. Mol. Biol. 40:65.

A POSSIBLE COMMON ROLE FOR DNA POLYMERASE I AND EXONUCLEASE V IN ESCHERICHIA COLI

Peter T. Emmerson and Peter Strike

Department of Biochemistry, University of Newcastle upon Tyne

Newcastle upon Tyne, England

In an attempt to construct the double mutant polA1 recB21, which is thought to be inviable (Monk and Kinross, 1972), we have transduced JG138 polA1 thyA with phage P1 grown on AB2470 recB21 thy$^+$ and selected thy$^+$ transductants. Among the transductants were some strains of the type polA1 recB21 sbcA (Strike and Emmerson, 1972), which had acquired an ATP-independent DNase characteristic of recB mutants suppressed by sbcA mutations (Barbour et al., 1970). Since it is extremely improbable that JG138 could have acquired both recB and sbcA mutations simultaneously by transduction it is likely that an intermediate strain polA1 recB21 was involved. On further examination of small transductants some apparently unsuppressed polA1 recB21 strains were also found. Fig. 1 shows the results of assaying lysates of one such strain, PE114, for DNA polymerase I, Exonuclease I and Exonuclease V activity. This strain lacks DNA polymerase I, Exonuclease V and the ATP-independent DNase characteristic of sbcA mutants. Moreover, since it has normal levels of Exonuclease I it does not carry an sbcB suppressor mutation which could also suppress recB (Kushner et al., 1971). Thus, PE114 does not appear to have an amber suppressor of polA1 or an indirect suppressor of recB such as sbcA or sbcB. Only about one per cent of the viable cells in a culture of PE114 are able to give rise to colonies compared with about ten per cent in the case of recB21 single mutants. We conclude therefore that polA1 recB21 double mutants are viable although very unhealthy and prone to reversion.

Studies with the temperature sensitive mutant polA12 have shown that polA12 recB21 double mutants are not viable at 42°C (Monk and Kinross, 1972). This result, which we have also obtained, may indi-

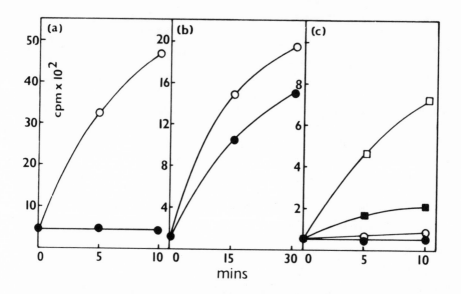

Fig. 1.   a) Cell cultures were grown to E_{650nm} = 0.6 in K medium,
washed in buffered saline and resuspended at a concentration of
approximately 3 x 10^9 cells/ml in ice-cold 10% sucrose, 0.1M Tris
(pH 8.5). Lysozyme and EDTA were added to a final concentration of
50µg/ml and 0.005M, respectively, and the mixture left on ice for
30 min. A mixture of Brij-58 and $MgSO_4$ at room temperature was then
added to give a final concentration of 5% and 0.05M respectively
and the suspension was centrifuged at 1500g for 30 min. The supernatant was taken and assayed for DNA polymerase I by adding
sonicated calf thymus DNA to a concentration of 50µg/ml and the four
deoxyribonucleoside triphosphates to a final concentration of
4nmoles/ml ATP, dGTP, dCTP and 2nmoles/ml ^{3}H-TTP. Samples were taken
at 0, 5 and 10 min, TCA precipitable counts collected on GFC filters,
and radioactivity determined in a scintillation counter. (o) AB1157
polA+; (•) PE114 polA1 recB21.

          b) Cells grown in K medium were harvested in log phase at
3-5 x 10^8 bacteria/ml from 40 ml of culture, and washed in buffered
saline. The cells were suspended in 0.4ml of chilled Tris-sucrose
(0.01M Tris, pH 8, 25% sucrose) and 40µl lysozyme (1mg/ml in 0.01M
Tris, pH 8) and 30µl EDTA (0.1M, pH 7.8) were added. The suspension was mixed well, stood on ice for 2 min, then 0.5ml cold 0.1M
MgSO4 (50mM final concentration) and 1ml 5% Brij-58 (0.01M Tris,
pH 7.2) were added. After mixing well the suspension was kept on

Legend to Fig. 1 (cont'd).

ice for 45 min, then centrifuged at 15×10^3g for 15 min. The supernatant was assayed for Exonuclease I by diluting immediately prior
to use with an equal volume of 0.05M Tris (pH 8.0), 0.25M $(NH_4)_2SO_4$,
and crystalline bovine serum albumin (1mg/ml). The reaction mixture
consisted of 0.2ml diluted supernatant, 0.4ml 0.1M glycine buffer
(pH 9.5), 10mM $MgCl_2$, 1.5mM β-mercaptoethanol, and 0.2ml heat
denatured ^{32}P-labelled E. coli DNA (5µg/ml based on OD_{260}). At
times 0, 15 and 30 min 0.2ml samples were taken into 0.25ml 10%
TCA, 0.1ml 2.5mg/ml carrier DNA was added, and the tube allowed to
stand for at least 20 min on ice. The precipitate was removed by
centrifugation at 12,000g for 10 min, the supernatant taken into
scintillation fluid, and the acid-soluble counts determined in a
scintillation counter. (o) PE114 polA1 recB21; (•) AB1157 wild-
type.

c) The cell lysis was carried out by the Brij-58 method as
described in part (b). The supernatant was assayed for Exonuclease
V by adding 0.3ml of extract to 0.1ml ^{32}P-labelled E. coli DNA
(5µg/ml), 0.1ml 10^{-3}M ATP in 0.01M Tris pH 8.0, and 0.2ml 0.01M
Tris, pH 8.0. At times 0, 5 and 10 min 0.2ml samples were taken
into 0.25ml 10% TCA, 0.1ml 2.5mg/ml carrier DNA was added and the
tube stood on ice for at least 20 min. After centrifugation, the
supernatant was transferred to scintillation fluid and the acid-
soluble counts determined in a scintillation counter. (■) PE108
polA1 recB21 sbcA50 + ATP; (□) PE108 - ATP; (•) PE114 polA1 recB21
+ ATP; (o) PE114 - ATP.

cate that the relevant activity of DNA polymerase I is more deficient
in polA12 strains at 42°C than it is normally in polA1 strains.   In
seeking reasons for the non-viability of some polA recB double
mutants it is natural to look for some function which can be carried
out by both gene products.   One property which DNA polymerase I and
Exonuclease V have in common is the ability to degrade DNA in the
5' to 3' direction.   In their 5' to 3' exonucleolytic modes both
enzymes degrade DNA to oligonucleotides and DNA polymerase I, at
least, is relatively unspecific in that it can degrade DNA containing mismatched bases and UV photoproducts (Kelley et al., 1969).
One function of the 5' to 3' exonucleolytic action of DNA polymerase I may be removal of the RNA which has provided 3' -OH primer
termini for new DNA synthesis (Kornberg, this volume; Wickner et al.,
1972; Sugino, Hirose and Okazaki, 1972).

A possible explanation for the non-viability of polA12 recB21
strains is that in recB$^+$ strains Exonuclease V can also remove the
priming RNA (Fig. 3). We have therefore examined the ability of
polA12 recB21 strains to join Okazaki pieces at 42°C. The results,
presented in Fig. 2, show that the double mutant is slightly slower
than polA12 single mutants at joining these pieces. This effect,
although reproducible, is very small and after periods of incubation
at 42°C of greater than 3 min no difference can be detected in this
system.   Thus, polA12 recB21 strains are capable of joining Okazaki
fragments but apparently at a slightly lower rate than polA12 single
mutants.   Strains carrying polA mutations are themselves much
slower than polA$^+$ strains in this respect (Kuempel and Veomett, 1970;
Okazaki, Arisawa and Sugino, 1971).   The slow joining of Okazaki
pieces in polA1 mutants is consistent with the idea that, in the
absence of DNA polymerase I, Exonuclease V or some other nuclease
can remove the priming RNA, but at a slower rate.   The fact that
Okazaki pieces are joined in polA12 recB double mutants at 42°C
(Fig. 2) could indicate that (1) there is a significant residual
level of DNA polymerase I and Exonuclease V, (2) that some other
enzyme is responsible for removing the RNA, or (3) that RNA is incorporated into the DNA in a form which is relatively alkali stable.

Aberrant control of the nuclease which removes the priming RNA
could be responsible for the non-viability of polA12 recA mutants.
This would be consistent with the observation that recA mutants
(Howard-Flanders and Boyce, 1966) and particularly polA12 recA
mutants (Monk and Kinross, 1972) break down their DNA continuously
during growth.

When polA12 recB21 strains are grown at 30°C and then transferred
to 42°C DNA synthesis, as measured by incorporation of ^{3}H-thymidine,
increases for over an hour despite the fact that the cells die (Monk
and Kinross, 1972).   Since the newly synthesized DNA may be defective

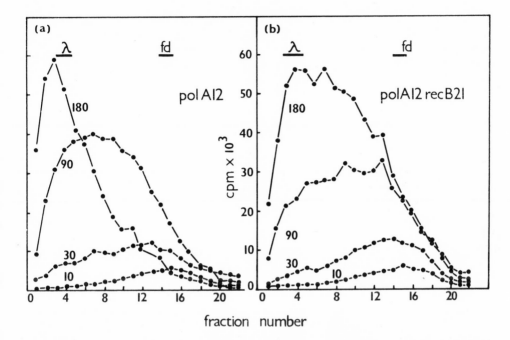

Fig. 2. Cultures of PE133, polA12 recB21 and PE134, polA12 recB+ were grown to E_{650nm} = 0.5 in K medium at 30°C. The cultures were shifted to 42°C for 30 min and then 21 ml of culture was pulse-labelled with 100μCi 3H-thymidine (23.7 Ci/m-mole, 1mCi/ml). Samples of 5ml were taken at 10, 30, 90 and 180 sec on to 3 g of ice + 0.5ml 0.25M KCN. Cells were collected by centrifugation for 5 min at 5 x 10^{3}g and 0°C, then gently resuspended in 1ml ice-cold 0.2N NaOH containing 20mM EDTA and 0.5% sodium dodecyl sarcosinate. The suspension was incubated at 37°C for 20 min, chilled on ice, and the insoluble material removed by centrifugation at 14 x 10^{3}g for 15 min. 0.1ml of this sample was then run in a 5ml 5-20% sucrose gradient in 0.1M NaOH, 0.9M NaCl, 1mM EDTA for 16 h at 22.5K in the SW 40 head of an MSE high-speed centrifuge. Ten drop samples were collected on to filter discs and acid insoluble counts were determined.

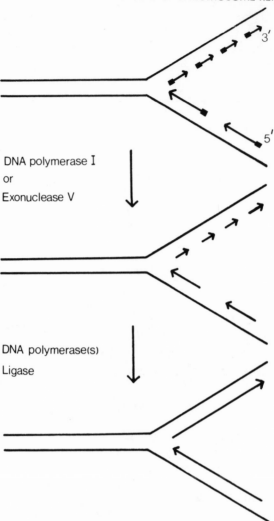

Fig. 3.   Illustration of the suggested role that might be shared by
DNA polymerase I and Exonuclease V.   Priming RNA is shown as short
thick lines.   Discontinuous synthesis is shown on both sides of the
replicating fork in conformity with the results of Bonhoeffer (this
volume; Olivera and Bonhoeffer, 1972; Herrmann, Huf and Bonhoeffer,
1972).   To account for the non-viability of some <u>polA</u> <u>recB</u> double
mutants it is suggested that the priming RNA might be removed by
either the 5' to 3' exonuclease activity of DNA polymerase I or by
Exonuclease V.   It is possible, however, that this RNA is not
removed by either of these enzymes but by some other nuclease.
Inability to terminate the nuclease activity when all of the RNA
is removed could result in some DNA degradation.   This might explain
the non-viability of <u>polA12</u> <u>recA</u> double mutants and the observation
that <u>recA</u> mutants degrade their DNA during growth.

in some way, we intend to examine it for anomalous behaviour in equilibrium density gradients.

The viability of polA1 recB21 double mutants is not inconsistent with the notion that the cell needs either DNA polymerase I or Exonuclease V in an exonucleolytic capacity, in view of the recent finding (Lehman and Chien, personal communication) that the polA1 amber fragment has substantial amounts of 5' to 3' exonuclease activity when assayed in vitro.

ACKNOWLEDGMENTS

We thank Marilyn Monk for kindly providing some of the strains and Christine McCarthy for technical assistance. This work was supported by a grant from the Medical Research Council.

REFERENCES

Barbour, S. D., H. Nagaishi, A. Templin and A. J. Clark. 1970. Proc. Nat. Acad. Sci. U.S.A. 67:128.

Herrmann, R., J. Huf and F. Bonhoeffer. 1972. Nature New Biology. 240:235.

Howard-Flanders, P. and R. P. Boyce. 1966. Radiation Res. Suppl. 6:156.

Kuempel, P. L. and G. E. Veomett. 1970. Biochem. Biophys. Res. Comm. 41:973.

Kushner, S. R., H. Nagaishi, A. Templin and A. J. Clark. 1971. Proc. Nat. Acad. Sci. U.S.A. 68:824.

Monk, M. and J. Kinross. 1972. J. Bacteriol. 109:971.

Okazaki, R., M. Arisawa and A. Sugino. 1971. Proc. Nat. Acad. Sci. U.S.A. 68:2954.

Olivera, B. M. and F. Bonhoeffer. 1972. Nature New Biology. 240:233.

Strike, P. and P. T. Emmerson. 1972. Molec. gen Genet. 116:177.

Sugino, A., A. Hirose and R. Okazaki. 1972. Proc. Nat. Acad. Sci. U.S.A. 69:1863.

Wickner, W., D. Brutlag, R. Schekman and A. Kornberg. 1972. Proc. Nat. Acad. Sci. U.S.A. 69:965.

THE JOINING OF DNA DUPLEXES AT THEIR BASE-PAIRED ENDS

Vittorio Sgaramella

Lab. CNR Genetica Biochimica ed Evoluzionistica

Pavia, Italy

J. P. Kennedy, Jr.

Lab. for Molecular Medicine, Genetics Dept.
Stanford University Medical School

Stanford, California, U.S.A.

SUMMARY

The enzymatic joining of two DNA duplexes can take place after their terminal single-stranded complementary sequences have formed hydrogen-bonded structures (cohesive joining). It is also possible to join DNA duplexes which have completely base-paired termini (terminal joining) in a reaction catalyzed by the T4 ligase. Fig. 1 summarizes the two modes of joining and this report describes some results obtained studying the terminal mechanism.

In the course of the work for the total synthesis of the structural gene for a yeast alanine transfer RNA, the extensive use of the T4 polynucleotide ligase, an enzyme which could seal nicks in DNA double helical molecules and thus join two DNA duplexes after they are annealed through their complementary single-stranded arms, has allowed the detection of an unpredicted ability of this ligase to act on DNA duplexes with completely base-paired ends. These can be obtained either through previous enzymatic joining of appropriately selected chemically synthesized oligonucleotides, or through the DNA polymerase-catalyzed "repair" of duplexes with protruding 5' ends and 3' hydroxyl bearing terminal nucleotides.

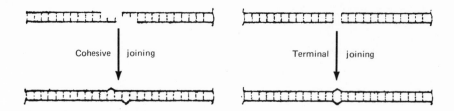

Fig. 1.   Alternative modes for the end-to-end joining of DNA molecules.   The term "cohesive" joining is proposed for the reaction (left half) involving the interaction of single-stranded complementary sequences (often called cohesive ends).   "Terminal" joining describes the alternative reaction, which takes place at completely base-paired, apposed termini (right half).   By analogy with chromosomal rearrangements, "molecular translocations" is proposed as an apt term for the products resulting from either mode of joining of different DNA molecules.

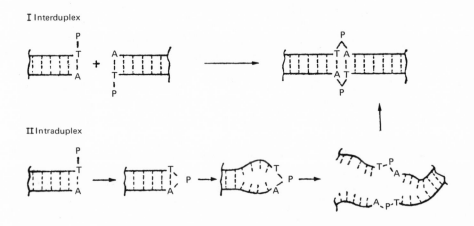

Fig. 2.   Alternative modes in the T4 ligase-catalyzed joining of duplexes at the base-paired ends.   In the intermolecular mechanism (I), two molecules of the duplex(es) are brought together at the terminal base pairs and the covalent joining then occurs.   In the intramolecular mechanism (II), the frayed end of a duplex may be covalently closed by the ligase.   The resulting hairpin structure could be extended by denaturation and then could form a duplex from two molecules.

It was first shown that the site of action of the T4 ligase
results from the apposition of two base-pair terminated duplexes
(intermolecular joining) better than from a single base-paired
terminus in which the two strands could be joined through a phosphodiester bond (intramolecular reaction). The alternative modes are
represented in Fig. 2. The experimental demonstration of the
occurrence of the intermolecular mode was sought in different ways:
(1) a duplex carrying only one base-paired terminus, when challenged
with the T4 ligase, gave rise to a dimeric product, and essentially
no intramolecularly joined hairpin-like structure of size "one"
could be detected; (2) a duplex carrying both termini completely
base-paired produced a continuous series of oligomeric compounds;
(3) two duplexes, each carrying only one base-paired end, and each
of them being specifically labelled with a different phosphorous
isotope (^{32}P and ^{33}P) at the 5' phosphoryl end, originated the
mixed or hetero-dimeric product, in addition to the two homodimers,
according to the scheme presented in Fig. 3. Fig. 4 shows joining
kinetics and fractionation patterns of the three reaction mixtures,
one containing one duplex only, the second the other duplex only,
and the third an equimolar mixture of the two duplexes. Fig. 5
gives the pattern of degradation of the joined products to 3'
mononucleotides (nearest neighbor analysis): it results that in
the homoaddition experiments, each label is transferred only to
one nucleotide, that facing it in the duplex structure. In the
heteroaddition, each label is transferred to both the 3' terminal
nucleotides, as to be expected from the formation of the dimeric
homoduplexes as well as the heterocompound (see scheme in Fig. 3).
These results prompted a research aiming at finding natural DNA's
endowed with completely base-paired ends. Among the various
possible candidates, the choice fell on the DNA of the Salmonella
typhimurium bacteriophage P22, because of two useful characteristics,
the circular permutation and the terminal permutation of its gene
sequence. In addition, P22 is a temperate phage capable of per-
forming both generalized and specialized transduction, and its DNA
--a population of different molecules as for the ends--can be
used for transfecting experiments (Sgaramella, Bursztin and
Lederberg, to be published). Using this natural substrate it has
been possible to demonstrate that the T4 ligase can catalyze the
oligomerization of P22 DNA in a way essentially similar to the
synthetic substrates. In addition it has been impossible to obtain
any effect on P22 DNA using the E. coli DNA ligase, even in reaction
mixtures containing, in addition to the bacterial ligase and its
co-factor DPN, the viral one but no ATP. From the results of these
experiments, obtained by sucrose zone sedimentation, it seems
reasonable to conclude that the joining of P22 DNA molecules takes
place at the intact base-paired termini, and leads to the formation
of linear dimers, trimers and higher oligomers--as expected given
the high initial concentration of "monomers," and as confirmed by
electron microscopy.

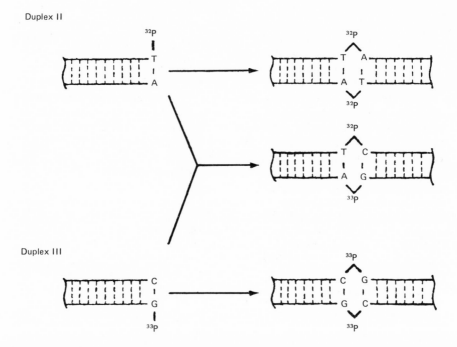

Fig. 3.   Test for interduplex joining in T4 ligase-catalyzed reaction.   One duplex contains A·T as the terminal base pair while the
second contains G·C as the terminal base pair.   The first duplex was
labeled at the 5'-T end with ^{32}P, and the second was labeled at the
5'-G end with ^{33}P.   Only the formation of the heteroduplex (middle)
should give, on degradation to 3'-nucleotides, ^{32}P radioactivity in
d–Cp and ^{33}P radioactivity in d–Ap.

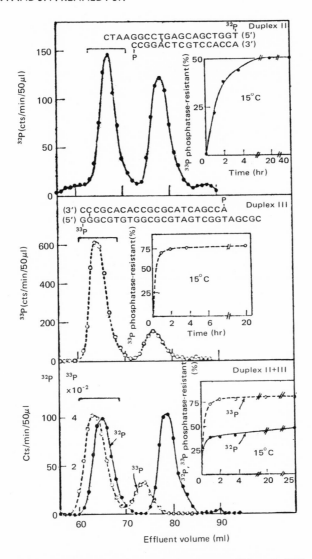

Fig. 4. Self and mixed joining of duplexes II and III. Three reaction mixtures were prepared, containing 2nmoles of duplex molecules/ml., 10mM-Tris-HCl (pH 7.6), 10 mM-MgCl$_2$, 10 mM-dithiothreitol, 60 µM-ATP and 280 units of T4 ligase per ml. The first tube (25 µl.) contained duplex II only, the second (25 µl.) duplex III, and the third (50 µl.) an equimolar mixture of the two. The joining was performed at 15°C, and was monitored with the phosphatase DEAE-cellulose paper assay. The time-courses of the reactions are given in the insets. The mixtures were fractionated by gel filtration through a Biogel-A (0.5 m) column (1.2 cm x 140 cm) equilibrated at 5°C with 0.05 M-triethylammonium bicarbonate.

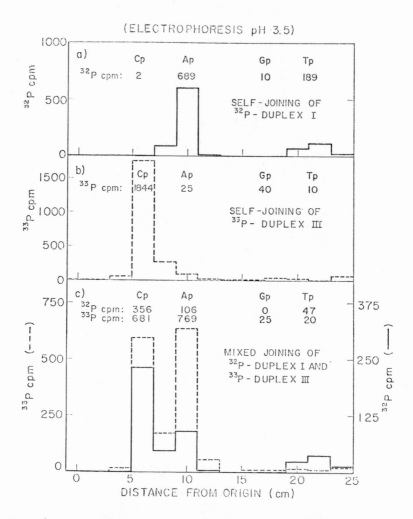

Fig. 5. 3'-nucleotides analysis of joined products of Fig. 4. The first peaks of the three fractionations were subjected to degradation to 3'-mononucleotides (nearest neighbor analysis) by the successive use of micrococcal nuclease and spleen phosphodiesterase. The digests were fractionated by paper electrophoresis and (not shown) paper chromatography. The strips were cut into 2 cm-wide pieces and counted by liquid scintillation spectrometry.

A natural extension of these experiments entailed the attempted covalent joining of two different natural DNA's through the terminal joining reaction. To this purpose, the linear form of Simian Virus 40 DNA, as obtained after nucleolytic cleavage with the restriction enzyme coded for by the E. coli resistance transfer factor R_1, was chosen since, by analogy with the products of the hemophilus influenzae restriction endonuclease, it was generally thought to have completely base-paired ends. In the attempts to join it with linear P22 DNA molecules no mixed product SV 40 DNA – P22 DNA was obtained, but SV 40 linear DNA could be extensively oligomerized by either the T4 or the E. coli ligase. It was therefore concluded that the termini of the R_1 endonuclease linearized SV40 DNA are cohesive. This finding prompted the successful joining by another group of the linear SV 40 DNA to the similarly linearized lambda dv DNA.

In conclusion, the T4 ligase has been shown to be able to join different DNA's of synthetic as well as natural origin endowed with completely base-paired ends. The joining occurs through the formation of the physiological phosphodiester bond. Using the natural P22 DNA as substrate, the ligase extracted from E. coli did not exhibit any "terminal" joining activity.

ACKNOWLEDGMENTS

The author is a researcher of the Consiglio Nazionale delle Ricerche, Lab. di Genetica Biochimica ed Evoluzionistica, Pavia, Italy, on leave at the Genetics Department of the Stanford University Medical School, Stanford, California. The experiments with the synthetic DNA's carry the obvious trademark of Dr. H. G. Khorana, the ones with P22 DNA were performed with the support and guidance of Dr. J. Lederberg: to both warm thanks are here expressed.

REFERENCES

Consult the following references for a more complete bibliography and for detailed description of the experiments.

Sgaramella, V. 1972. Il Farmaco Internl. Ed. 27:809.

Sgaramella, V. 1972. Proc. Nat. Acad. Sci., U.S.A. 69:3389.

Sgaramella, V. and H. G. Khorana. 1972. J. Mol. Biol. 72:493.

THE ATTACHMENT OF THE BACTERIAL CHROMOSOME TO THE CELL MEMBRANE

P. Dworsky[1] and M. Schaechter
[1]Institut für Allgemeine Biochemie
Universität Wien, Vienna, Austria
Department of Molecular Biology and Microbiology
Tufts University School of Medicine

We would like to pose three questions:
  1) Is the bacterial chromosome attached to the membrane?
  2) If so, is it attached at a specific portion of the membrane?
  3) Is it attached by a specific region of the DNA?

While definitive answers cannot be provided for these questions, we think that the available evidence suggest useful strategies for future work.

The notion that the chromosome of bacteria is attached to the membrane was originally proposed by Jacob, Brenner and Cuzin, (1) as a necessary element of the replicon model. They postulated that a chromosome-membrane complex plays the role of a primitive mitotic apparatus, and directs the partition of daughter chromosomes. This postulate relied to a large extent on electron microscopic studies by A. Ryer (2). In ultrathin sections of gram positive bacteria, the nuclear region is seen in juxtaposition with invaginations of the cell membrane, the mesosomes. Ryter showed that when the mesosome evaginates as a consequence of placing _Bacillus subtilis_ in hypertonic sucrose, the DNA no longer occupies a central position in the cell but becomes marginal. It appears to be pulled to the edge of the cell, conceivably by its attachment to the mesosome. Similar evidence is not easy to obtain in gram negative organisms such as _Escherichia coli_ where mesosomes are less evident.

Much work has been done with various cell fractionation techniques. The methodology used is not fully understood and these data must be interpreted with caution. Cell fractionation with bacteria is particularly difficult because they do not possess organelles bound by limiting membranes. The fractions that are usually obtained are ill defined and may result from different types of artifacts of preparation. In fact it is difficult to rule out artifactual associations that may result from spurious attachments occurring in vitro. Moreover, it is nearly impossible to determine which artifacts occurs at very instant of cell breakage, when cell constituents are still present at or near intracellular concentrations. Thus, finding that DNA does not readily associate with the membrane in vitro does not by itself, prove that DNA-membrane fractions represent structural associations present within the living cell.

Despite these gloomy prospects, a considerable amount of useful work has been done on the isolation of DNA-membrane complexes. In general, the techniques used are of one of two kinds: lysates are obtained by gently breaking cells with detergents or osmotic shock and fractionating them by centrifugation at moderate speeds (for examples, see 3,4,5). The pelleted membrane fraction contains DNA. Such fractions can also be retrieved by isopycnic centrifugation (6). The alternative method, which we have used in our work (7), consists of adding to lysates a suspension of crystals made by mixing magnesium and the detergent, Sarkosyl (sodium lauroyl sarcosinate). The surface of these crystals is hydrophobic and, possibly for this reason, certain portions of the membrane appear to adhere selectively to them. With this technique, the pieces of membrane are likely to remain in a two dimensional configuration and not form vesicles which may entrain DNA. Practically all the DNA is recovered in a band formed by centrifuging the crystals at low speeds. We have called this the M-band. M-bands obtained from a variety of organisms are viscous, indicating that the DNA has expanded and is not in the compact state characteristic of nuclear bodies obtained by Stonington and Pettijohn (8). M-bands do not represent the state of condensation of DNA inside the cells.

The argument that M-bands consist of a specific DNA-membrane complex is based largely on the observation that such complexes are not formed when native DNA of various sources is mixed with suspensions of membranes. In addition, when DNA is sheared a substantial amount of it is not found in the M-band.

Our second question is whether the DNA is attached to a specific portion of the membrane. The proportion of membrane that is found in M-bands varies with the conditions of preparation. It can be decreased by a variety of manipulations to as little as 4% of the total cell membrane, as measured by its phospholipid content (9). If the amount of membrane is decreased further, the DNA content in M-bands also begins to decrease, which suggests that as little as 4% of the membrane can support all the DNA. Work done in this laboratory (9) indicates that the portion of the membrane

to which the DNA is attached has different chemical or physical
properties. As reported elsewhere (9), the portion of the membrane
found in the M-band shows, upon re-banding, an unusually high
affinity for Mg-Sarkosyl crystals. The phospholipid composition of
M-bands from E. coli and Bacillus megaterium is different than that
of the rest of the membrane (Table 1). In both cases the membranes
M-bands contain less phosphatidylglycerol and more phosphatidylethanolamine than the rest. This finding goes along with the notion
that phosphatidylethanolamine in E. coli might play a structural
role (10) while phosphatidylglycerol, some of which is unstable
(11), may be involved in various metabolic processes. M. J. Daniels
(12) has modified this technique to obtain nine distinct membrane
fractions which differ a great deal in phospholipid composition.
DNA is found in only one or possibly two of these fractions. The
association of DNA with one of these membrane fractions is maintained throughout further purification by isopycnic centrifugation.
The Daniels technique is the only one which we are aware of for
fractionating the bacterial cell membrane into several fragments
of different chemical composition (other than the fractionation
of the inner and outer membranes of gram-negative bacteria).

Is the chromosome attached at the special region? The replicon model proposed that it is attached at a specific point, the
origin of replication. There is evidence from the work of Sueoka
and colleagues (5, 13) for the fact that the origin of replication
in Bacillus subtilis is attached to the membrane. The question
arises whether the DNA is attached at other points as well. We
have tried to answer this question with experiments based on the
following argument: the number of the attachment points of the
DNA to the membrane may be estimated by determining the number of
double strand breaks necessary to remove a given amount of DNA
from the membrane. Thus, if the DNA were attached at a single
point, an average of two breaks would suffice to remove a large
proportion of it, while if it were attached at many points, many
breaks would be necessary. Double strand breaks were produced by
X-irradiation of frozen cells. This procedure produces double
strand breaks and in addition, causes single strand breaks and
many other kinds of chemical effects. The number of breaks was
calculated by determining the molecular weight of DNA from irradi-
ate cells in neutral sucrose gradients. As seen in Fig. 1, there
is a monotonic relationship between the extent of irradiation and
the molecular weight of the DNA, as well as the amount of DNA
released from the M-band. This permits one to calculate the
number of breaks necessary to release a given amount of DNA from
the M-band. When 50% of the DNA is released, its molecular weight
is 1×10^8 daltons. From this value we have calculated that this
corresponds to about 20 attachment points per chromosome. The
details of the calculations will be presented elsewhere.

We have attempted to understand the possible meaning of this

Table 1.   Phospholipid Composition of M-bands

Organism	Fraction	Phosphati-dylethanol-amine	Phosphati-dylglycerol	Cardiolipin	Percent total lipid in M-bands
E. coli B/r (4 determinations)	Whole cells	56 ± 4	27 ± 5	14 ± 2	
	M-band	69 ± 5	19 ± 2	11 ± 1	49
E. coli A x 14 (4 determinations)	Whole cells	64 ± 5	16 ± 3	22 ± 7	
	M-band	78 ± 3	10 ± 2	12 ± 1	52
B. megaterium KM (6 determinations)	Whole cells	34 ± 2	41 ± 1	25 ± 3	
	Top fraction	31 ± 2	42 ± 2	25 ± 4	
	M-band	70 ± 2	6 ± 1	22 ± 2	15

Reprinted from ref. (9) by permission of the publishers.

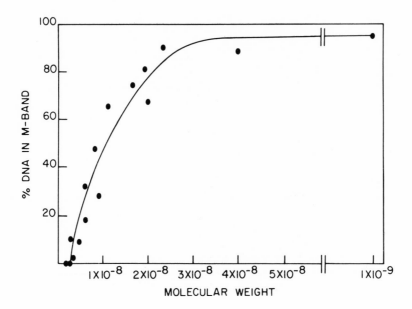

Fig. 1. Amount of DNA in the M-band vs. molecular weight after
X-irradiation. ³H-thymidine labeled cells were converted to
spheroplasts, frozen and irradiated for different times. After
thawing each sample fractionated by the M-band procedure (7, 9).

rather large number of attachment points by studying the effects
of various antibiotics. Of the various drugs tested, rifampicin
was the one that had a marked effect (Table 2, Fig. 2). Treatment
with this drug resulted in 4-5 fold decrease in the apparent number of attachment points. Other antibiotics did not have a
measurable effect. This result permits us to postulate the existence of two kinds of chromosome attachment points: those that
are resistant to rifampicin treatment, and those that are sensitive
to it. While we have no further information that would allow us
to speculate about the nature or function of these two kinds of
points, the following guesses may be in order: The rifampicin
sensitive points may be involved in maintaining the compact structure of the nucleoids. This is a possibility because when cells
are treated with the rifampicin, their chromosomes cannot be isolated in a compact configuration obtained by the Stonington-
Pettijohn fractionation (8). Instead, the DNA is viscous and
expanded.

     The effects of rifampicin concern RNA polymerase since the
structure of the nucleoid is not affected by treatment of rifampicin

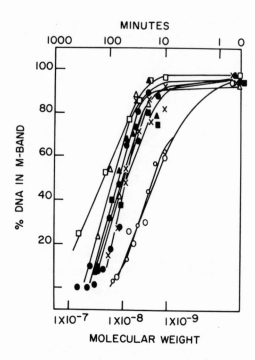

Fig. 2. Amount of DNA in the M-band vs. time of X-irradiation in antibiotic treated E. coli D-10. The proportion of DNA in M-bands was determined after different times of X-irradiation as in figure 1. Unless stated otherwise, treatment of cultures with antibiotics was for 30 min.
●, untreated control; x, actinomycin; Δ, tetracycline; ▲, puromycin; ■, erythromycin;   , chloramphenicol; o, rifampicin (7 min treatment); O, rifampicin (40 min treatment).

resistant RNA polymerase mutants. We do not know as yet if this action of the drug involves the known catalytic properties of the enzyme or some other, possibly structural, property of RNA polymerase.

The other type of attachment, that resistant to rifampicin, can be demonstrated even when the chromosome is fully expanded. This leads credence then to the idea that these points represent a structurally and physiologically different mode of attachment of the chromosome to the membrane. The number of rifampicin resistant attachment sites is interesting because it corresponds

Table 2

Number of attachment points per chromosome in drug-treated cultures

number of attachment points per chromosome

antibiotic	random spacing	equal distances
chloramphenicol	45	30
tetracyclin	23	15
puromycin	23	15
none	19	12.5
erythromycin	18	12
actinomycin D	17	11
rifampicin	.5	3

Two models were used, one with attachment points at
equal distances, the other with random spacings. The
number of attachment points was calculated using the
data in Fig. 2 by comparing values from drug treated
cultures to those of the control.

to that expected for the number of origins for replication per
chromosome in relatively fast growing cells (13, 14).

We suggest that in order to find mutants in the membrane
attachment of the origin of replication, candidates should be
fractionated after rifampicin treatment.

REFERENCES

Jacob, F., S. Brenner and F. Cuzin. (1963). Cold Spring Harbor
    Symp. Quant. Biol. 28: 329-348.
Ryter, A. (1960). Bacteriol. Rev. 32: 39-57.
Ganesan, A. T. and J. Lederberg. (1965). Biochem. Biophys. Res.
    Comm. 18: 824-835.

Smith, D. W. and P. C. Hanawalt. (1967). Biochim. Biophys. Acta 149: 519–531.

Sueoka, N. and W. G. Quinn. (1968). Cold Spring Harbor Symp. Quant. Biol. 33: 695–705.

Ivarie, R. and J. J. Pene. (1970). J. Bacteriol. 104: 839–850.

Tremblay, G., M. J. Daniels and M. Schaechter. (1969). J. Mol. Biol. 40: 65–76.

Stonington, O. S. and D. W. Pettijohn. (1971). Proc. Natl. Acad. Sci. USA. 68: 6–9.

Ballesta, J. P., E. Cundliffe, M. J. Daniels, J. L. Silverstein, M. M. Susskind and M. Schaechter. (1972). J. Bacteriol. 112: 195–199.

Ballesta, J. P. and M. Schaechter. (1971). J. Bacteriol. 107: 251–258.

Kenfer, J. and E. P. Kennedy. (1963). J. Biol. Chem. 238: 2919–2922.

Daniels, M. J. (1971). Biochem. J. 122: 197–207.

O'Sullivan, M. A. and N. Sueoka. (1972). J. Mol. Biol. 69: 237.

Helmstetter, C. E. and S. Cooper. (1968). J. Mol. Biol. 31: 507–518.

REPLICATION OF THE SINGLE-STRANDED DNA OF BACTERIOPHAGE φX174 IN NUCLEOTIDE-PERMEABLE CELLS

Ulrich Hess, Hans-Peter Vosberg, Hildegard Dürwald, Otto Schrecker and Hartmut Hoffmann-Berling

Max-Planck-Institute für medizinische Forschung

Heidelberg, West Germany

The replication of φX174 is initiated by conversion of single-stranded ring DNA (SS) into a double-stranded replicative form (RF) DNA molecule. The fact that this conversion of SS to RF is insensitive to high concentration of chloramphenicol suggests involvement of pre-existing host enzyme(s) in this step. The product of a viral gene is required only for the next stage in which the RF replicates to yield further RF (Sinsheimer, 1968).

This replication of RF to RF occurs in a rolling circle process. It involves displacement of the viral template strand from its complement through replication followed by discontinuous synthesis on a largely single-stranded viral DNA (Dressler and Wolfson, 1970; Schroeder and Kaerner, 1972). The evidence for discontinuous replication of the displaced viral strand is based on results obtained by DNA pulse-labelling in ether-permeabilized cells, a system in which the resolution of DNA pulse-labelling is especially high (Vosberg and Hoffmann-Berling, 1971). The results indicate formation of the complementary daughter strand through the joining of newly polymerized short DNA pieces which sediment at 5 to 6s in alkali (Dürwald and Hoffmann-Berling, 1971) and accordingly have 10 to 30% of the mean length of Okazaki's 10s-pieces of E. coli or T4 phage DNA (Okazaki et al., 1968) and 5 to 10% of the unit length of φX DNA (5,700 nucleotides). Pieces of nascent DNA of only 5 to 6s in alkali are observed also in the replication of E. coli DNA in ether-treated cells under certain conditions (Geider and Hoffmann-Berling, 1971) and in vivo in the replication of SV40 virus DNA (Fareed and Salzman, 1972).

71

In an attempt to clarify the role of these short pieces in DNA replication we have studied the question whether the complementary strand is made from short piece-precursors also in the conversion of SS to RF. The following results indicate that this is actually the case.

The phage was added to intact cells in the presence of caffeine which at high concentration blocks synthesis of DNA, RNA and protein in E. coli without interfering with the adsorption of φX. Following treatment of the cell-phage complexes with ether and incubation in the absence of caffeine with 4 deoxynucleoside triphosphates (dNTP) and rATP for 15 min at 35° C single-stranded DNA is converted to RF equivalent to the number of added infectious particles (5 per cell). About 80% of this RF sediments as the supercoiled form I (21 s at neutral pH; about 45 s at pH 12.5), the remainder as the open circular form II (16 s) (Fig. 1). Host DNA synthesis was suppressed with mitomycin C (Lindqvist and Sinsheimer, 1967; Vosberg and Hoffmann-Berling, 1971); so > 90% of the subcellular DNA synthesis is virus-directed in the following experiments.

DEMONSTRATION OF NASCENT SHORT CHAINS

The velocity sedimentation profiles in Fig. 2 were prepared to characterize the state of the nascent complementary DNA 3 min after start of DNA synthesis when half of the final amount of DNA is synthesized in the ether-treated cells. The uniform ^{3}H labelled new DNA (Fig. 2a) sediments at neutral pH as a broad band at velocities between 24.5 s (the velocity of SS) and 16 s suggesting that the new DNA is present in intermediates of SS to RF conversion mixed, possibly, with RFII and RFI. At alkaline pH (Fig. 2b) part of the incorporated ^{3}H label sediments as a continuous spectrum of pieces at velocities up to 14 s (the rate of unit-length linear φX DNA) indicating that, in fact, complementary strands in all stages of growth are present. Radioactivity at the bottom of this gradient is denatured RFI. On the other hand, DNA pulse-labelled at 3 min sediments predominantly as short pieces at 5 to 6s in alkali (Fig. 2c). After a chase with excess cold dNTP the label of this DNA sediments as long units including rings (present in RFI at the bottom of the tube; Fig. 2d).

The previous studies on the replication of RF to RF (Dürwald and Hoffmann-Berling, 1971) had shown that 5 to 6s-pieces of new complementary DNA are rapidly joined in the presence of an approximately physiological, 20 µM concentration of the dNTP and only slowly or not at all in the presence of 1 µM dNTP. Fig. 3 shows the velocity sedimentation profile in alkali of DNA after labelling of ether-treated cell-phage complexes by incubation for 3 min with tritiated dNTP at 1 µM concentration. When immediately

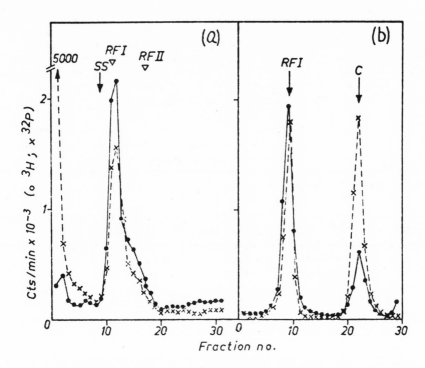

Fig. 1.   Characterization of the final product of the SS to RF
conversion in ether-treated cells.   E. coli H512 (uvrA⁻ endI⁻)
cells, grown and treated with mitomycin C as previously described
(Dürwald and Hoffmann-Berling, 1971) were suspended in 0.1 vol.
of growth medium containing 15 mg/ml. caffeine at 37°C and
infected with 5 infectious φX174 particles per cell.   After
allowing 2 min. for adsorption the cells were diluted into 0.4 vol.
of the same warm caffeine containing medium and aerated for further
5 min.   After sedimentation the cells were treated with ether and
stored frozen (Dürwald and Hoffmann-Berling, 1971).   Viral DNA
extracted from such cells is single-stranded (Hess et al., 1973).
The mixture for DNA synthesis (2 ml.) contained 5×10^9 cells/ml
in "basic medium solution" (80 mM KCl, 40 mM tris HCl pH 7.6, 7 mM
$MgCl_2$, 2 mM EDTA, 0.5 M sucrose), 1 mM rATP and 4 dNTP (20 μM)
including [^{3}H] dTTP of 0.04 Ci/mmole.   After incubation for 15
min at 35°C the cells were diluted into 10 ml. of borate-EDTA
solution (0.05 M sodium tetraborate; 0.05 M EDTA of pH 9.1) to
elute non-infectious phage from the cells (Newbold and Sinsheimer,
1970) and sedimented for 7 min at 7000 g at 0°C in two Servall
centrifuge tubes through underlayers of 25% sucrose prepared in
borate-EDTA.   DNA was extracted by lysis with 0.2 mg/ml. lysozyme
in the presence of EDTA, followed by overnight digestion at 37°C

Legend to Fig. 1 (cont'd).

with 100 μg/ml. pronase in the presence of 1% Sarcosyl NL97. The
extracts were studied by velocity sedimentation through sucrose
gradients (20°C; 5 to 20%) (Dürwald and Hoffmann-Berling, 1971)
prepared in a Spinco SW27 rotor. The bottom of the emptied tubes
was washed and the wash combined with the first fraction of the
gradient. Radioactivity was counted after precipitation with 5%
trichloroacetic acid on nitrocellulose filters.

(a) Sedimentation at neutral pH (12 H at 22,000 revs./min).
The 24.5 s-marker (SS) was run in a separate gradient. Parental
labelled viral DNA at the bottom of the gradient derives from
non-infectious particles which had failed to replicate and which
were not completely destroyed by the Sarcosyl-pronase treatment.

(b) Sedimentation at pH 12.5 (12 h at 22,000 revs./min)
following additional extraction of the DNA with phenol. C.
indicates the position of circular single strands of φX DNA.
Sedimentation is from right to left in these and the following
gradients. (x) Parental ^{32}P; (•) incorporated ^{3}H.

Fig. 2. Velocity sedimentation analysis at neutral and alkaline
pH of intermediates of SS to RF conversion.

(a) and (b) Uniform labelling experiment. After incubation of
the ether-treated cells for 3 min under conditions as described in
Fig. 1 ([^{3}H] dTTP of 5 Ci/mmole) DNA was extracted with pronase
and Sarcosyl followed by phenol treatment. Half of the extract was
sedimented through a neutral sucrose gradient as described (Fig.
1 (a)) and half through an alkaline sucrose gradient (SW27 rotor;
20°C; 15 h; 26,500 revs./min).

(c) Pulse labelling experiment. After incubation for 3 min
in the presence of cold dNTP (1 ml.) under conditions as in (a, b)
[^{3}H] dTTP, [^{3}H] dCTP and [^{3}H] dATP (0.2 mCi each, dissolved in 0.4

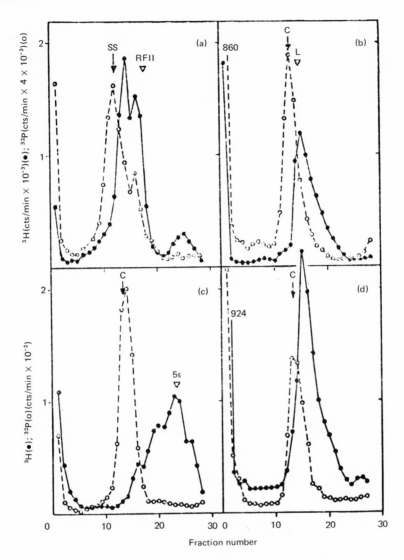

ml. basic medium; 1 mM rATP) were added. DNA synthesis was
interrupted 4 sec later with borate-EDTA. DNA was extracted as in
(a, b) and sedimented through alkaline sucrose as in (b). C and
L indicate the positions of circular and linear strands, respectively, of φX SS.

(d) Pulse-chase experiment. Pulse-labelling was carried out
as in (c), four cold dNTP (0.5 mM each) were then added and the
incubation was continued for a further 10 min. Sedimentation
through alkaline sucrose was as in (b). (o) Parental ^{32}P; (●)
incorporated ^{3}H.

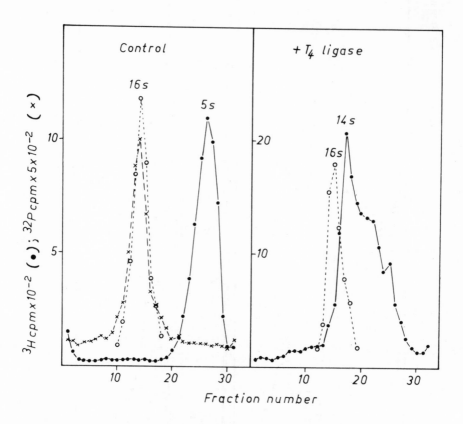

Fig. 3. Joining of 5-6 s pieces <u>in vitro</u>. The reaction mixture
for DNA synthesis contained [³H] dTTP, [³H] dATP, [³H] dGTP
(9 Ci/mmole each) and dCTP (1 μM each). After incubation for 3.5
min at 35°C DNA was extracted with Sarcosyl-pronase and phenol
and sedimented through a neutral sucrose gradient. Intermediates
of the SS to RF conversion were collected from the 24.5 to 16 s
region, pooled, dialysed versus 0.15 M tris HCl (pH 7.6); 0.2 mM
EDTA and in 0.85 ml. brought to 1 mM rATP, 10 mM MgCl₂ and 10 mM
2-mercaptoethanol with stock solutions. 0.05 ml. T4-ligase was
added (kindly provided by Dr. W. Knopf) and the mixture (1.0 ml.)
was incubated for 20 min at 25°C followed by the addition of 0.1
ml. of 0.5 M EDTA and re-extraction of the DNA with phenol. The
DNA was finally analyzed by sedimentation through an alkaline
sucrose gradient as described in Fig. 2 (b). - (a) control without
ligase treatment; (b) ligase-treated sample. (●) Incorporated ³H;
(x) ³²P-labelled parental DNA; (o) added [¹⁴C] SS.

added to an alkaline sucrose gradient this DNA sediments entirely
as short pieces at about 5 to 6s. When treated with T4-ligase before sedimentation this DNA sediments essentially as long units
including such of unit length of φX DNA. Sedimentation of the
parental DNA at the rate of circular strands (16 s) indicates that
this DNA is not discontinuous.

We take these results to indicate that the complementary strand
is discontinuous in the nascent state and that the short pieces can
be joined in vitro by ligase. In an early stage of the SS to RF
conversion this is not possible except after incubation of the
extracted DNA with a DNA polymerase (polymerase II was used) and
4 dNTP (Vosberg, unpublished) before the ligase treatment suggesting that the short pieces are separated by gaps in this stage of
replication. This finding supports a concept of repeated initiation
and argues against endonucleolytic scission as the reason for the
occurrence of short pieces.

EVIDENCE FOR TWO POLYNUCLEOTIDE JOINING PROCESSES INVOLVED IN THE FORMATION OF RFI

E. coli DNA ligase requires NAD as a cofactor and is not stimulated by deoxynucleotide. The enzyme combines with the AMP-
moiety of NAD to form an active complex which can be discharged
with nicotinamide mononucleotide (NMN) to the free, inactive enzyme.
Several authors have studied the influence of NMN on DNA synthesis
in subcellular E. coli systems including gently lysed (Olivera and
Bonhoeffer, 1972) and ether-treated (Geider, 1972) cells and from
the results have concluded that adenylated ligase is present in
such crude systems and involved in the conversion of RFII to RFI
and the joining of Okazaki-pieces of E. coli DNA.

If this is true and if the NAD-dependent ligase is not
influenced by deoxynucleotides, why were the 5 to 6s-pieces not
joined in the cells under the conditions of the experiment in Fig. 3?
To study this problem ether-permeabilized cell-phage complexes
containing such pieces in a ^{3}H-labelled form were incubated with
either NMN or NAD in the presence of 4 dNTP. As shown in Fig. 4 the
short pieces were as efficiently joined in the presence of NMN as in
the presence of NAD. The primary difference in the products
resulting from the two treatments resides in the formation of RFI.
In the presence of NMN it was not formed (absence of radioactivity
at the bottom of the alkaline sucrose gradient; Fig. 4c) whereas in
the presence of NAD it was a major product (Fig. 4b).

These results suggest that there are two joining reactions
involved in the formation of RFI and that these are separable by
NMN: One process which is insensitive to NMN is responsible for

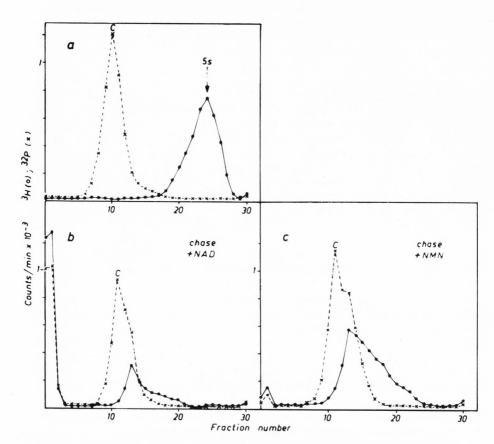

Fig. 4.   Joining of 5-6 s pieces in the presence of NMN.   After
incubation for 3.5 min at 35°C the reaction mixture for DNA
synthesis (1.5 ml.; 1 μM dNTP; [^{3}H] dTTP of 19.3 Ci/mmole) was
divided into three equal parts.   The cells were centrifuged at room
temperature for 2 min at 12,000 g in an Eppendorf centrifuge,
resuspended and either (a) directly lysed in 0.5 ml. borate-EDTA or
(b) incubated for further 7 min at 35°C in 0.5 ml. basic medium
containing 1 mM rATP, 4 cold dNTP (20 μM) and 0.2 mM NAD or (c)
incubated as in (b) in the presence of 0.2 mM NMN replacing NAD.
The cells of (b) and (c) were then lysed as in (a).   DNA was
extracted from the three samples (Sarcosyl-pronase) and studied by
sedimentation through alkaline sucrose gradients as described in
Fig. 2 (b).   (x) Parental ^{32}P label; (●) incorporated ^{3}H.

the joining of short pieces, the other which is sensitive to NMN is responsible for ring closure.

Similar results were recently obtained in studies on E. coli DNA replication in ether-treated cells (Dürwald, unpublished). In this system the mean size of the new discontinuous DNA depends on the dNTP concentration applied in the assay mixture, in analogy with the results presented here: in the presence of 1 μM dNTP the mean size of the discontinuous fraction of the new DNA (about 50% of the new DNA) is 5 to 6s; in the presence of 20 μM dNTP it is about 10 s resembling Okazaki-pieces. When cells containing radioactive 5 to 6s-pieces are incubated with NMN in the presence of cold dNTP the label of the 5 to 6s-pieces is chased into Okazaki-type pieces (about 13 s); when they are incubated with NAD in the presence of dNTP the label of the short pieces is chased into continuous DNA (>30s). The results suggest that the new, discontinuous E. coli DNA is joined in two steps: one gives rise to Okazaki pieces, the other to continuous DNA.

On the basis of these data and in view of the fact that φX makes use of host replication factors for the conversion of SS to RF (Schekman et al., 1972; Wickner, 1972) and largely also for the replication of RF to RF it is tempting to propose the following unifying concept of DNA replication in E. coli. This concept takes into account the results obtained with non-infected cells as well as those obtained with φX infected cells: The primary products of discontinuous replication are 5 to 6s-pieces of new DNA and this is true regardless of the nature of the template. In the presence of a physiological concentration of the dNTP these pieces are rapidly joined together to form longer units, possibly by an enzyme which is not the NAD-dependent ligase. These longer units are in the case of E. coli DNA Okazaki-pieces (10s) and in the case of φX DNA unit length linear chains (14s).

Okazaki-pieces of E. coli DNA are known to be separated by gaps immediately after their formation (resulting possibly from the removal of RNA primer (Sugino, Hirose and Okazaki, 1972)) which in a subsequent step are filled by repair-like synthesis (Kuempel and Veomett, 1970; Okazaki, Arisawa and Sugino, 1971). Conceivably this gap filling process requires time and accordingly stabilizes the new, discontinuous DNA for a while at the 10 s-level. This lag in the formation of a continuous DNA together with the direct flow of deoxynucleotides into Okazaki-pieces through their repair-like covalent extension might provide an explanation why 10 s-pieces, as distinct from 5 to 6s-pieces, are so prominent after pulse-labelling in E. coli DNA under conditions of physiological dNTP concentration.

The final step in the formation of a continuous chain through discontinuous synthesis would be the joining of Okazaki-pieces to

pre-existing DNA through the NAD-dependent ligase and, in the case
of φX DNA, the closure of a ring through this enzyme.

REQUIREMENT FOR rATP

The replication of φX SS in ether-treated cells depends
strictly on the presence of rATP (Fig. 5) and, after transfer of
the cells to an rATP-free mixture of dNTP, comes to a rapid halt
(Fig. 6).

Two explanations are conceivable to account for this requirement for rATP throughout the replication process:

(1) rATP plays a role not only in chain initiation as suggested
by in vitro results of Schekman and coworkers (1972) but also in
chain elongation. Repeated initiation in a process of discontinuous
chain growth could be a function which requires rATP.

(2) Replication of φX SS in cells might be coupled to uncoating
of the adsorbed viral DNA and penetration through the bacterial
envelopes (see the subsequent section). Penetration rather than
chain elongation might require rATP.

Attempts to decide between these alternatives by polyethylene-
glycol 6,000 co-precipitation of φX SS with protein out of a DNA
synthesizing system of soluble E. coli enzymes (developed by
Müller and similar to the system of Wickner and coworkers (1972))
have failed to decide between these alternatives. It is true,
after resuspension of the DNA-protein co-precipitate in an rATP-
free mixture of dNTP an initiated replication of the single strands
continues, although at a reduced rate compared to a non-precipitated
control. On the other hand, Dau (unpublished) has not been able to
identify short pieces of precursor DNA in this system. So it is
not clear whether the lack of DNA penetration or a lack of repeated
initiation accounts for the occurrence of rATP-independent DNA
synthesis in this system. Work is in progress to further elucidate
the problem.

CHARACTERIZATION OF THE REPLICATION COMPLEX OF φX SS

Studies on the site of φX SS replication in cells have led
Knippers and coworkers (1969) to conclude that the viral DNA
penetrates into the cytoplasm before being replicated whereas
Francke and Ray (1971) postulate that this replication starts while
part of the DNA molecule is still outside the cell confined within
the adsorbed viral coat.

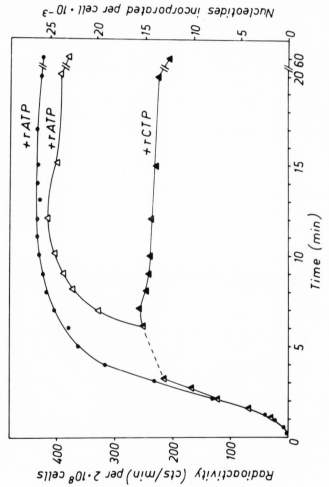

Fig. 6. Interruption of replication by removal of rATP. DNA synthesis (1.0 ml.) was carried out as in the preceding experiment and
interrupted after 3 min with 3 ml. of iced basic medium. The cells
were pelleted, resuspended and supplied with 4 dNTP including [3H]
dTTP to yield the former concentrations and specific radioactivity.
The cells were then divided into two parts which received either
(●) 1 mM rATP or (o) 1 mM rCTP. DNA synthesis was followed as in
the preceding experiment.

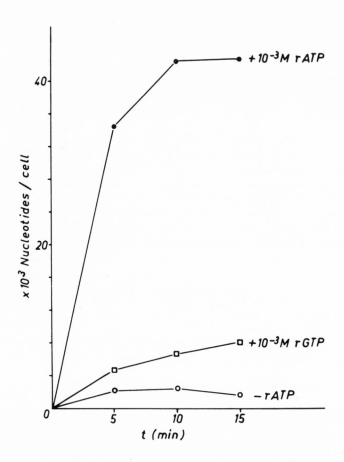

Fig. 5. Requirement of replication synthesis for rATP. DNA synthesis was carried out essentially as described in Fig. 1 (0.5 ml.;
20 μM dNTP; [^{3}H] dTTP of 0.48 Ci/mmole). At the times indicated
0.1 ml. samples were withdrawn from the mixture, precipitated with
5% TCA, washed on nitrocellulose filters and counted in toluene-PBD
scintillator. The counting efficiency for tritium is 16% under
these conditions. (●) Control; (o) rATP omitted; (□) 1mM rGTP substituted for rATP.

When ether-treated cells are gently lysed after lysozyme-EDTA treatment with 1% Brij 58 (a non-ionic detergent) more than 90% of the parental ^{32}P-labelled viral DNA pellets with the fast-sedimenting fraction of the cells in a low-speed run for 30 min at 14,000 g at 4° C in a Servall centrifuge. Furthermore, when limited DNA synthesis was allowed to occur in the cells before lysis 90% of the nascent DNA is also found in this fast-sedimenting cell-fraction suggesting that the viral DNA replicates in connection with a large structure. Although there is no definite proof we consider this structure to be the cell's membrane and accordingly are inclined to believe that the viral DNA penetrates into the cell while being replicated.

Detailed information regarding these large complexes was obtained after disintegration of the cell-phage complexes with sarcosyl NL97 (an anionic detergent) followed by short digestion of the lysate with pronase and sedimentation through a borate-sucrose gradient of pH 9.1 (Newbold and Sinsheimer, 1971). Under these conditions the parental ^{32}P label sediments essentially as complexes at about 50 s and 40 s (free SS sediments at about 14.5 s and intact phage at 114 s) suggesting that the viral DNA is not freed from its association with fast-sedimenting material by this rigorous lysis-treatment. These 50 s- and 40 s-complexes do not form on addition of SS to lysates of the protein concentration used for sedimentation analysis and they are not defective phages as shown by their low content of viral coat protein sulfur after infection with ^{35}S-protein, ^{32}P-DNA double-labelled ϕX. The ratio ^{35}S-counts to ^{32}P-counts in the experiment of Fig. 7a was 0.1 in the complexes as compared to 2.6 in the original phage stock.

When studied by electrophoresis in SDS-containing poly-acrylamide gels the two main proteins of these complexes have molecular weights of about 87,000 and 22,000 dalton. Three minor proteins, also observed, have M.W. between 60,000 and 75,000 dalton. The M.W. of the largest of the 4 proteins detectable in ϕX particles is 48,000 dalton and only a minor protein of the phage has M.W. near 20,000 dalton (Burgess and Denhardt, 1969). It thus appears that most of the protein of these complexes is derived from the host. It is interesting to note that E. coli unwinding protein, which might be required for the replication of ϕX SS to RF, has M.W. 22,000 dalton (Sigal et al., 1972).

When limited DNA synthesis occurred in the permeable cells before lysis most of the parental labelled viral DNA sedimented at rates intermediate between 50 s and the rate of RFII (Fig. 7b) and when extensive DNA synthesis occurred most of this DNA sedimented to the positions of free RFII and RFI together with the newly synthe-sized DNA. This is in accordance with the results shown in Fig. 1a. These data, especially those obtained at an intermediate stage of

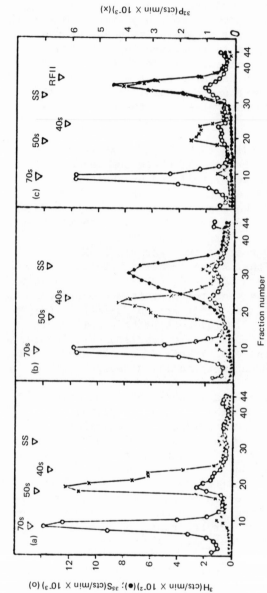

Fig. 7. Velocity sedimentation of replication complexes. 32P, 35S double labelled φX was used for infection. Samples containing 10¹⁰ ether-treated cells were either (a) not incubated with triphosphates or (b) incubated with 1 μM deoxynucleosidetriphosphates ([³H] dTTP of 7 Ci/mmole) for 8 min or (c) with 20 μM deoxynucleosidetriphosphates ([³H] dTTP of 0.7 Ci/mmole) for 3.5 min. 2,200 and 10,500 deoxynucleotides, respectively, were incorporated per cell. The cells were washed 3 times with borate-EDTA, lysed with lysozyme and 1% Sarcosyl and incubated for 2 h at 37°C with 100 ug/ml. pronase. Sedimentation was through borate-EDTA sucrose gradients. (o) 35S; (x) 32P; (●) ³H. The peak of 35S at 70 s which is probably a composite profile of empty viral coats (72 s) and defective phage particles, served as a sedimentation marker.

the SS to RF, suggest that the 50 s- and 40 s-particles are
replication complexes and that the viral DNA is not immediately
released from its complexes when replication starts.

Complexes with a sedimentation velocity near 50 s, considerable
stability in sarcosyl and a protein composition similar to that of
the complexes described here form also on addition of ϕX SS to a
DNA synthesizing extract of E. coli of the kind mentioned before.
In this case, too, the DNA is released from its complex after
replication. High capacity for complex formation and SS to RF
conversion (10 SS/ cell equivalent extract in our hands) make
these extracts a promising starting material for the characterization
of the complex proteins.

Part of this work has been published (Hess, Dürwald and
Hoffmann-Berling, 1973).

REFERENCES

Burgess, A. B. and D. T. Denhardt. 1969. J. Mol. Biol. 44:377.

Dressler, D. and J. Wolfson. 1970. Proc. Nat. Acad. Sci., U.S.A.
67:456.

Dürwald, H. and H. Hoffmann-Berling. 1971. J. Mol. Biol. 58:755.

Fareed, G. C. and N. P. Salzman. 1972. Nature New Biol. 238:274.

Francke, B. and D. S. Ray. 1971. Virology. 44:168.

Geider, K. and H. Hoffmann-Berling. 1971. Europ. J. Biochem.
21:374.

Geider, K. 1972. Europ. J. Biochem. 27:554.

Hess, U., H. Dürwald and H. Hoffmann-Berling. 1973. J. Mol. Biol.
73:407.

Knippers, R., W. O. Salivar, J. E. Newbold and R. L. Sinsheimer.
1969. J. Mol. Biol. 39:641.

Kuempel, P. L. and G. E. Veomett. 1970. Biochem. Biophys. Res.
Comm. 41:973.

Lindqvist, B. H. and R. L. Sinsheimer. 1967. J. Mol. Biol. 30:69.

Newbold, J. E. and R. L. Sinsheimer. 1970. J. Mol. Biol. 49:49.

Okazaki, R., T. Okazaki, K. Sakabe, K. Sugimoto and A. Sugino. 1968. Proc. Nat. Acad. Sci., U.S.A. 59:598.

Okazaki, R., M. Arisawa and A. Sugino. 1971. Proc. Nat. Acad. Sci., U.S.A. 68:2954.

Olivera, B.M. and F. Bonhoeffer. 1972. Proc. Nat. Acad. Sci., U.S.A. 69:25.

Schekman, R., W. Wickner, O. Westergaard, D. Brutlag, K. Geider, L. Bertsch and A. Kornberg. 1972. Proc. Nat. Acad. Sci., U.S.A. 69:2691.

Schroeder, C. H. and H. C. Kaerner. 1972. J. Mol. Biol. 71:351.

Sigal, N., H. Delius, Th. Kornberg, M. L. Gefter and B. Alberts. 1972. Proc. Nat. Acad. Sci., U.S.A. 69:3537.

Sinsheimer, R. L. 1968. Progr. Nucl. Acid Res. and Mol. Biol. 8:115.

Sugino, A., S. Hirose and R. Okazaki. 1972. Proc. Nat. Acad. Sci., U.S.A. 69:1863.

Vosberg, H. P. and H. Hoffmann-Berling. 1971. J. Mol. Biol. 58:739.

Wickner, R. B., M. Wright, S. Wickner and J. Hurwitz. 1972. Proc. Nat. Acad. Sci., U.S.A. 69:3233.

Wickner, W., D. Brutlag, R. Schekman and A. Kornberg. 1972. Proc. Nat. Acad. Sci., U.S.A. 68:2826.

REPLICATION OF BACTERIOPHAGE ɸX174 REPLICATIVE FORM DNA

IN VIVO

Claus H. Schröder and Hans-Christian Kaerner

Max-Planck-Institut für Medizinische Forschung

Abteilung für Molekulare Biologie; Heidelberg, W. Germany

SUMMARY

The replication of the circular double-stranded bacteriophage ɸX174 replicative form DNA was studied by structural analysis of pulse-labeled replicative intermediates. Evidence is presented that ɸX replicative form replicates according to a rolling circle model proposed by Dressler & Wolfson (1970). Replication involves continuous elongation of the viral (= positive) strand component of replicative form resulting in the displacement of a single-stranded tail of increasing length. Replicative intermediates sedimenting at 27 to 28 s are found to contain linear viral strands of approximately double ɸX unit length. The synthesis of the new complementary (= negative) strand on the single-stranded tail appears to be initiated with considerable delay and converts the tail to double-stranded DNA. Before the new negative strand is completed, the replicative intermediates split into (I) a complete RF molecule containing the "old" negative and the "new" positive strand and (II) a linear partially double-stranded "tail" consisting of the complete "old" positive strand and a fragment of the "new" negative strand.

The second part of this study is concerned with the fate during RF replication of these fragments of the rolling circles. The RF II molecules containing the "old" negative strands appear to go into further replication rounds repeatedly. Some of the "tails" were found in the infected cells in their original linear form. "Gapped" RF II molecules which have been described earlier by Schekman and coworkers (1971) are supposed to originate from the tails of rolling circle intermediates by circularization of their positive strand components. Evidence is provided by our experiments that even late during RF replication the gaps exist in the

negative strands of RF II rather exclusively. Appropriate chase
experiments indicated that the "tails" finally are converted to
RF I molecules. Progeny RF I molecules could not be observed to
start new replication rounds under our conditions although we cannot exclude that this might happen to some minor extent.

The results presented suggest that the first negative strands
rather than the parental positive strands persist as master templates during ϕX RF replication.

1. INTRODUCTION

Three processes can be distinguished during the multiplication
of the single-stranded circular DNA of the bacteriophage ϕX174
(Sinsheimer, 1968): (1) Upon infection the single-stranded DNA
circle is converted to a double-stranded circle, the parental replicative form (RF). (2) The parental RF replicates semiconservatively resulting in the formation of some 20 - 50 progeny RF molecules. These are found in infected cells either as supercoiled
RF I in which both strands are covalently closed or as relaxed RF
II with at least one single strand break. (3) The RF molecules
participate in the synthesis of single-stranded viral DNA in that
the positive strands are displaced from the RF duplexes by concomitant synthesis of new positive strands on the persisting circular negative templates.

Two different rolling circle mechanisms (Gilbert & Dressler,
1968) are discussed to govern the replication of the double-
stranded ϕX replicative form. Knippers, Whalley & Sinsheimer
(1969) suggested that the positive strand of the RF remains as a
closed ring during replication, whereas the negative strand is
nicked and elongated in the 5' → 3' direction and the new positive
strand is formed on the old negative strand. In contrast, Dressler
& Wolfson (1970) proposed a rolling circle model, in which the
positive strand of RF is opened and elongated on the persisting
circular negative strand as a master template.

In order to distinguish between those models we have pulse-
labeled replicating intermediates of RF with tritiated thymidine
and elucidated their structure by various centrifugation techniques.
Evidence could be provided that ϕX RF replicates according to the
rolling circle mechanism proposed by Dressler & Wolfson (1970).

The sedimentation rate of replicating intermediates is considerably altered between 16 s and about 28 s in the course of the
replication process. Replicating RF molecules are found to contain growing linear positive strands longer, and negative strands
shorter than ϕX unit length. The results presented here further
suggest that the synthesis of positive and negative strands is not

synchronous: the elongation of the positive strands partially precedes the synthesis of the new negative strands. This leads initially to RF molecules with single-stranded tails. The new negative strands are initiated and polymerized on the positive strand tails with considerable delay. Before the new negative strands have been completed the replicating intermediates split into their circular parts and the linear parts ("tails"). The circular parts are complete RF II molecules containing the circular old negative strands and the new linear positive strands. The "tails", on the other hand, consist of the linear old positive strands and incomplete new negative strands.

From these findings the question arises, which of these two fragments of the rolling circle enter further rounds of replication.

We have studied this problem by introducing different radio-active label into the positive and the negative strand component of parental RF molecules in order to follow separately the fate of the fragments of the rolling circles during replication. Those daughter RF II molecules stemming from replicating parental RF which contain the first strands are found to participate in repli-cation continuously whereas the pathway of the corresponding "tails" appeared more complicated. Evidence is provided that the tails which contain the parental positive strands develop to RF I mole-cules. In the course of this process the circularization of the positive strands appears to precede the completion and circulari-zation of the negative strand components of the duplexes. This leads to "gapped" RF II molecules which had been detected prev-iously by Schekman et al. (1971). From our result that the gaps proved to exist in the complementary strand component of these duplexes exclusively it might be concluded that "gapped" RF II molecules could indeed originate from the tails of rolling circle intermediates.

2. MATERIALS AND METHODS

(a) Bacteria and viruses

Escherichia coli H514 (F$^-$, uvrA, thy, arg, endI, su$_{amber}$) ser-ved as the host strain and was infected with ϕX174 wt.

(b) Media

NP medium has been described (Kaerner, 1970) and was supplied with 10 µg/ml. each of thymine and arginine for the growth of E. coli H514.

(c) Mitomycin C treatment

Before ϕX infection the H514 cells were treated with mitomycin

C according to Lindqvist & Sinsheimer (1967) in order to preferentially suppress the synthesis of host cell DNA.

(d) Bacterial growth and ϕX infection

H514 was grown at 37°C in 200 ml. of NP medium to 4 to 5.5 x 10^8
cells/ml. The culture was then concentrated by centrifugation and
resuspensin of the cells in 20 ml. of fresh medium. Following mitomycin treatment the cells were transferred to another 20 ml. of
fresh medium. 100 µg/ml. chloramphenicol was added to the culture
and after 1 min the bacteria were infected at 37°C with 4 plaque-
forming-units of ^{32}P-labeled ϕX per cell.

Under these conditions the infection process stops after the
formation of the parental RF (Tessman, 1966; Sinsheimer, Hutchinson
& Lindqvist, 1967). The following experimental steps of concentrating the culture and temperature shift could thus be carried out
under good control so that the cells could not switch early into
phase II or III of ϕX DNA replication, which would complicate the
interpretation of the results. After 5 min of infection in the
presence of chloramphenicol the bacteria were pelleted and without
further washing resuspended in 5 ml. thymine- and chloramphenicol-
free NP medium. Aeration was continued at 27°C in order to reduce
the rate of RF replication (Knippers, 1969). Concentrating the
infected cells was necessary to allow sufficiently hot pulse-
labelling of ϕX RF replication intermediates. The procedure described leads to a transfer of about 10 µg of chloramphenicol/ ml.
to the final 5 ml. of culture. This appeared to have some inhibitory effect upon ϕX phage formation, presumably on the synthesis of
ϕX single-stranded DNA. Phage ϕX RF replication, however, should
not be influenced at this low chloramphenicol concentration (Sinsheimer, Starman, Nagler & Guthrie, 1962). The infectious cycle
has been checked in control experiments in which the cells were
either washed free of chloramphenicol before concentrating them
in 5 ml. NP medium or by suspending them in 20 ml. of medium in
order to reduce further the concentration of chloramphenicol. In
either case after 60 min aeration at 27°C progeny phage appeared
in the cells, reaching a burst size of about 30 to 40 phage particles per infected cell within the following 60 min.

(e) Pulse-labelling of phage ϕX-infected cells

After resuspension in 5 ml. of thymine- and chloramphenicol-
free NP medium the cells were vigorously aerated for 9 min at 27°C.
At that time 2 to 3 mCi of [^{3}H]thymidine (spec. act. 40 to 60 Ci/
m-mole) was added to the cells and 30 sec later the culture was
poured into 20 ml. of acetone cooled to -70°C, in order to definitely stop all further DNA synthesis. From the results of Dressler
& Wolfson (1970) obtained at 37°C, it can be estimated that a 30-
sec pulse under the conditions described is shorter than the

replication time of one RF molecule. It should be mentioned here that a [^{3}H]thymidine pulse given earlier than 7 min after removal of chloramphenicol yielded no incorporation of label into ϕX-specific DNA. This could indicate that the initiation is a time-limiting step in the RF replication process.

(f) Growth of host bacteria and ϕX infection in
    presence of chloramphenicol and [^{3}H] thymidine

H514 was grown and treated with mitomycin C as described in section (d) of methods.

Chloramphenicol (120 μg/ml.) was added to the 20 ml. culture together with 0.1 μg/ml. [^{3}H] thymidine, specific activity 50 Ci/m-mole. After 1 min the cells were infected with a multiplicity of 4 plaque-forming-units of ^{32}P-labelled ϕX per bacterium and aerated for 8 min at 37°C. Then cold thymidine was added to give a concentration of 400 μg/ml. Another 2 min later the bacteria were pelleted in a Servall centrifuge and resuspended without further washing in 10 ml. NP medium at 27°C, containing 400 μg/ml. cold thymidine. The concentrated culture was then vigorously aerated at 27°C for 10 min. At that time the culture was divided: One half (5 ml.) was poured into 15 ml. of acetone which was pre-cooled to -70°C. The cells were harvested and extracted as described below. To the remaining 5 ml. culture 30 μg/ml. chloramphenicol was added and after further 20 min aeration at 27°C the cells were treated with acetone and harvested as described.

(g) Extensive labelling of ϕX RF II

H514 cells were treated with mitomycin and infected with 4 plaque-forming-units ϕX per bacterium at a concentration of 1 x 10^9 cells/ml. in presence of 0.1 μg/ml. [^{3}H] thymidine, specific activity 15 Ci/m-mole. After 9 min aeration at 37°C the culture was poured into -70°C pre-cooled acetone and the cells were harvested by centrifugation.

(h) Extraction of DNA

The acetone-treated bacteria were pelleted, washed with 0.1 M NaCl, 0.05 M Na-tetraborate, 0.005 M-EDTA pH 8.1 and suspended at 0°C in lysis buffer (0.01 M-Tris-HCl pH 8.1, 0.005 M-EDTA, 200 μg/ml. lysozyme and 20 μg/ml. pancreatic ribonuclease) at a concentration of 1x10^{10} cells per ml. buffer solution. After 10 min incubation at 0°C the lysate was brought to 0.5 M-NaCl and lysis was completed by addition of 0.4% sarcosyl. The lysate was deproteinized by digestion for 4 h at 37°C with 200 μg/ml. of pronase.

(i) E. coli exonuclease III digestion of ϕX RF II

Purified E. coli exonuclease III preparations (Schröder, 1970)

containing 100 units/ml. (Richardson, Lehman & Kornberg, 1964) were
used. Assay mixtures were 0.066 M-Tris-HCl pH 8.1, 0.0066 M-MgCl$_2$
and 0.005 M mercaptoethanol and contained 66 units exonuclease III
per ml. and about 0.1 µg/ml. of ϕX RF II. The assays were incubated
for 1 h at 37°C. The reaction was stopped by addition of 0.01
M-Na$_4$EDTA.

(j) <u>Filling of gaps in ϕX RF II molecules by E. coli
    DNA polymerase II</u>

Purified DNA polymerase II was a kind gift of Dr. H.P. Vosberg.
α^{32}P-dTTP was prepared by H.F. Lauppe and had a specific activity
of 10 Ci/m-mole. Reaction mixtures (0.3 ml.) were 0.066 M-Tris-HCl
pH 8.0, 0.008 M-MgCl$_2$, 3 x 10^{-5} M each of dATP, dCTP, dGTP and α
^{32}P-dTTP and contained about 0.05 µg of ^{3}H-labeled ϕX RF II and 0.6
units (Reed et al., 1972) DNA polymerase II. After 30 min at 40°C
the reaction was stopped by addition of 0.01 M-Na$_4$EDTA.

(k) <u>Centrifugation techniques</u>

Neutral sucrose gradients (7 - 20% w/v) contained 0.01 M-Tris-
HCl buffer pH 7.6; 0.3 M-NaCl and 0.005 M-EDTA and were centrifuged
15 h at 24,500 rev./min and 10°C in a Spinco SW 27 rotor. 5- 20%
alkaline (pH 12.6) sucrose gradients were 0.25 M-NaOH, 0.3 M-NaCl and
0.005 M-EDTA and were run either in a Spinco SW 41 rotor for 15 h at
33,000 rev./min and 8°C or in a SW27 rotor for 15 h at 26,000 rev./
min and 20°C.

Viral strand and complementary strand components of ϕX RF II and
of replicating intermediates were separated by banding in alkaline
(pH 12.6) CsCl equilibrium density gradients (Vinograd et al., 1963).
The average density of the gradients was adjusted to 1.755 g/cm^3.
In neutral equilibrium density gradients serving for the separation
of single-stranded from double-stranded DNA, the CsCl density was
adjusted to 1.71 g/cm^3. The CsCl gradients were spun for 40 h at
40,000 rev./min and 20°C in a Spinco type 50 angle head rotor. Radioactivity in the gradient fractions was measured in a Packard Tri-carb
scintillation spectrophotometer. All samples were counted to 10%
standard deviation by appropriate counting times.

(l) <u>Materials</u>

Chloramphenicol ("Paraxin") was a product of Boehringer, Mann-
heim. Mitomycin C was purchased from Kyowa Hakko Kogyo Co., Tokyo.
Mitomycin C was purchased from Kyowa Hakko Kogyo Co., Tokyo. Lysozyme, pronase and ribonuclease, 3 times crystallized, free of DNase,
were bought from Serva-Entwicklungslabor, Heidelberg. Before use,
the pronase was autodigested 3 h at 37°C, RNase stock solutions were
heated to 80°C for 10 min. [^{3}H] thymidine (tp Ci/m-mole) was bought
from New England Nuclear Corp., Boston, Mass./USA. α ^{32}P-dTTP (spec.

activity 20 Ci/m-mole) was prepared by H.F. Lauppe. E. coli DNA
polymerase II was kindly provided by Dr. H.P. Vosberg.

3.   RESULTS

(a) Sedimentation of pulse-labeled phage φX DNA
    at neutral pH

    Phage φX DNA was pulse-labeled during the period of RF repli-
cation at 27°C and extracted as described in section (e) and (h) of
Methods. The sedimentation pattern in a neutral sucrose gradient
of the extract is plotted in Figure 1. Pulse-labeled material was

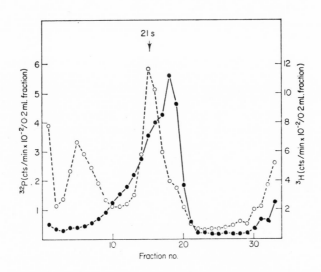

FIG. 1. Sedimentation of pulse-labeled φX DNA in a neutral 7 to 20%
sucrose gradient. H514 cells were infected with ^{32}P-labeled phage
φX, pulse-labeled and extracted as described in Materials and Methods.
The extract was sedimented through 6 parallel 7 to 20% neutral sucrose gradients for 15 h at 24,500 rev./min, at 10°C in an SW27 Spinco
rotor. Distribution of radioactivity was measured in 0.2 ml. portions
of the combined gradient fractions. Sedimentation is from right to
left. -- o -- o --, [32]P (cts/min); — ● — ● —, [3]H (cts/min).

observed sedimenting as a peak of 17 s with considerable tailing
up to about 28 s, as referred to 21 s as the established sedimentation coefficient of the super-coiled replicative form I (RFI) of
ϕX DNA (Burton & Sinsheimer, 1965). Parental ^{32}P-labeled phage ϕX
DNA sedimented as 21 s peak with a 16 to 17 s shoulder corresponding
to the sedimentation rates of RFI and RFII. In addition a 35 s peak
of undefined parental material was observed. The latter, however,
seemed not to be associated with the pulse-label and was not further
investigated in this study. The incorporation under the given conditions of pulse-label into E. coli DNA has been controlled and was
found to account for about 0,5% of the total [^{3}H] thymidine label
only. 16 s to 28 s gradient fractions (Fig. 1) associated with
pulse label were supposed to contain various ϕX RF replicative intermediates and were submitted separately to further structural analysis.

(b) Velocity sedimentation of alkali-denatured phage ϕX RF
    replicative intermediates at pH 12.6

Upon alkali denaturation the strands of RFII and of replicative
intermediates are separated. Resulting single-stranded ϕX DNA rings
sediment at 16 s, linear strands of ϕX unit length at 14 s at pH 12.6
(Sinsheimer, 1968). The supercoiled RFI does not undergo strand
separation by denaturation and sediments at about 50 s at pH 12.6.
Figure 2 shows a set of alkaline sucrose gradients of different
fractions recovered from the neutral sucrose gradient shown in
Figure 1. Referring to the indicated position in the plots of circular ϕX DNA reference revealed that replicative intermediates sedimenting faster than RFII, beside a spectrum of shorter material,
apparently contain strands longer than ϕX unit size. The faster the
corresponding native structures sedimented at neutral pH, the longer
single-strand components were observed in alkaline gradients. Pulse-
labeled material from the 26 to 28 s region of the neutral sucrose
gradient proved to contain strands sedimenting at 18 s, which,
according to the results of Studier (1965) is approximately the
sedimentation coefficient of linear single-stranded DNA of double
ϕX unit length.

(c) CsCl equilibrium density centrifugation of denatured
    phage ϕX RF replicative intermediates

In order to investigate the distribution of incorporated [^{3}H]
thymidine pulse label between positive and negative strand components of replicative intermediates, various fractions of the neutral
sucrose gradient of Figure 1 were banded in alkaline CsCl (pH 12.6)
density gradients. At pH 12.5 the established density of the positive strand is 1.765 g/cm^3, that of the negative strand 1.756 g/cm^3
(Burton & Sinsheimer, 1965; Siegel & Hayashi, 1967; Sinsheimer,
1959). The density difference between the two strands is supposed
to originate from the higher thymine and guanine contents of the
positive strand (Vinograd et al., 1963). The denatured super-coiled

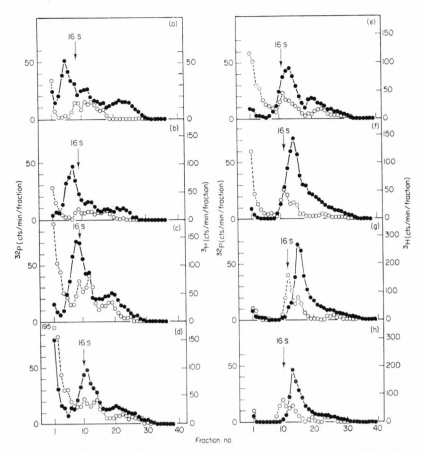

FIG. 2. Alkaline velocity sedimentation analysis of denatured ⌀X RF replicative intermediates. (a) to (h) show the sedimentation pattern of denatured replicative intermediates contained in the fractions 10, 12, 14, 15, 16, 17, 18 and 19 of the neutral sucrose gradient of Fig. 1. The position in the gradients of circular single-strandedphage ⌀X DNA (arrows 16 s) was determined by addition to some of the samples of ^{14}C-labeled ⌀X DNA, extracted from phage particles (not plotted). Sedimentation is from right to left. -- o -- o --, ^{32}P(cts/min); — • — • —, ^{3}H (cts/min).

RFI bands at 1.774 g/cm^3 (Knippers, Komano & Sinsheimer, 1968). Figure 3 shows that the ^{3}H label is associated with positive and negative ⌀X DNA strands in different ratios; fast sedimenting replicative intermediates (23 to 27 s) evidently contained more pulse label in their positive strand component than in the negative strand (Fig. 3(a) and (b)) whereas in structures sedimenting with 17 to 20 s at pH 7 approximately equal amounts of pulse label were found in each of the strands (Fig. 3(e) and (f)).

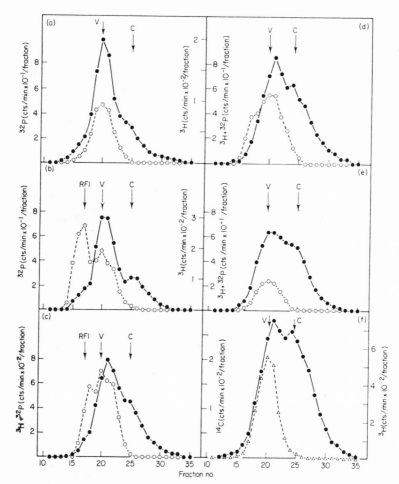

FIG. 3. Alkaline CsCl equilibrium density centrifugation of RF replicative intermediates. Plots (a) to (f) show the alkaline CsCl density pattern of the fractions 10, 14, 15, 16, 18 and 19 of the neutral sucrose gradient of Fig. 1. ^{14}C-labeled phage ϕX DNA was added to the gradients as a reference to mark the position of the positive type DNA. In plot (f) the distribution of the ^{14}C-labeled phage ϕX DNA reference is shown and the parental ^{32}P distribution is omitted. Density increases from right to left. -- o o --, ^{32}P (cts/min); — • — • —, ^{3}H(cts/min); -- ▽ -- ▽ --, ^{14}C (cts/min); V, positive strands; C, negative strands of phage ϕX DNA.

Because of the lower content of thymine of the negative ϕX DNA strand (Sinsheimer, 1959), the amount of newly synthesized negative strand is about 30% higher than represented by the incorporation of [^{3}H] thymine. In the gradients of Figure 3 this correction has not been made. Taking into account this correction the relative amounts

of positive and negative strands synthesized during the pulse in the
whole cells can be estimated from the data of Figure 3. Integration
of ^{3}H label in positive and negative regions in Figure 3 (a) to (f)
and in the alkaline density profiles of the fractions 11, 12 and 13
of the neutral sucrose gradient of Figure 1 which are not shown demonstrates that the pulse label has been incorporated symmetrically
into both strands. Only a negligible amount of pulse label is found
in RFI (Fig. 3 (b) and (c)) which, according to the results of Dressler & Wolfson (1970), can be supposed to be associated with negative
strands. These results suggest that essentially all the DNA synthesis occurring during the pulse is indeed RF replication.†

Replicative intermediates sedimenting at 21 to 23 s (Fig. 3 (c)
and (d)) showed a less unequivocal distribution of pulse label: a
significant part of the newly synthesized DNA banded close to, but
not exactly at the density of the positive strands. One possible
explanation for this finding would be the following: as it has been
pointed out above, the positive strand is believed to have a higher
density than the negative strand because of its higher thymine and
guanine contents. Growing viral DNA strands of different lengths
could slightly differ in their density from complete strands if the
"heavy" bases were not distributed at random along the polynucleotide chain.

(d) Nature of phage ϕX DNA strands longer, and of strands
     shorter than phage ϕX unit length

As described in section (b) above, replicative intermediates of
ϕX RF proved to contain DNA strands longer and shorter than ϕX unit
length (Fig. 2). In order to decide between the two proposed rolling
circle models for the ϕX RF replication the nature of the elongated
strands and of the short pieces had to be studied in detail. For
this purpose fractions containing either pulse-labeled strands of
approximately double ϕX unit length (18 s) and shorter strands (8
to 12 s) were separately recovered from alkaline sucrose gradients
of replicative intermediates and banded in CsCl equilibrium density
gradients at pH 12.6. The results of this centrifugation are shown
in diagram in Figure 4. As judged from their density the elongated

---

†A low level of early single-strand production accounting for 0.5
to 1% of the total phage-specific DNA synthesis was observed upon phage M13 infection of non-mitomycin-treated cells (Forsheit, Ray &
Lica, 1971). A similar synthesis (if any) would not interferw with
this interpretation of the above results. Thus pulse-labeled intermediates analysed in Figures 2 and 3 can be taken to be involved in
RF replication exclusively.

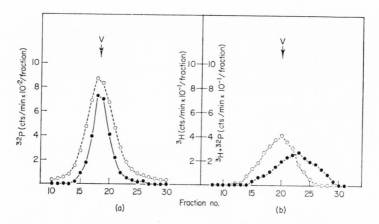

FIG. 4. <u>Banding in alkaline CsCl equilibrium density gradients of</u>
<u>single-strand components of replicative intermediates.</u> (a) DNA
chains of about double phage ϕX unit length (18 s); (b) shorter
pieces (8 to 12 s) were recovered from an alkaline sucrose gradient
similar to that shown in Fig. 2(a) and centrifuged to equilibrium in
alkaline CsCl density gradients. To the sample of gradient (a) ^{32}P-
labeled phage ϕX DNA was added as a reference because the sample contained only negligible amounts of original viral ^{32}P label. Radioactivity in the gradient fractions was counted to a standard deviation of 10%. Density increases from right to left. -- o -- o --,
^{32}P(cts/min); — • — • —, ^{3}H(cts/min); V, positive strands.

DNA strands evidently are of the positive type (Fig. 4(a)).

As it was expected no sharp banding could be achieved with
small DNA pieces. They formed a rather broad band, roughly symmetrical to the position of the density of negative strands (Fig.
4(b)). We would conclude from the density distribution of the short

DNA components of replicative intermediates that they are of the negative strand type of ϕX DNA. The low amount of this material left after the preceding steps of isolation did not permit an additional confirmation of this fact by hybridization experiments.

(e) CsCl buoyand density centrifugation of native RF replicative intermediates

The asymmetric distribution between positive and negative strand components of [^{3}H] thymidine pulse label in replicative intermediates at different stages suggested that the two strands are synthesized asynchronously during the replication process. This would mean that partially single-stranded intermediates were involved in RF replication, as had been originally proposed by Gilbert & Dressler (1968) in their rolling circle model. We have investigated replicative intermediates with regard to this point by banding them in neutral CsCl density gradients. Single-stranded regions could be expected to cause a density shift of the molecules from the density of double-stranded DNA versus the density of single strands. Figure 5 shows the density profiles of pulse-labeled intermediates of increasing sedimentation rates recovered from a neutral sucrose gradient similar to that presented in Figure 1. Starting from complete RF II (Fig. 5(f)) a continuously increasing density shift versus single-strand density of pulse-labeled DNA could be observed, corresponding to increasing sedimentation coefficients of the replicating structures (Fig. 5(b) to (e)).

Most of the parental ^{32}P-label present in the gradients of Figure 5 apparently is part of RFI or of non-replicating RFII, overlapping the replicating intermediates in the original neutral sucrose gradient (Fig. 1). Part, however, of the parental DNA appeared to be shifted together with the replicating structures (Fig. 5(b), which could be an indication that parental viral DNA might be a structural component of these molecules. The maximum density shift observed (Fig. 5(b)) can be calculated to account for about 25% of the difference between ϕX double-strand and single-strand density. Assuming that the single-strand regions of replicating RF correlate to peeled off single-stranded positive tails, this would mean that the initiation on the tails of the new negative strand occurs late in the replication process.

The results presented in sections (a) - (d) appear to reflect the following processes occurring in the course of RF replication; presumably early in replication the continuous elongation of the positive strand of RF II around the negative strand template results in displacement of a single-stranded tail of increasing length. It is likely that this tail is responsible for a quick increase of the sedimentation rate of the replicating molecule (Figs. 1 and 2). The most rapidly sedimenting structures (27 to 28 s) in our experiments proved to contain (1) the longest viral

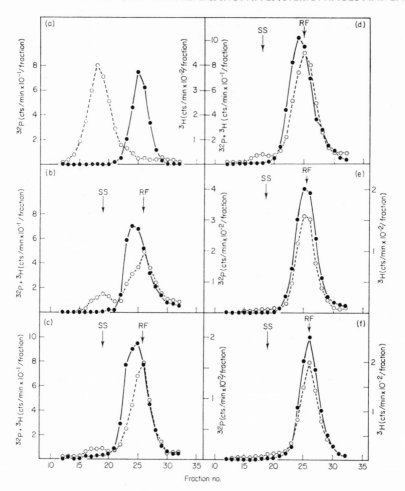

FIG. 5. Distribution in neutral CsCl equilibrium density gradients of RF replicative intermediates. Samples of replicative intermediates, sedimenting at about 27 s (b), 25 s (c), 22 s (d), 20 s (e) and 17 s (f), were centrifuged to equilibrium in neutral CsCl density gradients. (a) shows a reference gradient containing single-stranded ^{32}P-labeled phage ϕX DNA extracted from phage particles and ^{3}H-labeled RFI. Density increases from right to left. -- o -- o --, ^{32}P(cts/min); — • — • —, ^{3}H(cts/min); SS, single strand.

strand components (Fig. 2 (a)), (2) the most extended single-stranded region (Fig. 5), and (3) a rather small ratio of pulsed label incorporated in their negative strand component (Fig. 3 (a)).

Proceeding backwards from 27 to 28 s in terms of the sedimentation rate, the ratio of newly synthesized negative strand component

in the intermediates increases as compared to the positive strand.
This finding could be explained reasonably by continuous conversion
of the single-stranded viral tail to double-stranded DNA. This
would imply that the "tailing" RF molecules in the sedimentation
pattern of a neutral sucrose gradient are superimposed by a population of replicative intermediates involved in the synthesis of
the new negative strands. Complete RF dimers should sediment at
about 21 s according to Studier (1965). Structures sedimenting at
21 s, however, evidently do not contain detectable amounts of positive strands of twice ϕX unit length. Hence one could conclude
that the replicative intermediate is nicked in the positive strand
before the new negative strand had been completed, presumably resulting in the formation of (1) a complete RF II molecule containing
the circular old negative and the linear new positive strand and
(2) a linear partially double-stranded tail containing the old
positive and a fragment of the new negative strand.

The fate of these two types of daughter molecules during further RF replication was investigated in another series of experiments which are described in the following sections.

(f) $[^3H]$ Thymidine labelling of the negative strands of
parental ϕX RF molecules

The technique of specific labelling of the parental negative
ϕX DNA strands made use of the fact that the process of ϕX DNA
replication stops after the formation of the parental RF when the
host cells are infected in the presence of 120 μg/ml. chloramphenicol (Tessman, 1966; Sinsheimer, Hutchinson & Lindqvist, 1967).
Normal semiconservative replication of the RF starts with some delay
after the removal of the drug. The lag period has been shown above
to be about 8 - 10 min if the cells are aerated at 27°C. H514 cells
were treated with mitomycin C, infected with ^{32}P-labeled ϕX phage
and labeled with $[^3H]$ thymidine in the presence of chloramphenicol
as described in section (f) of Methods. The ^3H label was diluted
out in two steps: (1) At 8 min after addition of $[^3H]$ thymidine,
i.e when chloramphenicol was still present, a 4000-fold excess of
cold thymidine was added to the culture. (2) Upon removal of the
chloramphenicol by centrifugation the pelleted cells were resuspended in fresh medium containing another 4000-fold excess of cold
thymidine as referred to the original concentration of $[^3H]$ thymidine.

ϕX DNA was extracted as described from equal portions of the
cells harvested (a) at 10 min after removal of chloramphenicol, i.e.
at the beginning of RF replication and (b) 30 min after the removal of the drug, i.e. after several rounds of replication could
be assumed to have occurred. In order to prevent the synthesis of
single-stranded ϕX DNA during the 10 min to 30 min period, 30 μg/
ml. chloramphenicol was added to the culture (Sinsheimer, Starman,
Nagler & Guthrie, 1962). The sedimentation pattern in 7 - 20%

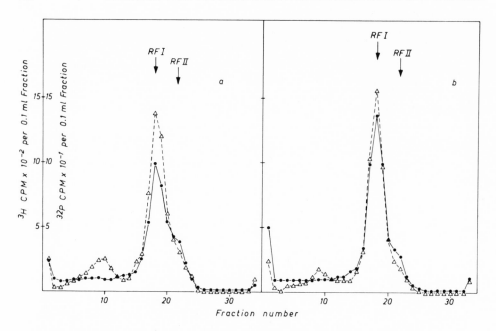

FIG. 6. Sedimentation in neutral sucrose gradients of øX DNA extracted early and late during øX RF replication. E. coli H512 cells were treated with mitomycin, infected with ^{32}P øX and labeled with [^{3}H] thymidine at 120 μg/ml. chloramphenicol as described in Methods. DNA was extracted from the bacteria (a) 10 min and (b) 30 min after removing the chloramphenicol, i.e. after a further 20 min period of RF replication at 27°C in the presence of cold thymidine. Each of the two DNA extracts were sedimented through 3 parallel 7 - 20% neutral sucrose gradients for 15 hr at 24,500 rev./min and 10°C in a SW27 Spinco rotor. The distribution of ^{32}P and ^{3}H radioactivity was measured in 0.1 ml. aliquots of the combined fractions of the corresponding gradients. Sedimentation is from right to left. Δ — Δ , ^{32}P (cts/min); ● — ●, ^{3}H (cts/min).

neutral sucrose gradients of ^{32}P-and ^{3}H-labeled øX DNA in the two extracts are plotted in Figure 6. At both times of extraction the majority of the parental viral ^{32}P and of the ^{3}H label was found to be associated with RF I which sediments with 21 s at neutral pH (Sinsheimer, 1968). RF II containing both labels appeared as a 16 -

17 s shoulder on the RF I peak, and accounted for about 10 - 15% of the total ^{3}H and ^{32}P label associated with RF II early in RF replication (Fig. 6 (a)) apparently had been transferred into RFI after 20 min of replication at 27°C (Fig. 6 (b)).

(g) The pathway of the negative strand component of parental RF and of early synthesized positive strands during RF replication

Three forms among the population of the RF replicative intermediates appeared most suitable to follow the fate of the replicating molecules and were isolated from the neutral sucrose gradients shown in Fig. 6: (1) Rolling circles sedimenting at 25 - 27 s which were shown in sections (b) - (e) to have positive strand tails up to a full ϕX DNA length (Fig. 2 (a) and (b)) and to contain short fragments of newly synthesized nagative strands (Figs. 2 (a), 4 and 5 (b)). (2) RF I and (3) RF II. For this purpose samples of fractions (14 + 15), 18, and (21 + 22) both of the gradients of Fig. 6 were brought to pH 12.6 and submitted to equilibrium density centrifugation in alkaline CsCl gradients. In order to achieve separation of the two strands of RF I, the relevant samples were heated to 95°C for 15 min before centrifugation. This treatment generates at least one single strand break per RF I molecule. To all gradients ^{14}C-labeled single-stranded ϕX DNA, isolated from phage particles, was added as a density marker.

Fig. 7, panels (a), (c) and (e), show the density distribution of ^{3}H-labeled ϕX DNA components of fast sedimenting (26 s) replicative intermediates, RF I and RF II, respectively, at the initial period of RF replication. Panels (b), (d) and (e) represent the corresponding density patterns obtained late in RF replication. For reasons to be discussed below the distribution in the gradients of the parental ^{32}P label is not shown in Fig. 7.

As expected the majority of [^{3}H] thymidine label was found to be associated with the negative strand components of all investigated forms of ϕX DNA early in RF replication (Fig. 7 (a), (c) and (e)). One surprising result, however, of these experiments was that some [^{3}H] thymidine was also present in positive strands, mainly of fast sedimenting rolling circles and of RF II, early during RF replication (Fig. 2(a) and (e)). This finding could be reasonably explained in two ways: Either RF replication was not completely blocked by 120 µg/ml. chloramphenicol, or part of the [^{3}H] thymidine was not diluted out even by the given high concentrations of cold thymidine. The former argument was ruled out by control experiments: Infected cells were treated with acetone and extracted (a) immediately after the removal of ^{3}H-thymidine and chloramphenicol, (b) after 5 min, (c) after 7 min and (d) after 9 min of further aeration at 27°C. ϕX DNA was recovered from neutral sucrose gradients of the cell extracts, denatured at pH 12.6 and

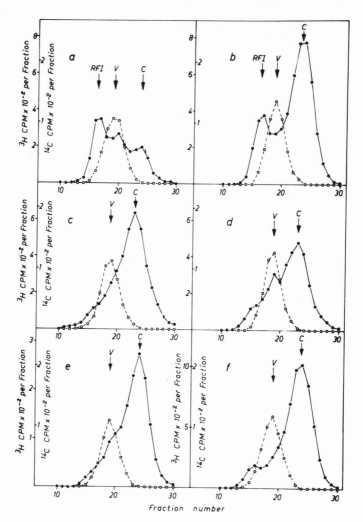

FIG. 7. <u>Alkaline CsCl equilibrium density centrifugation of different forms of ∅X RF replication intermediates.</u> ∅X DNA contained in different fractions of the neutral sucrose gradients of Figure 6(a) and (b) was banded in alkaline CsCl density gradients for 40 hr at 40,000 rev./min and 20°C in a type 50 angle head Spinco rotor. Panels (a), (c) and (e) show the alkaline density patterns of the fractions (14 + 15), (18) and (21 + 22) of Figure 6(a); panels (b), (d) and (f) represent the density patterns of the corresponding fractions of Figure 6(b). In order to separate the strands of RFI, fractions (18) both of Figure 6(a) and (b) were heated to 95°C for 10 min before centrifugation. ^{14}C labeled ∅X DNA extracted from ∅X phage particles was added to all gradients as a density marker. Density is increasing from right to left. • — • —, ^{3}H (cts/min); o -- o, ^{14}C (cts/min). V, positive strands; C, negative strands.

banded in alkaline CsCl density gradients. In each case no ^{3}H-label
was detected in positive strands. In additional controls no increase of the total ^{3}H label incorporated into ϕX DNA could be
observed between 10 min and 30 min after the removal of chloramphenicol.

From these results it could be concluded that the incorporation
of [^{3}H] thymidine into positive DNA strands of replicative intermediates occurred soon after the 8-10 min lag period described
above, using up a residual precursor pool which had persisted in
the infected cells. From Fig. 6 the total ^{3}H-label incorporated
into ϕX RF and the total parental ^{32}P-label present in the cell
extracts at both times of extraction can be estimated to account
for about 1.5 x 10[5] ^{3}H-CPM and 1.5 x 10[4] ^{32}P-CPM, respectively.
Fig. 7 (a) shows that early in RF replication about 900 ^{3}H-CPM,
i.e. 0.6% of the total ^{3}H-label, was associated with negative
strands of rolling circles sedimenting at 25 - 27 s (fractions 14
+ 15 of Fig. 6 (a)). Since the amount of ^{3}H-label in newly synthesized negative strands being part of these structures can be
neglected as compared to their extensively labeled complete "old"
negative strand components, the amount of 900 ^{3}H-CPM can be taken
to represent a ratio of 0.6% of replicative intermediates as referred to the total population of replicating and non-replicating
RF DNA. If these rolling circles were replicating parental RF they
should consequently contain also 0.6% of the total parental [32]P
label, i.e. about 90 ^{32}P-CPM. That low amount of ^{32}P radioactivity,
however, could not be detected in the gradient of Fig. 7 (a). It
was not possible therefore to observe the parental positive DNA
strands associated early during RF replication with fast sedimenting rolling circles. The participation in replication of the
parental viral DNA became evident, however, from the structural
analysis of the tails of early rolling circle intermediates which
will be be discussed in sections (h) and (i) of the Results.

The limited incorporation of [^{3}H] thymidine of high specific
activity into positive DNA strands of early replicative intermediates
appears to provide an opportunity for investigating the fate of the
positive strand components of RF in the subsequent replication
rounds.

After 20 min of RF replication at 27°C, the ^{3}H-labeled positive
strands associated early with fast sedimenting rolling circles and
with RF II (Fig. 7 (a) and (e)) appeared to have been chased out of
these structures and proved to be now constituent of progeny RF I
(Fig. 7 (d). On the other hand the amount of fast sedimenting (25 -
27 s) rolling circles containing the parental negative strands had
increased about 5-fold as compared to Figure 7 (a) as judged from
the amount of ^{3}H label present in negative strands in Fig. 7 (b).
In the case that these replicating molecules contained ^{32}P-labeled
parental positive DNA strands, this ^{32}P-label should account for

about 450 ^{32}P-CPM which should have been detected in the gradient of Figure 7 (b), in contrast to the situation illustrated in Figure 7 (a). No significant amounts of ^{32}P label were found, however, to be associated with positive strands of these replicative inter- mediates. From the fact that none of the progeny RF I containing early synthesized positive strands was involved late in replication, we conclude that late entering of progeny RF I molecules into re- plication rounds is a rare event. Taking into account these con- siderations, those of the parental negative strands observed in late rolling circles could be assumed to have passed several replication rounds during the 20 min chase period.

The above results would suggest the following pathway of early synthesized positive strands and presumably also of the parental positive strands: Starting as components of the circular part of rolling circles, flowing into the tails in the following replication round according to the proposed rolling circle mechanism, they would finally end up in the RF I pool.

In the subsequent sections experiments concerning the fate of the "tails" of rolling circles will be described.

(h) Evidence for the existence among RF II type ϕ DNA of linear duplexes considered to be "tails" of rolling circles

Treatment of ϕX RF II molecules with E. coli exonuclease III results in the 3' → 5' digestion of the linear components of the duplexes, whereas the circular strands remain intact (Burton & Sinsheimer, 1963). If the 3'hydroxyl end of a linear DNA chain is not hydrogen-bonded to a complementary nucleotide, it is not at- tacked by the enzyme.

Whereas ϕX RF II sediments at 16 - 17 s, its circular strand components remaining after extensive exonuclease III digestion sedi- ment at 24 s in neutral sucrose gradients. Complete linear ϕX single- stranded DNA also sediments at 24 s at neutral pH. Upon alkali de- naturation at pH 12.6 linear single-stranded ϕX DNA sediments at 14 s, ϕX DNA rings at 16 s in alkaline sucrose gradients (Sinsheimer, 1968). Linear "tails" derived from rolling circles could be ex- pected to sediment at neutral pH at a rate similar to that of cir- cular RF II (Schröder, 1972).

In order to check for linear "tails" among the population of ϕX RF II, H514 cells were infected with ^{32}P-labeled ϕX phage and pulse-labeled with [^{3}H] thymidine early during the period of RF re- plication as described in Methods. Pulse-labeled RF II was isolated from the 16 - 17 s-region of neutral sucrose gradients of the ex- tracted DNA.

One aliquot of the RF II sample was re-sedimented through a neutral sucrose gradient, another aliquot, after denaturation by

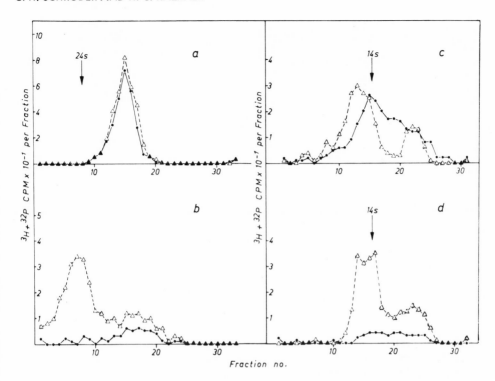

FIG. 8.   Digestion by exonuclease III of ^{3}H pulse-labeled RF II type
DNA containing ^{32}P parental viral strands. Neutral sedimentation
patterns of purified RF II in 5 - 20% neutral sucrose gradients before and after extensive digestion with E. coli exonuclease III are
diagrammed in (a) and (b) respectively. Plots (c) and (d) show the
corresponsing alkaline sedimentation profiles of RF II samples. The
arrows in the diagrams indicate the position in the gradients of
added linear ^{14}C-labeled phage ϕX DNA. Sedimentation is from right
to left. Δ -- Δ, [32]P (cts/min); $\bullet$ -- $\bullet$, ^{3}H pulse label (cts/min).

alkali at pH 12.6, was subjected to alkaline sedimentation analyses.
The radioactivity patterns of these gradients are plotted in Figure
8 (a) and (c) respectively. The alkaline sedimentation profile in
Figure 8 (c) displayed the heterogeneity of the population of the
RF II type DNA which, at neutral pH, had sedimented as a rather
homogeneous peak (Fig. 8 (a)). The ^{32}P-labeled parental DNA components of the RF duplexes proved to exist in their circular and
their linear form in an approximately equimolar ratio. The short
pieces of parental DNA present in the gradient of Figure 8 (c) had
been earlier shown to stem from RF II type duplexes having more than
one single strand break (Schröder, 1972). The ^{3}H-pulse-labeled DNA
was exclusively linear and showed a wide distribution of lengths.
The panels of Figure 8 (b) and (d) represent the neutral and alkaline sedimentation patterns of samples of the same RF II fraction

which had been extensively digested with exonuclease III prior to centrifugation. As expected the ^{32}P-labeled parental DNA components now sedimented at 24 s at pH 7 (Fig. 8 (b)), indicating that they had been liberated from the RF II duplexes by the hydrolysis of the pulse-labeled linear negative strands. The sedimentation through an alkaline sucrose gradient of the digest showed that, beside the circles and the shorter pieces, the linear positive parental DNA strands had not been hydrolysed by exonuclease III. This finding suggested that the latter fraction of parental positive DNA strands were originally part of linear duplexes, containing incomplete negative strands which did not match the 3'hydroxyl ends of the positive strands. This fact is of some importance with respect to the site of initiation on the positive strand template of the negative strand. As judged by their structure and considering the pathway of the positive strand component of rolling circles described in section (g) of the Results we would suggest that these duplexes represented the linear form of the tails of rolling circles. Furthermore the analysis of these intermediates displayed the participation of the parental DNA early in the replication process.

(i) <u>Circular "gapped" RF II can be evisaged to arise from linear tails</u>

Schekman et al. (1971) had detected that a considerable fraction of RF II molecules isolated from ϕX-infected bacteria have a gap in their linear DNA component. In the following experiments we tried to determine whether the gaps in RF II were (a) in the positive strand, (b) in the negative strand or (c) in both strands.

For this purpose RF II was extensively labeled during RF replication with ^{3}H thymidine in ϕX-infected cells as described in Methods. After purification by repeated sedimentation through neutral sucrose gradients RF II samples were incubated with <u>E. coli</u> DNA polymerase II together with deoxyribonucleoside triphosphates including α^{32}P-dTTP as described. After the reaction was stopped by EDTA the incubation mixtures were brought to pH 12.6 by NaOH and spun through 5 - 20% alkaline sucrose gradients. The radioactivity pattern of one of these gradients is presented in Figure 9 (a). Beside a discrete 9 s-peak and short pieces which were neglected in this study ^{32}P-labeled material was found cosedimenting with the ^{3}H-labeled linear component of RF II at 14 s.

The 14 s-DNA was recovered from the gradient and banded in an alkaline CsCl equilibrium density gradient in order to identify the nature of the DNA strands having been completed by the action of DNA polymerase II. The diagram of the density distribution shown in Figure 9 (b) demonstrates that ^{32}P dTMP had been incorporated into negative ϕX DNA strands almost exclusively. These results clearly indicated that the gaps had existed in the negative strands of RF II type molecules.

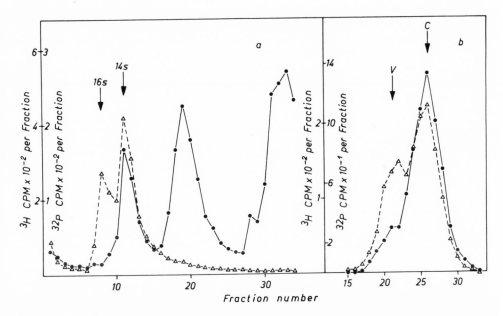

FIG. 9. **Filling of gaps in the linear component of RF II type DNA.**
∅X DNA was extensively labeled with ³H thymidine during RF replication.
RF II type DNA was isolated from neutral sucrose gradients of DNA
extracts and treated with E. coli DNA polymerase as described in
Methods, section (j). (a): After terminating the reaction with 0.1
M-EDTA the assay mixture was brought to pH 12.6 and spun through a
5 - 20% alkaline sucrose gradient together with circular ¹⁴C-labeled
phage ∅X DNA as a marker (arrow at 16 s). (b): Linear ∅X DNA strands
sedimenting at 14 s were recovered from the gradient shown in plot
(a) and subjected to alkaline CsCl equilibrium density centrifugation.
Added ¹⁴C-labeled phage ∅X DNA served as a density marker (arrow at
"v"). ∇ -- ∇ , ³H (cts/min); •——• ³²P (cts/min).

We could not distinguish, however, by this technique the relative amounts of circular and linear "gapped" RF II. Schekman et
al. (1971) have provided evidence that about 30% of the total RF
II population late in infection were of the circular gapped form.

Considering the presented results we are inclined to assume that
circular gapped RF II molecules might also represent physiological
intermediates of the RF replication process, arising from linear
"tails" of rolling circle structures by circularization of their
complete viral strand component. After filling the gaps, the negative strands were ready to be closed by DNA ligase resulting in
the formation of RF I. These two reaction steps were performed by
Schekman and his coworkers (1971) in vitro serving for the indenti-
fication of circular gapped RF II.

4. DISCUSSION

Progeny RF has been shown to increase linearly with time in ϕX
infected cells during the period of semiconservative replication of
ϕX replicative form DNA (Sinsheimer, 1968). This implies that only
a few RF molecules are replicating. Knippers & Sinsheimer (1968)
suggested that RF replication takes place at a limited number of
special sites of the bacterial membrane and that only RF molecules
containing the parental viral DNA strands would continuously replicate. Recent results of Baas & Jansz (1972a, b), and of Iwaya
& Denhardt (1973a, b), however, provided evidence that the negative
strands of parental RF rather than the parental positive strands
persisted throughout RF multiplication as master templates. This
would mean that only those of the daughter RF molecules containing
the parental negative strands would enter new replication rounds.

The special type of rolling circle mechanism derived from the
results of the first part of this study (sections (a) - (e)) implies
that in the course of the replication process the original positive
strand of an RF molecule is driven by chain elongation into the tail
of the rolling circle, whereas its negative strand remains in the
circular part of the replicative intermediate. This would mean that
one early event in RF replication is the introduction of a nick into
the circular positive strand of the parental RF. Evidence that this
is the case comes from experiments of Francke and Ray (1971).

Support for the asynchronous formation of progeny positive and
negative strands comes from several laboratories: In the case of
the filamentous DNA phage M13, Hohn, Lechner & Marvin (1971) suggested that a partially single-stranded intermediate is involved
in RF replication. Salstrom & Pratt (1971) confirmed the view that
the switch from M13 RF replication to the production of progeny
single-strand viral DNA is mediated by a late phage protein, asso-
ciating with the growing single-stranded tails of RF and thus

preventing the tails from being converted to double-stranded DNA. A similar mechanism could be envisaged for the replication of ϕX DNA assuming the gene D protein to be the inhibitor of the further synthesis of complementary strands on pre-formed viral strand tails of RF molecules.

Geider, Lechner & Hoffmann-Berling (1972) have density-labeled ϕX RF replicative intermediates with 5-bromodeoxyuridine triphosphate in ether-treated nucleotide triphosphate-permeable cells. In these experiments the single-strand component of newly formed RF completely labeled with bromouracil first was revealed to be of the negative type.

The role of parental viral DNA in RF replication is difficult to interpret from the results of the pulse-labeling experiments of sections (a) - (e). Evidently only a few of the RF molecules containing the parental viral strand participate in replication, as can be derived from most of the gradient profiles. Some circumstantial evidence about the presence of parental positive strands in early replicative intermediates may be deduced from the alkaline sucrose gradients shown in Fig. 2 (c) to (h). These intermediates contained few ^{32}P labeled parental DNA appearing either as circles or in the linear form. Part of these strands might be considered as components of the linear or the circular form of tails of rolling circles. This suggestion was confirmed in the experiments of sections (h) and (i) of the Results and will be discussed below.

In the second part of this study we have investigated the pathway of the two daughter molecules arising from rolling circles. For this purpose the host bacteria were infected with ^{32}P ϕX phage and the parental negative ϕX DNA strands were labeled with ^{3}H thymidine in the presence of 120 μg/ml. of chloramphenicol. We could not avoid, however, some limited ^{3}H-labeling of early synthesized new positive strands occurring after a lag period of initiation of RF replication. This fact, on the other hand, provided an experimental advantage: It appeared not to be possible to recognize the ^{32}P-parental ϕX DNA as constituent of early fast sedimenting rolling circle intermediates because of the small ratio of molecules being in that state (Fig. 7 (a)).

The latter fact could be explained in the following way: The infectious cycle in the experiments described here was synchronized in so far as it was limited by high concentrations of chloramphenicol to the formation of the parental RF, the majority of which proved to be RF I. The initiation of replication rounds evidently took considerable time as judged from the lag period of incorporation of [^{3}H] thymidine pulse label following after the removal of chloramphenicol from the infected bacteria. Hence it appears likely that the parental RF molecules enter replication rounds sporadically. This could be the reason why only few fast sedimenting rolling

circles containing parental viral DNA strands are found. From the
fact that early in replication the tails of rolling circles contained detectable amounts of parental positive strands one can
assume that the conversion of the partially double-stranded tails
into mature RF is a slow process.

^{3}H-labeled early progeny viral strands could be assumed to be
contained under the given conditions in the circular part of fast
sedimenting rolling circles, and - if the circular parts would
further replicate - to flow into the tails in the subsequent cycle.
They would mimic with that respect the role in the first replication
round of the parental viral strands.

The presented chase experiments (Fig. 7) provided evidence that
after 20 min of RF replication at 27°C the parental negative strands
were still involved in replication whereas the early formed positive strands had become part of RF I during that time. The latter
fact could be explained in two ways: 1) by conversion into RF I
of the circular parts of the rolling circles or 2) after a subsequent replication round of the circular parts after the splitting
of the rolling circles - by conversion of the new tails into RF I.
Considering that progeny RF I evidently did not participate in late
replication to a detectable extent (Fig. 7(b)), the parental negative strands would be diluted out of replication if the circular
parts of the rolling circles were directly converted into RF I.
Since this evidently did not occur we conclude that the circular
parts replicated further and that early synthesized positive strands
had been chased into RF I as part of the tails in a subsequent replication round. Fast sedimenting rolling circle intermediates
isolated late in RF replication did not contain detectable amounts
of parental viral (^{32}P) DNA (Fig. 7(b)). This finding suggested
that also the parental positive strands had been diluted out of the
replication process and that the parental negative strands found in
these replicative intermediates were involved repeatedly in replication rounds.

As a consequence of the delayed formation of the progeny negative strands linear and circular intermediate duplex structures of
the tails could be expected to exist among the population of replicative intermediates. Two types of molecules could be identified
among RF II type ϕX DNA which could be supposed to represent such
intermediates. Both of them proved to contain complete positive
strands, either in the linear or in the circular form, and incomplete negative strands. Early during RF replication the linear
as well as the circular duplex form of the tails proved to contain parental positive strands, (Fig. 8), as this has been shown
above in the pulse labeling experiments (Fig. 2). In contrast
the sucrose gradient shown in Fig. 6(b) indicates that late in
RF replication only small amounts of parental viral DNA were left
associated with RF II type DNA. These findings also could be an

indication that the parental viral strands were involved early in
replication only.

The complete viral strand components of the linear tails could
be expected to circularize before the complementary strands were
completed.†) This would consequently lead to RF II molecules with
gaps in their negative strands. Gapped RF II which has been detected by Schekman and coworkers (1971) has been shown by our experiments to contain incomplete negative øX DNA strands rather
exclusively (Fig. 9). We would emphasize from this result that
"gapped" RF II represents an intermediate form of the tail-type
progeny RF which for unknown reasons appears to be relatively
stable. Geider, Lechner & Hoffmann-Berling (1972) have shown by
in vitro gap filling experiments using [^{3}H] bromouridinetriphos-
phate replacing TTP that the gaps would be about 500 nucleotides
long. Recent results of Lauppe and Schröder in our laboratory indicated that the gaps comprize a special region of the negative
strands (Lauppe & Schröder, manuscript in preparation).

The above results are summarized in a tentative model for the
replication of øX174 replicative form DNA which is shown in Figure
10, starting from a parental RF molecule. From this scheme it
appears likely that the circular part of rolling circles is favoured
to remain involved continuously in replication: They might be considered as "initiated" RF molecules in that they have a nick in
their viral strands (Francke & Ray, 1971 and 1972). The tails
apparently have to pass a longer way before they could enter into
new replication rounds. One major conclusion of the proposed replication model is that the parental negative strands rather than
the infecting viral DNA govern øX RF replication. This suggestion
is consistent with the conclusion from genetic experiments of Baas
& Jansz (1972 a, b) and is strongly supported by the recent results
of Iwaya and Denhardt (1973 (a) and (b)).

The mechanism of the synthesis of progeny complementary strands
on the viral strand tails of rolling circle intermediates remains
to be clarified. From our results of exonuclease III digestion of

---

†) Circulation could be imagined for example to be accomplished
by DNA ligase if the viral strands were nicked in the course of
the replication process in their double-stranded hair pin region
(Fiers & Sinsheimer, 1965). Presumably by intramolecular base
pairing the ends of the viral strand would be fixed in such a
way that they provide a good substrate for the sealing enzyme.
A similar mechanism has been porposed by Knippers et al. (1969)
for the formation of circular øX progeny DNA and has been supported by the results of Schekman & Ray (1971).

the linear duplex form of the tails it appears that the very initiation site, or one of the initiation sites in the case of discontinuous formation of the negative strands, does not coincide
with the 3'hydroxyl end of linear positive strands. This fact
would provide an important aspect with respect to the hypothesis
that the parental negative strand synthesized on the circular infecting parental viral DNA might be formed discontinuously (Hess,
Dürwald & Hoffmann-Berling, 1972).

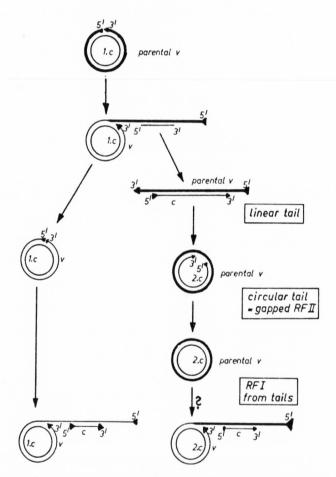

FIG. 10. A proposed model for the in vivo replication of ϕX
replicative DNA.
    A hypothetic scheme of RF replication is shown starting from
a parental RF molecule. Details are given in the text of the
discussion.
c = negative (complementary) strand; v = positive (viral) strand.

We gratefully acknowledge the skillful technical assistance
of Miss Renate Wolf and Miss Edith Erben. Miss Maria Kremser
kindly provided purified øX phage.

This work is being published in two parts elsewhere: The first
part appeared in J.Mol.Biol. (1972) 71, 351. The second part has
been submitted in 1972 to the J.Mol.Biol.

REFERENCES

1. Baas, P.D. & Jansz, H.S. (1972a) J.Mol.Biol. 63: 557.
2. Baas, P.D. & Jansz, H.S. (1972b) J.Mol.Biol. 63: 569.
3. Burton, A. & Sinsheimer, R.L. (1963) Science 142: 962.
4. Burton, A. & Sinsheimer, R.L. (1965) J.Mol.Biol. 14: 327.
5. Dressler, D. & Wolfson, J. (1970) Proc.Nat.Acad.Sci., Wash.
    67: 456.
6. Fiers, W. & Sinsheimer, R.L. (1965) J.Mol.Biol. 5: 424.
7. Forsheit, A.B., Ray, D.S. & Lica, L. (1971) J.Mol.Biol. 57: 117.
8. Francke, B. & Ray, D.S. (1971) J.Mol.Biol. 61: 565.
9. Francke, B. & Ray, D.S. (1971) J.Mol.Biol. 61: 565.
10. Geider, K., Lechner, H. & Hoffmann-Berling, H. (1972) J.Mol.
    Biol. 69: 333.
11. Gilbert, W. & Dressler, D. (1968) Cold Spr.Harb.Symp.Quant.
    Biol. 33: 473.
12. Hess, U., Durwald, H. & Hoffmann-Berling, H. (1972) J.Mol.Biol.
    In press.
13. Hohn, B., Lechner, H. & Marvin, D.A. (1971) J.Mol.Biol. 56: 143.
14. Kaerner, H.C. (1970) J.Mol.Biol. 53: 515.
15. Knippers, R. (1969) Habilitationsschrift. Heidelberg.
16. Knippers, R., Komano, T. & Sinsheimer, R.L. (1968) Proc.Nat.
    Acad.Sci., Wash. 59: 577.
17. Knippers, R., Razin, A., Davis, R. & Sinsheimer, R.L. (1969)
    J.Mol.Biol. 45: 237.
18. Knippers, R. & Sinsheimer, R.L. (1968) J.Mol.Biol. 34: 17.
19. Lindqvist, B.H. & Sinsheimer, R.L. (1967) J.Mol.Biol. 30: 69.
20. Reed, B., Wickner, W., Ginsberg, B., Berkower, I. & Hurwitz,
    J. (1972) J.Biol.Chem. 247: 489.
21. Richardson, C.C., Lehman, I.R. & Kornberg, A. (1964) J.Biol.
    Chem. 239: 251.
22. Salstrom, J.S. & Pratt, D. (1971) J.Mol.Biol. 61: 489.
23. Schekman, R.W., Iwaya, M., Bromstrup, K. & Denhardt, D.T. (1971)
    J.Mol.Biol. 57:177.
24. Schekman, R.W. & Ray, D.S. (1971) Nature 231: 170.
25. Schröder, C.H. (1970) Theses (Diploma). Faculty of Biology,
    University of Heidelberg.
26. Schröder, C.H. (1972) Doctoral Dissertation. Faculty of Biology,
    University of Heidelberg.
27. Siegel, J.E.D. & Hayashi, N. (1967) J.Mol.Biol. 27: 443.

28. Sinsheimer, R.L. (1968) Progr.Nucl.Acid Res. and Mol.Biol.
     8: 115.
29. Sinsheimer, R.L., Hutchinson, C.A. & Lindqvist, B. (1967)
     In The Molecular Biology of Viruses, p 175. New York: Academic
     Press.
30. Sinsheimer, R.L., Starman, B., Nagler, C. & Guthrie, S. (1962)
     J.Mol.Biol. 4: 142.
31. Studier, W.F. (1965) J.Mol.Biol. 11: 373.
32. Tessman, E.S. (1966) J.Mol.Biol. 17: 218.
33. Vinograd, J., Morris, J., Davidson, N. & Dove, W.F. (1963)
     Proc.Nat.Acad.Sci., Wash. 49: 12.

ALTERNATE PATHWAYS OF BACTERIOPHAGE M 13 DNA REPLICATION

Walter L. Staudenbauer, William L. Olsen and
Peter Hans Hofschneider

Max-Planck-Institut für Biochemie

8033 Martinsried bei München

The replication of bacteriophage M 13 provides an excellent system for studying different types of DNA synthesis. The replication of M 13 DNA occurs in three temporally separated steps (Marvin & Hohn, 1969). First the infecting single stranded phage DNA is converted into a double stranded replicative form (RF) by the synthesis of a complementary strand. Early in the infection the replicative forms multiply by symmetric semi-conservative replication to form a pool of replicative form molecules. Later double strand synthesis stops and single stranded viral DNA is produced by an asymmetric replication process in which the viral strand of the replicative form is displaced as a new one is synthesized (Ray, 1969). These processes depend largely on host proteins. Only two phage genes are involved in DNA replication (Pratt, 1969): Gene-2 is required for double strand replication as well as for single strand synthesis possibly by coding for a "nickase" (Lin & Pratt, 1972; Tseng & Marvin, 1972; Fidanian & Ray, 1972). The gene-5 protein is involved only in single stranded DNA synthesis (Salstrom & Pratt, 1972).

Hohn, Lechner & Marvin (1971) have proposed a model for M 13 DNA synthesis according to which the early RF replication and the late single strand synthesis both occur by the same asymmetric mechanism. Early in the infection double strand replication is predominant as progeny single strands are immediately converted into double strands by the same host enzyme(s) that convert the infecting viral single strands into the parental replicative form molecules. Accumulation of gene-5 protein is thought to cause a switch from double strand replication to single strand synthesis by blocking the synthesis of complementary strands.

117

It has been pointed out recently (Gross, 1971) that an obvious similarity exists between M 13 single strand DNA replication and DNA transfer in conjugation: each appears to occur by an asymmetric type of DNA synthesis involving the continuous displacement of one strand from a "rolling circle" type intermediate without concomitant synthesis of a complementary strand. As has been shown previously transfer replication differs from vegetative replication of the chromosome by not requiring the product of the dnaB gene (Marinus & Adelberg, 1970; Vapnek & Rupp, 1971). Thus one might expect that M 13 is capable of replicating in a dnaB host in the absence of a functional dnaB gene product. However infection of the temperature-sensitive E. coli mutant HfrH 165/70 (dnaB) with M 13 is abortive at the restrictive temperature (Primrose, Brown & Dowell, 1968). Since a localization of the block in M 13 replication may contribute information both to the replicative process of single stranded DNA phages and to the biochemical lesion in the dnaB mutant, abortive M 13 infection at the restrictive temperature was studied in more detail.

FORMATION OF PARENTAL RF

To determine if the parental single stranded DNA penetrates and is converted to RF at the restrictive temperature, E. coli HfrH 165/70 (dnaB) cells were infected in the presence of chloramphenicol at 41°C and 34°C with ^{32}P-labeled M 13. Chloramphenicol was added to inhibit the synthesis of phage-specific proteins required for RF replication. It should be noted that chloramphenicol does not prevent the SS → RF conversion (Pratt & Erdahl, 1968). 15 min after infection the cells were harvested and lysed, and the products analysed by sedimentation through a CsCl gradient. As shown in Fig. 1A nearly all the parental single strands were converted to replicative forms at both the permissive and restrictive temperatures. Thus M 13 DNA can enter the cell at 41°C and the presence of the dnaB mutation does not inhibit the conversion of viral DNA into the double stranded RF.

Although single strands are effectively converted into RF at 41°C, it is possible that this does not occur by the normal rifampicin-sensitive mechanism (Brutlag, Schekman & Kornberg, 1971). Recently it has been shown that an alternate pathway for the conversion of single stranded phage DNA into RF exists in E. coli which is not inhibited by rifampicin (Schekman et al., 1972; Wickner et al., 1972). However, when cells of HfrH 165/70 were treated with 200 µg/ml of rifampicin for 10 min and then infected with ^{32}P-labeled M 13, the antibiotic quantitatively inhibited the SS → RF conversion at both 34 and 41°C (Fig. 1B).

Attachment of the parental RF to the cell membrane appears to be a necessary condition for further in vivo replication (Stauden-

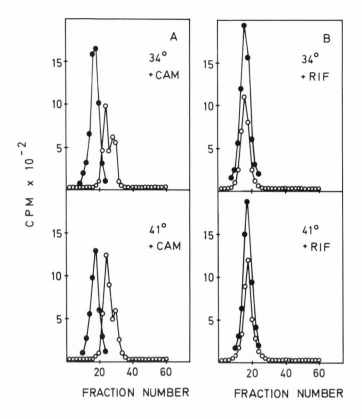

Fig. 1. Formation of parental RF. HfrH 165/70 (thy⁻, thi⁻, dnaB) was grown to 3 x 10⁸ cells per ml in 20 ml of Hanawalt medium at 34°C and chloramphenicol (100 μg/ml) or rifampicin (200 μg/ml) was added. The cultures were then divided into 10-ml subcultures, one of which was shifted to 41°C and one left at 34°C. The cultures were then incubated for 10 min, followed by infection with ³²P-labeled M 13 (2.2 x 10⁻⁶ cpm/PFU) at a MOI of 100. After 15 min of further incubation, the cultures were quickly cooled by placing them in a dry ice-ethanol bath, and KCN was added to 0.02 M. The cells were harvested by low speed centrifugation, washed twice with a Waring Blendor to remove F-pili and adsorbed phage, and lysed by treatment with 100 μg/ml of lysozyme for 30 min in an ice bath, followed by addition of 0.5% Sarcosyl. Samples of the lysates were mixed with ³H-labeled M 13 single strands and analysed by centrifugation in a linear neutral CsCl gradient (Staudenbauer & Hofschneider, 1971) in a SW 50.1 rotor at 45,000 rpm for 2 hrs at 15°C. Fractions were collected from the bottom of the tube and assayed for acid-precipitable radioactivity. Sedimentation in this and all subsequent gradients is from right to left.

o——o, ³²P-labeled parental DNA; o——o, M 13 single strands

bauer & Hofschneider, 1971; Forsheit & Ray, 1971). When the RFs formed at both 41° and 34°C were analysed after gentle lysis of the cells in a 5-20% sucrose gradient containing a cushion of 60% sucrose it was found that at both temperatures the parental RF was associated with the fast sedimenting cell membrane fraction (Olsen, Staudenbauer & Hofschneider, 1972).

INHIBITION OF RF REPLICATION

To investigate if the parental RF, once formed, could undergo subsequent replication at 41°C, cells were infected in the presence of chloramphenicol at the permissive temperature. After an adsorption period, the chloramphenicol was washed away and the cells were shifted to 34 and 41°C. Samples taken after 10 and 20 min of further incubation were pulse-labeled with ^{3}H-thymidine. Analysis of the cell lysates by sucrose gradient centrifugation (Fig. 2) indicated that replication of the parental RF did not occur at 41°C and apparently depends on the dnaB function. This inhibition of RF replication is not restricted to the parental RF, since it is also observed after a temperature-shift 15 min after infection, when a sizeable number of progeny RF molecules have already been formed.

SYNTHESIS OF PROGENY SINGLE STRANDS

If the cells were shifted to the restrictive temperature late in the infection, i.e. at the time of single strand synthesis, incorporation of a ^{3}H-thymidine pulse label into phage-specific DNA was observed. To decide whether this incorporation represents normal single strand synthesis a pulse-chase experiment was performed after shifting the cells to the restrictive temperature 90 min after infection. As shown in Fig. 3A most of the label incorporated during a 1-min pulse sedimented on a sucrose gradient as open circular RF II and supercoiled RF I. After a 5-min chase with unlabeled thymidine (Fig. 3B) the RF II has almost disappeared, whereas more than 50% of the label sediments in the position of single strands. During a 10 min chase part of the labeled single strands has become incorporated into mature phage particles (Fig. 3C). After a 20 min chase at 41°C more than 80% of the pulse-label has been chased into viral single strands and further into phage particles with a concomitant decrease of labeled RF I (Fig. 3D). From these results it is concluded that the normal mechanism of M 13 single strand synthesis is maintained at the restrictive temperature in the dnaB host.

During single strand synthesis a significant amount of label incorporated into phage-specific DNA during a short pulse is attached to the cell membrane (Forsheit & Ray, 1971). The

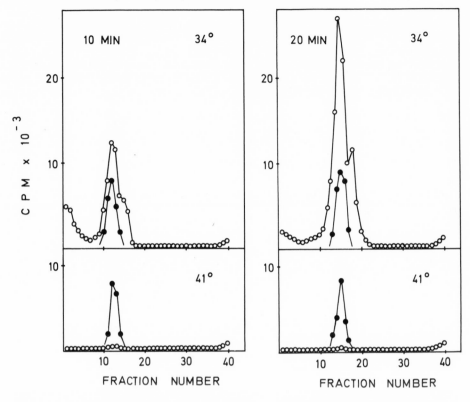

Fig. 2. Replication of parental RF. Exponentially growing HfrH 165/70 (2 x 10^8 cells per ml) incubated at 34° were treated with chloramphenicol (100 µg/ml) for 5 min, then infected with M 13 (MOI = 100). After 10 min the culture was divided in half; the infected cells were harvested by centrifugation at room temperature and washed twice with H medium without glucose or thymidine. After washing, the cells were resuspended in warmed H medium containing glucose and thymidine and immediately placed in water baths at 34 and 41°. After 10 and 20 min of incubation 2-ml samples of each culture were pulse-labeled (at 34 and 41°) with ^{3}H thymidine (10 µCi/ml) for 1 min. After the pulse, incorporation was stopped by addition of KCN (0.02 M), and the sample was immediately placed in a dry ice-ethanol bath. The cells were lysed by incubation with 5 mM EDTA and (100 µg/ml) of lysozyme for 20 min, followed by addition of 0.5% Sarkosyl. ^{32}P-Labeled M 13 RF I DNA was added as a marker, and the entire viscous lysate was chilled and layered on a 5-20% sucrose gradient (1 M NaCl) and centrifuged 15 hr at 25,000 rpm in a SW 27 rotor at 4°. Fractions (1 ml) were collected from the top of the tube by pumping 60% sucrose into the botton of the gradient, and were assayed for Cl_3CCOOH-precipitable radioactivity. (A) Pulse-labeled 10 min after removal of chloramphenicol. (B) Pulse-labeled 20 min after removal of chloramphenicol. o———o, pulse-labeled ^{3}H DNA; o———o, ^{32}P-labeled M 13 RF I.

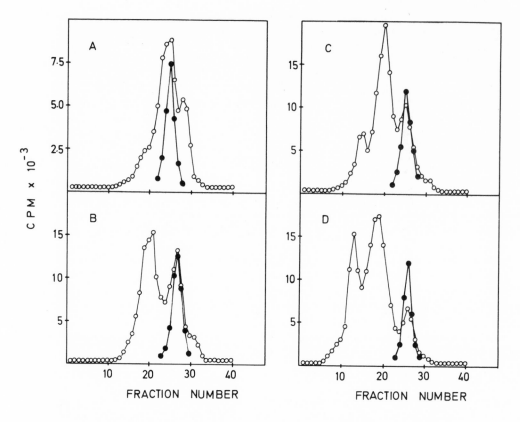

Fig. 3.   Sedimentation patterns from a pulse-chase experiment.
E. coli HfrH 165/70 was grown at 34°C to a density of 10^8/ml and
the culture was infected with M 13 at a multiplicity of 50 and
aeration continued for 90 min at 34°C.   0.5 ml aliquots were then
shifted to 41°C and after a pre-incubation of 10-min pulse-labeled
with 10 μCi ^{3}H thymidine for 1 min.   The pulse was stopped by
freezing the sample in acetone-solid CO_2 or followed by a chase
with 5 mg/ml unlabeled thymidine for the times indicated.   Cells
were lysed as described in Fig. 2 and the total lysates layered
on 5-20% (w/w) sucrose gradients in 17 ml Tris-EDTA-NaCl buffer
containing 1 M NaCl.   Centrifugation was carried out in a Spinco
SW-27.1 rotor at 25,000 rev./min for 16 h at 4°C.   40 fractions
were collected from the bottom of the tube and assayed for acid-
precipitable radioactivity, ^{32}P-labeled replicative form I was
used as sedimentation marker.   (A) Pulse-labeled DNA; (B) pulse-
labeled DNA after a 5-min chase; (C) pulse-labeled DNA after a
10-min chase; (D) pulse-labeled DNA after a 20-min chase.
o———o ^{3}H pulse-labeled DNA; o———o ^{32}P-labeled replicative
form I.

difficulty of studying the function of the cell membrane in phage
DNA synthesis in the presence of host chromosome replication can
be avoided by using a dnaB host in which bacterial DNA replication
can be turned off without blocking phage DNA synthesis. When M 13
infected dnaB cells were pulse labeled for 1 min after a shift to
41°C, up to 50% of the labeled phage-specific DNA was found to be
attached to the membrane after sucrose gradient centrifugation
(Staudenbauer & Hofschneider, 1972b).

EFFECTS OF ANTIBIOTICS ON SINGLE STRAND SYNTHESIS

It has been shown previously that chloramphenicol and rifampicin have different effects on M 13 single strand synthesis.
Chloramphenicol leads to a gradual decrease in the synthesis of
phage-specific DNA during which the newly synthesized single
strands are converted into double strands (Ray, 1970). On the
other hand, addition of rifampicin results in the immediate
cessation of phage DNA synthesis probably by directly interfering
with the single strand synthesis (Staudenbauer & Hofschneider,
1972a). Single strand synthesis in E. coli HfrH 165/70 at 41°C
was investigated to see whether it shows the same inhibition by
these antibiotics. Since under these conditions no bacterial DNA
is synthesized, phage DNA synthesis can be measured directly by
incorporation of ^{3}H-thymidine into acid-insoluble material (Fig. 4).
Rifampicin stopped phage DNA synthesis within 3 min, whereas in the
presence of chloramphenicol phage DNA synthesis continued for at
least 10 min with an even slightly increased rate during the first
5 min.

The radioactive label incorporated during a 3 min pulse in the
presence of chloramphenicol (50 µg/ml) was found predominantly in
RF I molecules (Fig. 5). To decide whether both strands were
synthesized under these conditions, the RF I molecules were nicked
by heating and subjected to equilibrium centrifugation in alkaline
CsCl. Under these conditions viral and complementary strands band
at different densities because of their different thymine and guanine contents (Ray, 1969). As shown in Fig. 6A less than 10% of
the radioactivity from RF I molecules labeled during single strand
synthesis at 41°C bands at the density of complementary strands.
On the other hand, a large amount of label incorporated in the
presence of chloramphenicol is found in complementary strands
(Fig. 6B). This demonstrates that not only the synthesis of progeny
single strands but also their conversion into double strands is
possible in the absence of the dnaB function.

INVOLVEMENT OF GENE-5 PROTEIN IN LATE M 13 DNA REPLICATION

So far we could show that the dnaB function is required for

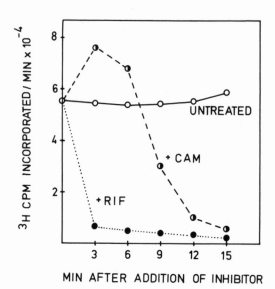

Fig. 4. Effect of antibiotics on single-strand synthesis. E. coli
HfrH 165/70 was grown at 34°C to a density of 10^8/ml and infected
with M 13 at a multiplicity of 50. 90-min after infection the culture was divided into three parts and shifted to 41°C. After a
pre-incubation of 10-min one part was left untreated, the other
parts were treated with rifampicin (200 μg/ml) and chlo3ampheni-
col (100 μg/ml), respectively. At 3-min intervals 1-ml aliquots
were removed and exposed to 10 μCi· ^{3}H thymidine for 1-min at 41°C.
Incorporation was terminated by freezing the samples in acetone-
solid CO_2. An equal volume of 10% trichloracetic acid was added
and the acid-precipitable radioactivity determined.
o———o untreated control; o·····o treated with rifampicin;
o-----o treated with chloramphenicol

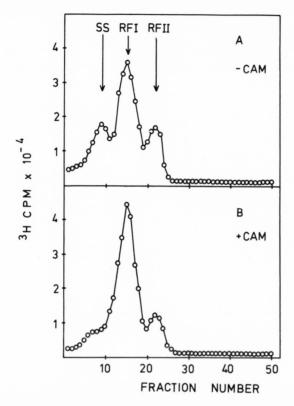

Fig. 5. Inhibition of single-strand synthesis by chloramphenicol.
E. coli HfrH 165/70 was grown at 34°C to a density of 10^8/ml and
infected with M 13 at a multiplicity of 50. Aeration at 34°C
was continued for 90-min. The culture was then divided into two
parts and shifted to 41°C. After a pre-incubation at 41°C for
10-min one part was treated with chloramphenicol (50 μg/ml), the
other part was left untreated. 5 min later 3 ml aliquots were
exposed to 50 μCi ^{3}H thymidine for 3-min respectively. Incorporation was stopped by freezing the samples in acetone-solid CO_2.
The cells were lysed as described in Fig. 2 and the lysates layered
on 5-20% (w/w) sucrose gradients in 35 ml Tris-EDTA-NaCl buffer
containing 1 M NaCl. Centrifugation was performed in a Spinco
SW-27 rotor at 25,000 rev./min for 16h at 4°C. 0.8 ml fractions
were collected from the bottom of the centrifuge tube and aliquots
assayed for acid-precipitable radioactivity. ^{32}P-labeled M 13
single strands were used as sedimentation marker. (A) 3-min
labeling in the absence of chloramphenicol; (B) 3-min labeling in
the presence of chloramphenicol.

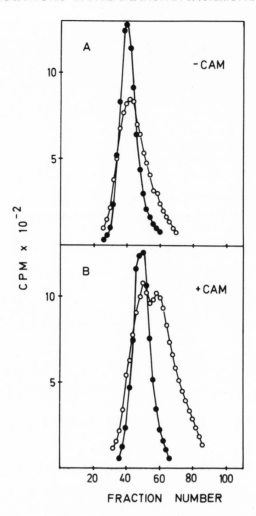

Fig. 6. Effect of chloramphenicol on the distribution of label between viral and complementary strands. The ^{3}H-labeled DNA sedimenting in the positions of replicative form I and replicative form II from the gradients described in Fig. 9 was pooled separately and dialysed against 0.05 M Tris—HCl pH 8.6—0.005 M EDTA. The material from the replicative form I positions was then heated to 100°C for 3 h. The boiled replicative form I samples and the replicative form II samples were combined with ^{32}P-labeled M 13 ssDNA and sedimented to equilibrium in alkaline CsCl gradients (Staudenbauer & Hofschneider, 1972b). Density increases to the left.
o———o ^{3}H-labeled DNA; o———o ^{32}P-labeled M 13 ssDNA.
(A) replicative form I from the gradient shown in Fig 5A,
(B) from Fig 5B.

early RF replication but dispensible for single strand synthesis
as well as for the conversion of single strands into double strands.
Therefore double strand replication early in M 13 infection appears
to occur by a pathway of DNA replication different from the RF
synthesis observed after addition of chloramphenicol late in M 13
infection. Conceivably this dnaB independent RF synthesis requires
some phage coded protein(s) not available in sufficient amounts
early in the infection. Since the gene-5 protein is the only phage
protein specifically required for late M 13 DNA synthesis we investigated whether this protein is involved in the switch from the
dnaB dependent to the dnaB independent mode of DNA synthesis.

HfrH 165/70 cells were infected with M 13 or an M 13 ts-5
mutant (Staudenbauer & Hofschneider, 1973) and shifted to 41°C
90 min after infection. 30 min after the shift the cultures were
pulse-labeled with ^{3}H-thymidine for 5 min and the labeled DNA
analysed by sucrose gradient centrifugation. As has been shown
previously (Salstrom & Pratt, 1971) in an M 13 ts-5 infected wild
type host a switch from single strand to double strand replication
is observed upon inactivation of the gene-5 protein (Fig. 7).
However in an M 13 ts-5 infected dnaB host phage-specific DNA
synthesis stops all together at the restrictive temperature
(Fig. 8). This demonstrates a definite requirement for gene-5
protein in the dnaB independent DNA replication late in M 13
infection. Conceivably the intracellular concentration of this
protein might control the switch from early to late M 13 DNA
synthesis.

DISCUSSION

We have shown that upon infection of a dnaB mutant with M 13
at the restrictive temperature the single stranded phage DNA
penetrates the cell and is converted in a rifampicin sensitive
step to the double stranded RF. However the subsequent semi-
conservative replication of the parental RF is blocked in the
absence of a functional dnaB gene product. RF replication is
also blocked at the restrictive temperature after progeny RF have
been formed at the permissive temperature. This result implies
that progeny RF replication also depends on the dnaB function.
However, once the progeny single strand synthesis has started at
the permissive temperature, it will continue after a temperature
shift. Addition of chloramphenicol causes the conversion of
newly synthesized progeny single strands into RF molecules.

It appears that the process of semi-conservative double
strand replication depends specifically on the dnaB function. On
the other hand, asymmetric synthesis, i.e. the displacement
synthesis of single stranded DNA on a double stranded template as
well as the synthesis of a complementary strand on a single

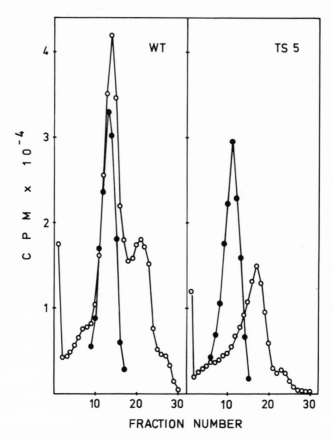

Fig. 7. Effect of blocking gene-5 function during single-stranded
DNA synthesis in E. coli HfrH 165/70 (dnaB⁺). E. coli Hfrh
165/70 (dnaB⁺) was grown in supplemented Hanawalt medium at 34°C
to a density of 10^8 cells/ml, shifted down to 32°C and infected
with M 13 wild type or ts 5 mutant phage at an m.o.i. of 25.   90
min after infection the cultures were transfered to 41°C. At 15
min and 40 min after the shift to the elevated temperature 2.5 ml
aliquots were pulse-labeled for 5 min with 50 μCi ³H thymidine.
Incorporation was stopped by freezing the samples in an acetone-
solid CO_2 bath. Cells were lysed and the labeled products
analysed by sucrose gradient centrifugation as described in Fig. 2.
³²P-labeled single-stranded DNA was used as sedimentation marker.
(A) M-13-infected cells, (B) M-13 ts 5-infected cells pulse-
labeled after 30 min at 41°C.
o———o ³H-pulse-labeled DNA; o———o ³²P-labeled single-stranded
DNA.

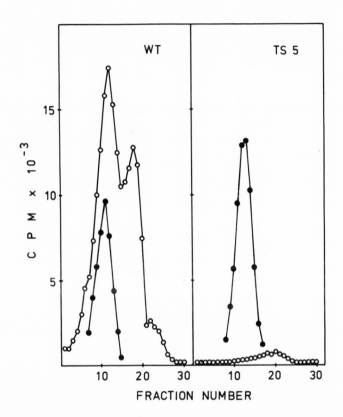

Fig. 8. Effect of blocking gene-5 function during single-stranded
DNA synthesis in E. coli Hfrh 165/70 (dnaB⁻). All procedures
were as described in Fig. 7 except that the isogenic dnaB⁻
host E. coli HfrH 165/70 (dnaB⁻) was used. (A) M-13-infected
cells, (B) M-13 ts 5-infected cells pulse-labeled after 30 min
at 41°C.
o———o ³H-pulse-labeled DNA; o———o ³²P-labeled single-stranded
DNA

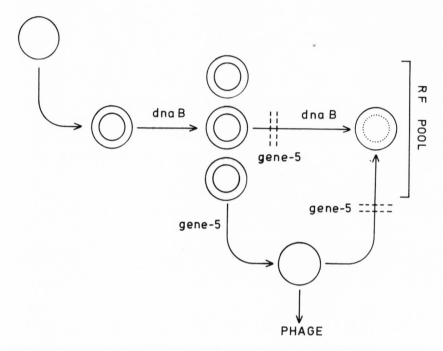

Fig. 9. Diagram of M 13 DNA replication. The parental viral
single strand is first converted to the parental RF by the dnaB
independent synthesis of a complementary strand (inner circle). In
the presence of the dnaB gene product the parental RF replicates
semi-conservatively establishing a pool of progeny RF molecules.
After accumulation of gene-5 protein, RF replication stops and
progeny single strands are synthesized in a dnaB independent
replication step. Under normal conditions these single strands
are extruded as progeny phage particles. However inactivation
of gene-5 protein results in the conversion of the progeny single
strands to RF molecules and to the resumption the dnaB dependent
RF replication.

stranded template, does not necessarily require the dnaB function.
In this respect M 13 RF replication resembles the vegetative replication of the E. coli chromosome and M 13 single strand synthesis
and the SS → RF conversion are similar to transfer replication
during conjugation.

On the basis of these data we suggest the following scheme
for M 13 DNA replication (Fig. 9): The infecting viral single
strand is converted to the parental RF which replicates using the
replication machinery of the host chormosome including the products of the dnaB and dnaE genes (Staudenbauer, Olsen & Hofschneider, 1973). Accumulation of progeny RF molecules leads to an
increase in the concentration of gene-5 protein which is required
for the replication machinery used in single strand synthesis
possibly by functioning as an unwinding protein. Also, due to its
strong binding to single stranded DNA gene-5 protein furthermore
acts as a repressor of complementary strand synthesis (Alberts,
Frey & Delius, 1972; Oey & Knippers, 1972).

Addition of chloramphenicol and simultaneous thermal inactivation of the dnaB gene product leads to a transient RF synthesis
in which in the absence of new protein synthesis the intracellular
concentration of gene-5 protein drops to a level which still
supports single strand synthesis but cannot preserve the newly
made viral DNA in a single stranded form. In the presence of a
functional dnaB gene product, depletion or inactivation of gene-5
protein results in a switch back to early dnaB dependent RF
replication. Thus the availability of gene-5 protein would control
which of the alternate pathways of M 13 DNA replication is operative.

REFERENCES

Alberts, B., Frey, L. & Delius, H. (1972). J. Mol. Biol. 68, 139.
Brutlag, D., Schekman, R. & Kornberg, A. (1971). Proc. Nat. Acad.
     Sci., Wash. 68, 2826.
Fidanian, H.M. & Ray, D.S. (1972). J. Mol. Biol. 72, 51.
Forsheit, A.B. & Ray, D.S. (1971). Virology 43, 647.
Gross, J.D. (1972). Curr. Top. Microbiol. Immunol. 57, 39.
Hohn, B., Lechner, H. & Marvin, D.A. (1971). J. Mol. Biol. 56, 143.
Lin, N.S.C. & Pratt, D. (1972). J. Mol. Biol. 72, 37.
Marinus, M.G. & Adelberg, E.A. (1970). J. Bacteriol. 104, 1266.
Marvin, D.A. & Hohn, B. (1969). Bact. Rev. 33, 172.
Oey, J.L. & Knippers, R. (1972). J. Mol. Biol. 68, 125.
Olsen, W.L., Staudenbauer, W.L. & Hofschneider, P.H. (1972).
     Proc. Natl. Acad. Sci., Wash. 69, 2570.
Pratt, D. (1969). Ann. Rev. Genetics 3, 343.
Pratt, D. & Erdahl, W.S. (1968). J. Mol. Biol. 37, 181.
Primrose, S.B., Brown, L.R. & Dowell, C.E. (1968). J. Virol. 2, 1308.

Ray, D.S. (1969). J. Mol. Biol. 43, 631.

Ray, D.S. (1970). J. Mol. Biol. 53, 239.

Salstrom, J.S. & Pratt, D. (1971). J. Mol. Biol. 61, 489.

Schekman, R., Wickner, W., Westergaard, O., Brutlag, D., Geider, K., Bertsch, L.L. & Kornberg, A. (1972). Proc. Nat. Acad. Sci., Wash. 69, 2691.

Staudenbauer, W.L. & Hofschneider, P.H. (1971). Biochem. Biophys. Res. Commun. 35, 354.

Staudenbauer, W.L. & Hofschneider, P.H. (1972a). Proc. Natl. Acad. Sci., Wash. 69, 1634.

Staudenbauer, W.L. & Hofschneider, P.H. (1972b). Europ. J. Biochem. 30, 403.

Staudenbauer, W.L. & Hofschneider, P.H. (1973). Europ. J. Biochem. 34, 569.

Staudenbauer, W.L., Olsen, W.L. & Hofschneider, P.H. (1973). Europ. J. Biochem. 32, 247.

Tseng, B.Y. & Marvin, D.A. (1972). J. Virol. 10, 384.

Vapnek, D. & Rupp, W.D. (1971). J. Mol. Biol. 60, 413.

Wickner, R.B., Wright, M., Wickner, S. & Hurwitz, J. (1972). Proc. Nat. Acad. Sci., Wash. 69, 3233.

DNA-UNWINDING PROTEINS AND THEIR ROLE IN THE REPLICATION OF DNA

Bruce Alberts

Department of Biochemical Sciences

Princeton University

The first "DNA-unwinding protein" to be characterized was the T4 bacteriophage gene 32-protein (Alberts and Frey, 1970). Since then, other proteins of similar type have been obtained from a variety of organisms. In this article, I would like to review some of their properties and discuss their possible roles in DNA replication.

ISOLATION OF DNA-UNWINDING PROTEINS

"DNA-unwinding proteins" share several common properties, even though their biological roles may be different. Of these, the most characteristic is that of binding tightly all along a single-stranded DNA chain, completely covering it to form an extended linear complex, resembling that diagrammed in Fig. 1A. Because these proteins have very little affinity for DNA in its double-helical form, the simple coupling of this protein-binding reaction to the DNA helix-coil transition greatly facilitates DNA denaturation, as schematically illustrated in Fig. 1B. Depending on protein concentration and ionic conditions, the DNA melting point depression can be 40° or more. In an analogous manner, histones and polyamines bind more tightly to double-helical than to single stranded DNA, and thereby _increase_ DNA-helix stability.

Selective binding to single-stranded DNA is clearly an essential property of any protein capable of destabilizing DNA helices. Therefore, to screen for DNA-unwinding proteins from any organism, we prepare a crude cellular extract, free it of endogenous DNA, and then pass it through a large column containing double-stranded DNA-cellulose (Alberts & Herrick, 1971). Most of the

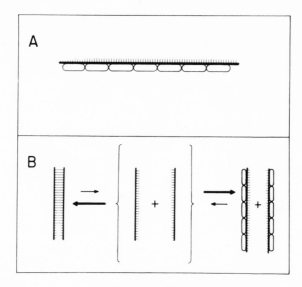

Fig. 1.   Schematic representation of DNA-unwinding protein interactions with DNA.   (A). Complex with single-stranded DNA.   (B).
Shift in helix-melting equilibrium caused by DNA-binding specificity.

proteins which bind to DNA in an extract are removed in this step.
The breakthrough (and/or a low-salt eluate) is then reapplied to a
single-stranded DNA-cellulose column.   Because of the previous
selection, only a few proteins are able to bind tightly to this
second column, and these are recovered by a series of high-salt
elution steps.   After further purification by standard chromatographic procedures (monitored by polyacrylamide gel electrophoresis), each pure protein is tested for its ability to denature
poly dAT.   This synthetic DNA serves as an especially sensitive
probe for DNA-unwinding proteins, being more easily denatured than
natural DNA's which contain G-C base-pairs.

PROPERTIES OF DNA-UNWINDING PROTEINS

    Proteins similar to the T4 bacteriophage gene 32-protein
include the E. coli DNA-unwinding protein (Sigal et al., 1972),
the filamentous bacteriophage gene 5 protein (Alberts, Frey &
Delius, 1972; Oey & Knippers, 1972) and a DNA-binding protein
isolated from calf thymus (G. Herrick & B.A., in preparation).

    A brief summary of some of the properties of these various
DNA-unwinding proteins is presented in Table 1.   The T4 gene 32
product is known from the work of J. Tomizawa, R. H. Epstein, and

TABLE 1

Properties of some DNA-unwinding proteins

Protein	Biological Role	Subunit Molecular Weight	Nucleotides Bound per Subunit	copies/cell	Isoelectric pH
T4 bacteriophage gene 32-protein	DNA replication and genetic recombination	35,000	10	10^4	5
E. coli	DNA synthesis?	22,000	8	8×10^2	6-7
Calf thymus	?	24,000	7	$>2 \times 10^6$	7-8
Ff bacteriophage gene 5-protein	Single-stranded DNA synthesis	10,000	4	1.5×10^5	8

their colleagues to be needed for both the replication and genetic recombination of T4 DNA, while the Ff gene 5 protein is needed for synthesis of the single-stranded viral DNA. The biological roles of the other two proteins are not known, although their stimulation of in vitro synthesis by DNA polymerases (Sigal et al., 1972; G. Herrick et al., in preparation) suggests that they may also function in DNA synthesis.

All of these proteins have been purified to electrophoretic homogeneity, and their subunit molecular weights have been determined by SDS-polyacrylamide gel electrophoresis (Table 1). The native molecules are often larger than these subunits; for example, gene 32 protein undergoes a concentration-dependent aggregation (Huberman, Kornberg & Alberts, 1971), with the dimer being preferred at low protein concentrations (Carroll, Neet & Goldthwait, 1972). The isoelectric points of these proteins, as determined by isoelectric focusing, are near neutrality (Table 1), and some of them carry a net negative charge at physiological pH. Nevertheless, they all appear to utilize a positively-charged site for DNA-binding, since their affinity for DNA is greatly reduced when electrostatic interactions are damped out by raising salt concentrations. The effect of salt is illustrated for 32-protein in Fig. 2, for which DNA affinity decreases gradually as the ionic strength is raised.

The number of these protein molecules per cell varies widely (Table 1); but when expressed as molecules per replication fork, the numbers become very similar for the E. coli and T4 proteins (Sigal et al., 1972).

Other important similarities between these proteins include:
1) At saturation, DNA-protein complexes are formed with a high protein to DNA ratio (weight ratio of 7:1 to 12:1; protein: DNA). The calculated number of DNA nucleotides per protein monomer thus ranges between 4 and 10, being greater for the larger proteins (Table 1).
2) Adjacent protein molecules in the DNA complex appear to contact each other. The favorable protein-protein interactions which ensue lead to the highly cooperative DNA-binding seen for the T4, E. coli, and bacteriophage Ff proteins. Such cooperativity may be detected as an increase in DNA-affinity as protein concentration increases. Moreover, since an incoming protein monomer strongly prefers a DNA-site next to a previously bound protein molecule rather than a region of DNA without protein, highly clustered binding occurs when DNA is in excess over protein. This effect is seen in Fig. 3 for 32-protein binding to both excess single-stranded fd and excess poly dT DNA's. Since in both cases the bulk of the radioactively-labeled protein sediments ahead of the main peak of nucleic acid, most of the 32-protein must be complexed to a small fraction of these DNA's; moreover, the exchange rate (the rate at which a

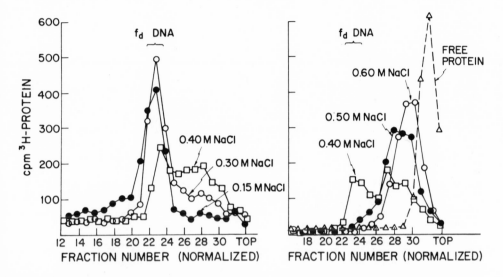

Fig. 2.   Salt-dependence of 32-protein binding to single-stranded
DNA as measured by sucrose-gradient sedimentation.   10 µg of single-
stranded bacteriophage fd DNA was mixed with 6 µg of ^{3}H-32-protein
in 0.2 ml of buffer containing the indicated NaCl concentration
plus 0.02 M Tris-HCl, pH 8.1, 0.5 mM Na$_3$EDTA, 0.15 M NaCl, 100 µg/
ml bovine serum albumin (BSA), 10% glycerol, 1 mM β-mercaptoethanol.
After 40 min. at 4°, the mixture was layered onto a 5 ml, 5-30%
sucrose gradient containing the same buffer.   Following centrifugation for 2 hrs at 46,000 rpm (4°), fractions were collected
and monitored for radioactivity by standard techniques.   The
brackets indicate the position of the narrow peak of fd DNA, determined by absorbance measurements.

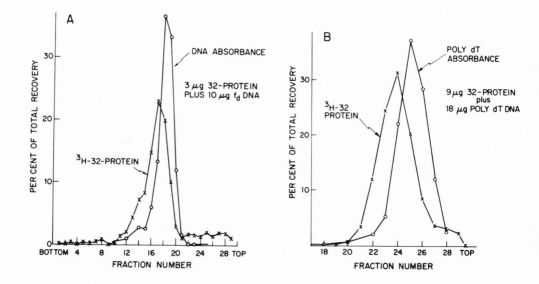

Fig. 3.   Clustered binding of 32-protein to excess single-stranded
DNA.   (A). A 40-fold excess of fd DNA (10 μg in 0.10 ml) was
quickly mixed with 0.10 ml containing 3 μg of 32-protein.   After
standing for 20 min at 4°C, the mixture (in 0.01 M Tris-HCl, pH 8.1,
0.02 M KCl, 0.01 M MgCl$_2$, 0.5 mM Na$_3$EDTA, 100 μg/ml BSA, 10% glycerol, 1 mM β-mercaptoethanol) was sedimented for 2.5 hrs at
46,000 rpm through a 5-30% sucrose gradient containing the same
buffer (Spinco SW 50 rotor, 4°).   Samples were then processed for
counting and absorbance measurements by standard techniques.   (B).
As above, except that 9 μg of 32-protein was mixed with 18 μg of
poly dT DNA, and the buffers contained 0.15 M KCl in place of Mg^{++}.
Sedimentation was for 3.5 hrs at 40,000 rpm.

bound cluster moves to a different DNA molecule) must be slow,
since discrete profiles are observed. Clustered binding has also
been visualized directly by electron microscopy (Delius, Mantell
& Alberts, 1972; Alberts, Frey & Delius, 1972; Sigal et al., 1972).
A possible exception is the calf protein, for which neither
clustered binding nor cooperativity has been detected.
3) DNA strands are held in an extended conformation by these proteins. This results in a slow sedimentation rate for each DNA-
protein complex relative to its mass. The exact conformation of
the complex appears to be quite distinctive for each protein.
Dr. Hajo Delius at the Cold Spring Harbor Laboratory has examined
single-stranded DNA saturated with unwinding proteins in the
electron microscope (after fixation with glutaraldehyde and spreading by the Kleinschmidt technique). With the circular bacteriophage fd DNA molecule (6600 nucleotides), he sees protein-thickened
circles which are 3.0 μm, 1.2 μm, and 2.4 μm long for the T4,
E. coli, and calf unwinding proteins, respectively (Delius et al.,
1972; Sigal et al., 1972; G. Herrick et al., in preparation). In
marked contrast, the Ff gene 5 protein appears to be bivalent; by
bringing two single-strands together, it collapses the fd DNA
circles into branched linear forms, in which the total single-
strand contour length appears to be 2.1 μm (Alberts et al., 1972).
4) For all four proteins, spectral measurements indicate that the
nucleotide bases in the DNA complex have nearly their maximum
absorbance at 260 nm. Hence, these bases appear to be held in an
unstacked conformation. Moreover, the bases appear not to be
covered by the protein [at least for 32-protein where very rapid,
second-order DNA-renaturation rates are observed for the DNA-protein complexes (Albert & Frey, 1970)]. The strong specificity of
these proteins for DNA single-strands probably reflects their
requirement for a particular DNA backbone conformation (which a
strand in the double-helix cannot attain), rather than a need to
bind near the hydrogen-bonding sites of the base-pairs.
5) These proteins have very little sequence or base specificity
for DNA binding. Although they preferentially denature those
regions of DNA-helix rich in A-T base pairs, this is only because
these helical regions are the weakest ones thermodynamically
(Delius et al., 1972; Alberts et al., 1972; Sigal et al., 1972).
6) While binding tightly to single-stranded DNA, the T4, E. coli,
and Ff proteins have only very weak affinity for single-stranded
RNA. This is determined either by their inability to substantially
depress the melting temperature of poly rAU, or by their failure
to cosediment with bacteriophage R17 RNA in sucrose gradients.
Data of the latter type is presented for 32-protein in Fig. 4.
(This protein can nevertheless be used in electron microscopy to
enhance visualization of RNA chains (Delius, Wesphal & Axelrod,
1973), apparently because its weak affinity for RNA is greatly
strengthened by the glutaraldehyde fixation used). In contrast,
the calf protein does bind significantly to RNA; after its purification it depresses the melting temperature (Tm) of poly rAU

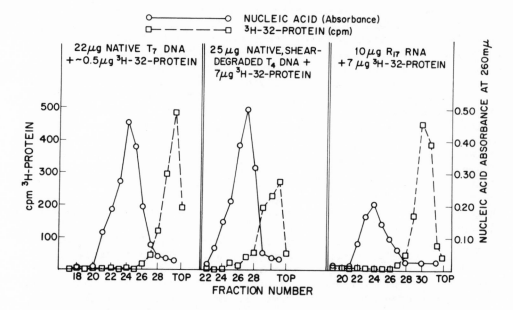

Fig. 4.   Nucleic acids that bind 32-protein poorly as judged by sucrose gradient sedimentation.   The indicated nucleic acids were mixed with ³H-32-protein and sedimented exactly as described for Fig. 2, except that 0.15 M KCl replaced the NaCl.

nearly as much as it depresses the Tm of the DNA analogue, poly dAT (G. Herrick & B. A., in preparation).   For this reason, it is not clear whether this protein functions in DNA or in RNA metabolism.

7) Three of the DNA-unwinding proteins, when added to single-stranded DNA template strands, markedly stimulate the rate of incorporation.   Sigal et al., 1972; G. Herrick et al., in preparation).   The apparent exception is the Ff gene 5 protein (Oey & Knippers, 1972), which may actually function in single-stranded viral DNA synthesis by inhibiting synthesis of the complementary DNA strand (Salstrom & Pratt, 1971).

Of special interest is the specificity of the polymerase stimulations by the E. coli and T4 unwinding proteins.   In comparable in vitro assays using the same DNA template, T. Kornberg and M. Gefter have observed marked stimulations of E. coli DNA polymerase II by the E. coli DNA-unwinding protein, and of T4 DNA polymerase by T4 32-protein.   But there was no effect of E. coli unwinding protein on the T4 polymerase, nor of T4 unwinding protein on the E. coli polymerase (Sigal et al., 1972).   This specificity could be due to the fact that the two DNA unwinding proteins hold the DNA single-strands of the template in quite different conformations; regardless of the explanation, the result strongly implies that these polymerases and unwinding proteins interact with each other in vivo, as well as in vitro.

DNA REPLICATION

What are the possible roles for DNA-unwinding proteins in DNA replication? Here it is best to restrict our attention to the T4 bacteriophage system, where there is strong genetic evidence that both the T4 DNA polymerase and DNA-unwinding protein are essential for the polymerization process.[1] Our current view of a replication fork is based upon work in many laboratories, and is schematically illustrated in Fig. 5. Because of the antiparallel orientation of strands in the DNA double-helix, a discontinuous mode of DNA synthesis is necessary on one template strand if all of the DNA is to be made with the T4 DNA polymerase, which (like all other such enzymes) catalyzes polymerization in the 5' to 3' chain direction only (Goulian, Lucas & Kornberg, 1968). Both biochemical analysis

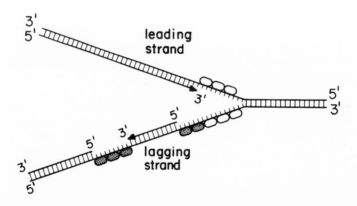

Fig. 5. Schematic representation of possible DNA polymerase and unwinding protein cooperation in the T4 DNA replication fork. The shaded 32-protein molecules are those required by the first two models discussed in the text, while the unshaded molecules are needed for Model 3 only. An important point omitted here is that Okazaki and his coworkers have recently shown that, at least in E. coli, the discontinuously synthesized DNA chains are primed by a special piece of RNA which appears to be synthesized de novo by a special, rifampicin-resistant enzyme (Sugino, Hirose & Okazaki, 1972). After it serves as a primer for DNA-polymerase, this RNA is erased and replaced by DNA.

of initial DNA products (Sugino & Okazaki, 1972) and electron microscopy of forks in similar systems (Inman & Schnos, 1971; Wolfson &
Dressler, 1972) indicate a length of about 10^3 nucleotides between,
adjacent polymerase starts on the "lagging side" of the fork.   On
the "leading side" of the fork, synthesis can be either continuous
(as indicated in Fig. 5) or discontinuous, perhaps depending on
physiological conditions.   However, the discontinuous synthesis on
the leading side is best viewed as a competition between polymerase
continuation and polymerase restart mechanisms (Olivera & Bonhoeffer, 1972); since such restarts do not appear to lead to gap formation in the daughter strand, this synthesis can be treated for our
purposes as continuous.

Note that there should be an average of one Okazaki-piece
length (about 10^3 nucleotides) of single-stranded DNA on the
lagging side of each fork according to Fig. 5.   As indicated in
Fig. 5 (shaded protein molecules), such DNA is almost certain to
be covered with tightly bound DNA-unwinding protein.   Since each
32-protein monomer binds 10 nucleotides, 100 molecules of 32-
protein would be required per fork to cover this DNA.   Genetic
experiments, in which 32-protein levels are varied by mixed
infections, suggest that 100-200 molecules of 32-protein are
indeed necessary to establish each T4 replication fork, in marked
contrast to the much smaller amounts of all of the other replication gene products required (Snustad, 1968; Sinha & Snustad, 1971;
reviewed in Alberts, 1971).

At least three types of models can now be proposed to explain
why 32-protein is essential for T4 DNA replication; these might be
viewed as "trivial", "modest", and "glamorous":
1) "Trivial." In this model, the only role of 32-protein in replication is to protect the exposed single-stranded DNA on the
lagging side from endonucleolytic degradation.   In fact, 32-
protein appears to reduce DNA degradation by nucleases both in vivo
(Kozinski & Felgenhauer, 1967; M. Curtis & B. A., in preparation)
and in vitro (Huang & Lehman, 1972; unpublished observations of
C. Manoil, this laboratory).   However, the specific stimulation of
T4 DNA polymerase by 32-protein is not explained by this hypothesis.   Also one would normally expect an organism to get rid of (or
modify) a deleterious nuclease rather than to make large quantities
of unwinding protein solely for its neutralization.
2) "Modest." This view explains the specific in vitro polymerase
stimulation observed by proposing that unwinding protein is
required in vivo to align the single-strands on the lagging side
of the fork in some favorable conformation which allows polymerase
to rapidly fill in these gaps (and perhaps is secondarily needed
for nuclease protection).

Note that no role is predicted for unwinding protein on the
leading side of the fork in either of the above two models.   If

either of these models is correct, one might therefore expect to observe two distinct rates of T4 DNA synthesis decay when the temperature is raised to inactivate a temperature-sensitive gene 32-protein: a fast decay (to roughly half of the normal rate) as lagging-side synthesis stops, followed by a slower decay during which leading-strand synthesis gradually ceases. In fact, after a shift from 25° (permissive temperature) to 42° (non-permissive temperature) of cells infected with T4 mutant ts P7 (temperature-sensitive for 32-protein), the rate of DNA synthesis drops to below 1% of normal within 2 minutes. One can calculate from this data that the replication fork moves less than 15% of a T4 genome length before stopping completely (M. Curtis & B. A., in preparation). Only a single rate of inactivation is observed, with no indication of a break in the decay curve at 50%. Thus, either 32-protein is needed for DNA synthesis on both sides of the fork (See Model 3, below), or the synthesis on the leading side is tightly linked to continued lagging-strand synthesis.

3) "Glamorous." In addition to the above postulated roles, 32-protein might be used to open up the DNA double-helix ahead of the replication fork, thus providing a single-stranded template strand for the leading strand molecule of DNA polymerase (indicated by non-shaded unwinding protein molecules in Fig. 5). Some such destabilization of the helix ahead of the fork seems essential, in view of the fact that double-helical templates cannot be copied by the T4 DNA polymerase in vitro (Goulian et al., 1968). An obvious difficulty with this proposal is that 32-protein is thought to be able to invade double-helical regions only transiently inside the cell (Delius et al., 1972). However, the pre-fork region may be much more susceptible to 32-protein opening than other helical regions, due to the fact that cooperative 32-protein binding can be nucleated on both the lagging strand 32-protein complexes (Fig. 5) and on the leading strand molecule of DNA polymerase (for which 32-protein has substantial affinity: Huberman et al., 1971). Other replication proteins could also be involved in helping to destabilize this pre-fork region (Alberts, 1971).

Another potential problem with this model is that, since replication procedes at about 10^3 nucleotides/second, 100 32-protein molecules must be added ahead of the fork each sec to maintain a steady state. Assuming that 32-protein is released as polymerase passes, and thereafter freely diffuses throughout the cell, its maximum rate of re-entry into the fork can be estimated from the rate at which pre-fork DNA sites and protein should collide. If 32-protein is assumed to have a bimolecular rate constant for collision with DNA sites of $k = 10^8$ M^{-1} sec^{-1} (about the theoretical naximum and equal to the association rate constant observed for the lactose repressor protein with its DNA operator in 0.1 M KCl; see Riggs, Bourgeois and Cohn, 1970), a rate of entry of 100/sec seems possible providing that a large fraction of collisions are effective.[2] Such a "diffusion-limited reaction"

might be reasonable for 32-protein binding to the single-stranded
DNA on the lagging side. However, only a very small fraction of
total collisions can be effective on the leading side, where
entry demands simultaneous helix-opening. Thus, to preserve this
model, one might imagine that the 32-protein molecules used to
unwind the helix ahead of the fork are directly reutilized without
leaving the DNA: either by invoking some unique 3-dimensional
convolution of template strands (Alberts, 1971); or by having these
protein molecules slide, pushed by the polymerization behind.

CONCLUSION

It is likely that many of the questions raised about DNA-
unwinding protein function in DNA replication will remain unan-
swered until much more is known about the detailed enzymology of
this process. Genetic analysis in the T4 bacteriophage system
(Epstein et al., 1963; Warner & Hobbs, 1967) makes it clear that
at least 4 proteins play essential roles in the replication process
in addition to DNA polymerase and unwinding protein. The same is
likely to be true for E. coli (Gross, 1972), although the known
unwinding protein has not yet been shown to be essential for E.
coli DNA replication (Sigal et al., 1972).

Our approach to the enzymology of DNA replication has been to
attempt to isolate all of the genetically identified proteins of
one DNA-replication apparatus. Once this is done, we hope to get
this structure to self-assemble in vitro by incubating intracellular concentrations of these proteins with DNA. With this goal in
mind, we have focused our attention on T4 bacteriophage replication, and developed an in vitro complementation assay which should
allow all of the known T4 replication proteins to be purified in
active form. Thus far, gene 45-protein and a tight complex of
gene 44 and gene 62 proteins have been purified to electrophoretic
homogeneity, and some of their physical properties determined
(Barry & Alberts, 1972; Barry et al., 1973). Attempts to reconstruct functional T4 replication forks in vitro are beginning;
indications are that this will be a major undertaking which will
take some time to accomplish.

FOOTNOTES

(1) Alternatively, it has been suggested that the abortive replication of gene 32-deficient mutants could be merely an indirect
effect of their recombination deficiency, which "prevents formation or restoration of circular replicons from infecting terminal-
ly redundant molecules or from DNA fragments, after intracellular
nucleases have cut potentially replicating circles" (Marsh,
Breschkin & Mosig, 1971). Support for this hypothesis came from

the fact that in 32-amber mutant infections some DNA is made, about half of the parental phage DNA being replicated once (Kozinski & Felgenhauer, 1967). This view is not, however, consistent with the rapid and complete shut-down of T4 DNA replication observed as soon as 32-protein is inactivated in a temperature-sensitive mutant infection (see below). Moreover, further studies now demonstrate that this initial 32^- replication, unlike normal replication, requires several proteins of the host DNA replication apparatus (Mosig, Bowden & Bock, 1972), possibly one of which is a substitute host DNA-unwinding protein.

(2) With a total of 10^4 molecules/cell, the free 32-protein concentration cannot exceed 10^{-5} M. Since the number of collisions/sec per DNA site equals k times the free 32-protein concentration, the maximum entry rate for 32-protein should be about 10^3/sec (diffusion-controlled binding). This must be an overestimate, since not all of the 32-protein is free; moreover the rate of DNA-fork movement is already maximal at early times after infection when 32-protein levels are still increasing (Werner, 1968).

ACKNOWLEDGEMENT

I would like to thank my coworkers, particularly Glenn Herrick, Linda Frey, and Nolan Sigal, whose efforts were crucial in the work on unwinding proteins, and Dr. Hajo Delius of Cold Spring Harbor Laboratory, whose electron micrographs have been inspiring. This work was supported by grants from the National Institutes of Health and the American Cancer Society.

REFERENCES

Alberts, B. 1971. On the Structure of the Replication Apparatus. In Nucleic Acid-Protein Interactions (Miami Winter Symposia, Vol. II, D. W. Ribbons, J. F. Woessner and J. Schultz, eds., p. 128-143, North Holland, Amsterdam.

Alberts, B. and L. Frey. 1970. T4 Bacteriophage gene 32: a structural protein in the replication and recombination of DNA. Nature. 227: 1313.

Alberts, B., L. Frey, and H. Delius. 1972. Isolation and characterization of gene 5-protein of filamentous bacterial viruses. J. Mol. Biol. 68: 139.

Alberts, B. and G. Herrick. 1971. In "Nucleic Acids," Methods in Enzymology, eds. Grossman, L. and Moldave, K. (Academic Press, N. Y.), Vol. XXII, pp. 198-217.

Barry, J. and B. Alberts. 1972. In vitro complementation as an as-
    say for new proteins required for bacteriophage T4 DNA repli-
    cation: purification of the complex specified by T4 genes 44
    and 62. Proc. Nat. Acad. Sci., USA. 69: 2712-

Barry, J. H. Hama-Inaba, L. Moran, J. Wiberg and B. Alberts. 1972.
    Proteins of the T4 bacteriophage replication apparatus. In
    DNA Synthesis In Vitro (R. D. Wells and R. B. Inman, eds.)
    University Park Press, Baltimore, Md.

Carroll, R. B., K. E. Neet and D. Goldthwait. 1972. Self-associa-
    tion of gene-32 protein of bacteriophage T4. Proc. Nat. Acad.
    Sci., USA. 69: 2741.

Delius, H., N. Mantell, and B. Alberts. 1972. Characterization by
    electron microscopy of the complex formed between T4 bacterio-
    phage gene 32-protein and DNA. J. Mol. Biol. 67: 341.

Delius, H., H. Westphal and N. Axelrod. 1973. Length measurements
    of RNA synthesized in vitro by E. coli RNA polymerase. J.
    Mol. Biol. 74: 677.

Epstein, R. H., A. Bolle, C. M. Steinberg, E. Kellenberger, E. Boy
    de la Tour, R. Chevallez, R. S. Edgar, M. Susman, G. H.
    Denhardt, and A. Lielausis. 1963. Physiological studies of
    conditional lethal mutants of bacteriophage T4D. Cold Spring
    Harbor Symp. Quant. Biol. 28: 375.

Goulian, M., Z. J. Lucas and A. Kornberg. 1968. Enzymatic syn-
    thesis of DNA. XXV. Purification and properties of DNA
    polymerase induced by infection with phage T4. J. Biol. Chem.
    243: 627.

Gross, J. D. 1972. DNA replication in bacteria. In Current Topics
    in Microbiology and Immunology. 57: 39. (Springer Verlag,
    Berlin, Heidelberg, New York).

Huang, W. M. and I. R. Lehman. 1972. On the exonuclease activity
    of phage T4 DNA polymerase. J. Biol. Chem. 247: 3139.

Huberman, J., A. Kornberg, and B. Alberts. 1971. The stimulation
    of T4 DNA polymerase by the protein product of T4 gene 32.
    J. Mol. Biol. 62: 39.

Inman, R., and M. Schnos. 1971. Structure of branch points in
    replicating DNA: Presence of single-strand connections in
    lambda DNA branch points. J. Mol. Biol. 56: 319.

Kozinski, A. W. and Z. Z. Felgenhauer. 1967. Molecular recombination in T4 bacteriophage DNA. II. Single-strand breaks and exposure of uncomplemented areas as a prerequisite for recombination. J. Virol. 1: 1193.

Marsh, R. C., A. Breschkin and G. Mosig. 1971. Origin and direction of bacteriophage T4 DNA replication. II. A gradient of marker frequencies in partially replicated T4 DNA as assayed by transformation. J. Mol. Biol. 60: 213.

Mosig, G., D. W. Bowden and S. Bock. 1972. E. coli DNA polymerase I and other host functions participate in T4 DNA replication and recombination. Nature New Biology. 240: 12.

Oey, J. L. and R. Knippers. 1972. Properties of the isolated gene 5 protein of bacteriophage fd. J. Mol. Biol. 68: 125.

Olivera, B. and F. Bonhoeffer. 1972. Discontinuous DNA replication in vitro. I. Two distinct size classes of intermediates. Nature New Biology. 240: 233.

Riggs, A., S. Bourgeois and M. Cohn. 1970. The lac repressor-operator interaction. III. Kinetic studies. J. Mol. Biol. 53: 401.

Salstrom, J. S. and D. Pratt. 1971. Role of Coliphage M13 gene 5 in single-stranded DNA production. J. Mol. Biol. 61: 489.

Sigal, N., H. Delius, T. Kornberg, M. Gefter and B. Alberts. 1972. A DNA-unwinding protein isolated from E. coli: its interaction with DNA and with DNA polymerases. Proc. Nat. Acad. Sci., USA. 69: 3537.

Sinha, N. K. and D. P. Snustad. 1971. DNA synthesis in bacteriophage T4-infected E. coli: evidence supporting a stoichiometric role for gene 32-product. J. Mol. Biol. 62: 267.

Snustad, D. P. 1968. Dominance interactions in E. coli cells mixedly infected with bacteriophage T4 wild-type and amber mutants and their possible implications as to type of gene-product function: catalytic vs. stoichiometric. Virology. 35: 550.

Sugino, A., S. Hirose and R. Okazaki. 1972. RNA-linked nascent DNA fragments in Escherichia coli. Proc. Nat. Acad. Sci., USA. 69: 1863.

Sugino, A. and R. Okazaki. 1972. Mechanism of DNA chain growth. VII. Direction and rate of growth of T4 nascent short DNA chains. J. Mol. Biol. 64: 61.

Warner, H. R. and M. D. Hobbs. 1967. Incorporation of uracil-C^{14}
     into nucleic acids in <u>Escherichia coli</u> infected with bacterio-
     phage T4 and T4 amber mutants. <u>Virology</u>. <u>33</u>: 376.

Werner, R. 1968. Initiation and propagation of growing points in
     the DNA of phage T4. <u>Cold Spring Harbor Symp. Quant. Biol.</u>
     <u>33</u>: 501.

Wolfson, J. and D. Dressler. 1972. Regions of single-stranded
     DNA in growing points of replicating bacteriophage T7 chromo-
     somes. <u>Proc. Nat. Acad. Sci., USA.</u> <u>69</u>: 2682.

THE ROLE OF THE HOST MEMBRANE IN THE MORPHOGENESIS OF THE HEAD OF

COLIPHAGE T4

Pamela J. Siegel and Moselio Schaechter

Department of Molecular Biology and Microbiology

Tufts University School of Medicine

Head morphogenesis in T4 represents an ideal system for studying the interaction of proteins with nucleic acid since many of the relevant components can be dissected genetically. T4 head morphogenesis has been studied by electron microscopic examination of cells infected with conditional lethal mutants under non-permissive conditions. However, the structures accumulated often are not normal intermediates. In vitro complementation (the formation of a head from its pieces) has not been achieved for T4. In fact, it has been carried out with good efficiency only with Tobacco Mosaic Virus. Here we describe a new approach for the examination of the early events of T4 head morphogenesis which conceivable could be used to study other viruses.

The method we have used is a modied M-band technique (for details, see (1) and Dworsky, P. and Schaechter, M., elsewhere in this volume). Briefly an M-band is formed by layering infected cells on a gradient and lysing them in the presence of sodium lauroyl sarcosinate (Sarkosyl) and $MgSO_4$. Sarkosyl and Mg++ form hydrophobic crystals to which the DNA-bearing membrane adheres preferentially. After centrifugation in the M-band is collected as a white dense band in the middle of the gradient. Cytoplasmic components and membrane components not associated with DNA are found in the top fraction above the M-band. Numerous controls have shown that mature T4 phage and purified T4 DNA are found in the top fraction only, and not in the M-band (1).

Parental and newly made T4 DNA are initially found in the M-band. Then, later in infection, the DNA detaches (figure 1). Late in infection, cells are very fragile and special techniques must be employed for handling them (1). It is necessary to use

149

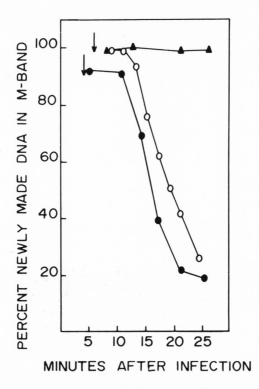

Fig. 1. Detachment of newly made DNA from M-band. At 4-1/2
minutes after infection at 37°C of <u>E. coli</u> B with phage T4D ac41
or phage T4 amH26, 15 µCi ^{3}H-thymidine/ml, was added for 45". Incorporation was terminated by resuspending infected cells in
broth containing 500 µg of thymidine/ml, (●-●). At 6 minutes after
infection of <u>E. coli</u> B with T4D ac41, 0.5 µCi/ml 2-^{14}C thymidine
was added (O-O). Four minutes after infection of <u>E. coli</u> B with
T4 tsA81 at 42C, 5 µCi/ml of ^{3}H thymidine was added, (▲-▲). Samples were taken at various times and placed on the top of gradients.
The percentage of DNA in the M-band was determined for each time
indicated.

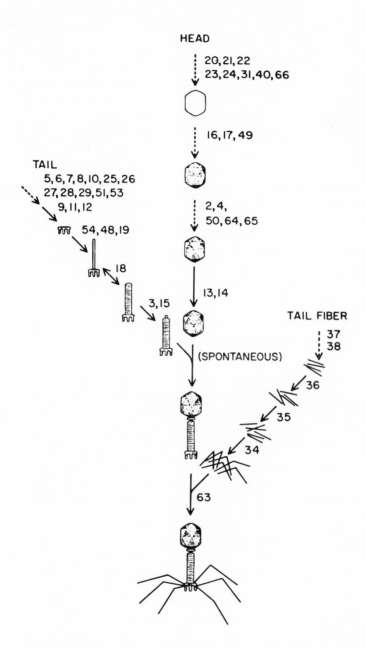

Fig. 2. The morphogenetic pathway of T4. Reprinted from Federation Proceedings 27: 1163, 1968 (3) by permission of the authors and the publisher.

lysis defective mutants of T4 (either e⁻, lysozymeless, or, t⁻ "membranase"-less) (8), or to carry out the conversion after the assembly of the major coat proteins, the products of genes 23, 24, 20, 22 (5). We know from our results that this cleavage occurs before mature DNA is cut from the replicative form, presumably by the action of P49 (6). Perhaps other proteins namely P16, P17, P4, and P65, also act before P49. Gene products P13 and P14 are involved in a late stage in head maturation. They are necessary to transform the head into the suitable substrate for the spontaneous addition of the tail, which can take place in vitro (7).

We have examined the effect of mutations in many late functions on the detachment of T4 DNA from the membrane. Gene products P20, P21, P22, P23, P24, P31, P16, P17, and P49 were found to be absolutely required for detachment (genes 40 and 66 were not tested). Gene products P2, P50, P64, P65 and P4 were not required. Mutants in gene 50 sometimes showed a slightly lower rate of DNA detachment, but this appeared to depend on the quality of the (amber) phage stock. Also, gene products involved in tail (P5) tail fiber (P37) or base plate (P12) morphogenesis were not necessary for DNA detachment (1).

It appears that from these data that protein cleavage is not required for DNA packaging and that gene products involved in head formation but not in other structural functions are required for detachment of DNA from the membrane. We confirmed the latter conclusion by temporal studies of the normal infection process.

The head does not detach in a fully mature state since its DNA is sensitive to pancreatic deoxyribonuclease (DNase) (figure 3). DNA of mature T4, on the other hand, is fully resistant to the action of this nuclease. DNase resistance is required rapidly, especially late in infection. A kinetic study suggests that the amount of structural proteins present in the cell determines the onset of DNA packaging and release from the membrane and, in addition, the requisition of resistance to DNase (1). The gene products involved in making DNA nuclease resistant are not known. Likely candidates are P13 and P14 since they appear to participate in a late stage of head morphogenesis (7).

Currently there is a controversy over whether initiation of T4 capsid formation takes place on the host membrane or directly on phage DNA and even whether T4 DNA participates at all in head maturation. We have approached this question by determining whether late proteins are bound to one or the other components. The distinction can be made in our methods by altering the amount of membrane in the M-band. As reported (11, and our unpublished data), mild treatment with Triton X-100 reduces the proportion of membrane material in the M-band from about 30%, the usual under the

ERRATUM

Part of the text of page 152 was accidentally omitted. The complete page should read as follows:

lysis defective mutants of T4 (either e⁻, lysozymeless, or t⁻ "membranase"-less) (8), or to carry out the conversion of cells to spheroplasts on the top of gradients prelayered with cyanide.

We have concluded that detachment of T4 DNA from the membrane occurs during the last stage of DNA encapsulation into a phage head, the cutting of DNA from a replicative concatemer. All the T4 DNA found in the top fraction (non-membrane bound) is of the size of mature T4 DNA, unlike the replicative DNA which is of heterogeneous size (unpublished data). As expected, detachment of progeny T4 DNA is the expression of late functions, and requires the synthesis of proteins made late in infection. During infection with T4 mutant in gene 55 (ts A81), DNA does not detach from the M-band (figure 1). Under non-permissive conditions, mutants in gene 55 synthesize DNA normally but do not make any late proteins (2). Late proteins are predominantly involved in phage morphogenesis (5).

We attempted to determine more precisely which step in morphogenesis is responsible for the detachment of phage DNA from the membrane. Figure 2 shows a diagram of the morphogenesis of T4. This is the work of Wood and his colleagues (3). As determined by complementation tests there are three independent branches of T4 morphogenesis: head, tail, and tail fiber assembly. These structures are assembled independently and then joined together (3). The products of gene 20, 21, 22, 23, 24, 31, 40, and 66 are involved in head formation. Genes 2, 4, 64, 65 participate in head morphogenesis but their function is not known with certainty. Genes 2, 50, and 64 are involved in the obligatory proteolytic cleavage of structural proteins (5, and our unpublished data). Five structural head proteins are known to be cleaved during morphogenesis: P22, P23, P24, and IPIII (5) and another protein of about 88,000 d (unpublished data). (The gene products of T4 are called by the number of their genes with the prefix P; IP denotes internal protein). The cleavage seems to occur after the assembly of the major coat proteins, the products of genes 23, 24, 20, 22, (5). We know from our results that this cleavage occurs before mature DNA is cut from the replicative form, presumably by the action of P49 (6). Perhaps other proteins, namely P16, P17, P4, and P65, also act before P49. Gene products P13 and P14 are involved in a late stage in head maturation. They are necessary to transform the head into the suitable substrate for the spontaneous addition of the tail, which can take place in vitro (7).

We have examined the effect of mutations in many late functions on the detachment of T4 DNA from the membrane. Gene products P20, P21, P22, P23, P24, P31, P16, P17, and P49 were found to be absolutely required for detachment (genes 40 and 66 were not tested). Gene products P2, P50, P64, P65, and P4 were not required. Mutants in gene 50 sometimes showed a slightly lower rate of DNA detachment, but this appeared to depend on the quality of the (amber) phage stock. Also, gene products involved in tail (P5), tail fiber (P37), or base plate (P12) morphogenesis were not necessary for DNA detachment (1).

It appears from these data that protein cleavage is not required for DNA packaging and that gene products involved in head formation but not in other structural functions are required for detachment of DNA from the membrane. We confirmed the latter conclusion by temporal studies of the normal infection process.

The head does not detach in a fully mature state since its DNA is sensitive to pancreatic deoxyribonuclease (DNase) (figure 3). DNA of mature T4, on the other hand, is fully resistant to the action of this nuclease. DNase resistance is required rapidly, especially late in infection. A kinetic study suggests that the amount of structural proteins present in the cell determines the onset of DNA packaging and release from the membrane and, in addition, the requisition of resistance to DNase (1). The gene products involved in making DNA nuclease resistant are not known. Likely candidates are P13 and P14 since they appear to participate in a late stage of head morphogenesis (7).

Currently there is a controversy over whether initiation of T4 capsid formation takes place on the host membrane or directly on phage DNA and even whether T4 DNA participates at all in head maturation. We have approached this question by determining whether late proteins are bound to one or the other components. The distinction can be made in our methods by altering the amount of membrane in the M-band. As reported (11, and our unpublished data), mild treatment with Triton X-100 reduces the proportion of membrane material in the M-band from about 30%, the usual under the

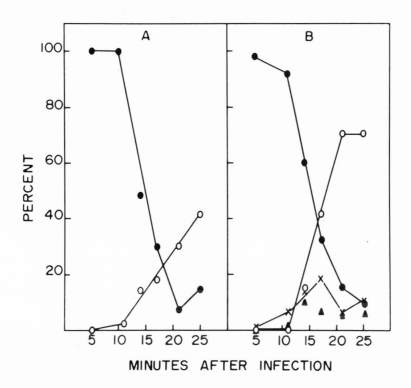

Fig. 3. DNase sensitivity of detached DNA in wild type infection.
At 4-1/2 minutes after infection of E. coli B with T4D ac41, at
37°C, 13 µCi/ml of ^{3}H thymidine was added. Incorporation was terminated at 5 min 15 sec by resuspending infected cells in broth containing 500 µg/ml thymidine. Samples were placed on gradients and
M-band determinations were made at the times indicated. The M-band
and top fraction from each sample were divided in two, one half was
treated with 50 µg/ml DNase and half served as a control, by addition
of 50 µg/ml bovine serum albumin.
(A) Percent of the DNA in the top fraction sensitive to DNase (●-●),
and percent of the DNA in the M-band resistant to DNase (o-o).
(B) DNase sensitive DNA in M-band/total DNA (o-o), DNase sensitive
DNA in top/total DNA (x-x), and DNase resistant DNA in M-band/total
DNA (▲-▲), DNase resistant DNA in top/total DNA (o-o).

conditions used here, to 5% of the total. We call these "minimal"
M-bands. Proteins that are found in "normal" M-bands (containing
30% or more membrane) but not in "minimal" M-bands can be assumed
to be bound to the membrane and not to the DNA. On the other hand,
treatment of M-bands with DNase releases proteins that are bound
to the DNA but not to the membrane.

After a short pulse (2 min. at 25C) with ^{14}C amino acids only
uncleaved P23 is found in the lysate or in "normal" M-bands. When
lysates were treated with Triton X-100 nearly all of P23 (about
90%) is found in the top fraction and is lost from the M-band. We
interpret these results to mean that during wild type infection
uncleaved P23 is first associated with the host cell membrane. If
the same experiment is performed but the label chased for 3 min,
both cleaved P23 (P23*) and P23 are found in the M-band. Some
P23* is found in "minimal" M-bands. It is directly associated with
the DNA in "normal" M-bands since DNase treatment releases P23* but
not P23 from these M-bands.

In an attempt to determine which proteins initiate capsid formation, we examined the proteins in "minimal" M-bands after a short
pulse of labeled amino acids. The first proteins found labeled in
this fraction are P20, and IP III, followed by P22 and small amounts
of P23 and protein X_1 (5). Also "normal" M-bands contain at least
3 proteins unique to the M-band (figure 4). T4 infected cells were
pulse labeled from 28-30 min after infection at 25°C with ^{14}C amino
acids. Samples were taken and M-bands made at 30, 32, and 35 min
after infection. These samples were subjected to SDS-acrylamide-
gel electrophoresis and subsequently to autoradiography according
to the method of Laemmli (5). As illustrated (figure 4) P20 and
IP III are the only unique proteins found in the M-band (we tenta-
tively called it "IP IV", It has a molecular weight of 88,000
daltons). On the other hand, P23, P24 (major structural proteins)
and P22 are equally distributed between the M-band and non-M band
membrane. Showe and Black (9), using a different approach, have
found that P22 and three internal proteins (IP I, IP II and IP III)
form the assembly core which serves as a nucleation center for
phage head formation. This core appears to be associated with DNA
in vivo.

We find that P20 plays a very early role in head morphogenesis
and appears to function as a capsid initiator protein.

"IP IV" which begins synthesis early and is cleaved late in
infection may also have a role in capsid initiation, it is particularly difficult to study since it is about the same size as major
early and late proteins. The cleavage product is however easily
detectable and has a molecular weight about half-way between that
of P18, 80,000 daltons (10) and P20, 63,000 daltons (approximately
71,500 daltons) (15).

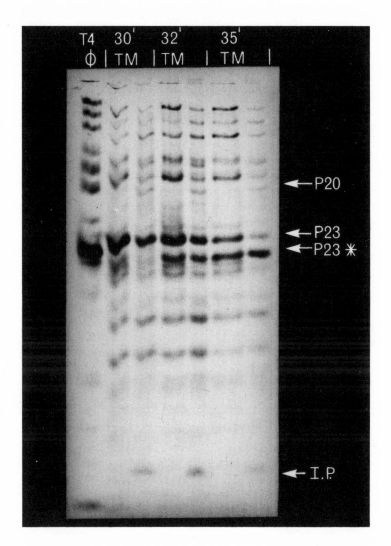

Fig. 4. Autoradiograms of SDS-acrylamide gels of M-bands and top
fractions from T4 infected cells. At 28 min after infection at 25C
of E. coli B/r with T4 amB5 (lysis⁻), a ^{14}C amino acid mixture was
added. At 30 min, incorporation of label was terminated by the
addition of unlabeled amino acids. Samples were taken at the times
indicated and M-bands were prepared.

    Preparation of samples and polyacrylamide gels, and electro-
phoresis were carried out according to Laemmli (5). Equal volumes
of each sample were placed onto the 10% gel. Gels were dried and
autoradiograms made.

We conclude then, that T4 DNA is always replicated in association with the host membrane. The DNA is released from the membrane in the form of a nearly complete but immature head. Although these studies are incomplete, we conclude that the following is a plausible sequence of events: P23, the major head protein, associates with the host membrane shortly after its synthesis. At the same time, "head initiator proteins" P20 and IP III and perhaps "IP IV" associate with the replicating DNA. P22 is perhaps a "linker" protein, being associated with the DNA and the membrane, and serves to bring the DNA-P20-IP III complex in association with the membrane bound P23. Cleavage of P23, P24, IP III, P22 and "IP IV" takes place normally as the head matures, then the DNA is cut and the head released to be completed in some unknown steps.

REFERENCES

Siegel, P. J. and Schaechter, M. 1973. J. Virol. 11: 359-367.
Pulitzer, J. F. 1970. J. Mol. Biol. 49: 473-488.
Wood, W. B., Edgar, R. S., King, J., Lielausis, I., Henninger, M. 1968. Fed. Proc. 27: 1160-1168.
Simon, L. S. 1972. Proc. Nat. Acad. Sci. USA. 69: 907-911.
Laemmli, U. K. 1970. Nature (London) 237: 680-685.
Frankel, F. R., Batcheler, M. and Clark, C. 1971. J. Mol. Biol. 62: 439-463.
Edgar, R. and Lielausis, I. 1968. J. Mol. Biol. 32: 263-276.
Josslin, R. 1970. Virology 40: 719-726.
Showe, M. K. and Black, L. W. 1973. Nature New Biology 242: 70-75.
King, J. and Laemmli, U. K. 1973. J. Mol. Biol. 75: 315-337.
Cundliffe, E. 1970. J. Mol. Biol. 52: 467-481.

BACTERIOPHAGE T7 DNA REPLICATION

IN VIVO AND IN VITRO

Rolf Knippers

Department of Microbiology, University of Konstanz

Konstanz, Germany

The genome of bacteriophage T7 is a linear double stranded DNA molecule (molecular weight: 25×10^6) with terminal redundancies of about 250 base pairs. The genetic information is sequentially expressed after infection: (1) An E. coli RNA polymerase transcribes the "pre-early" genes of T7 occupying the leftmost 17% of the genetic map. These genes (among others) code for a viral specific RNA

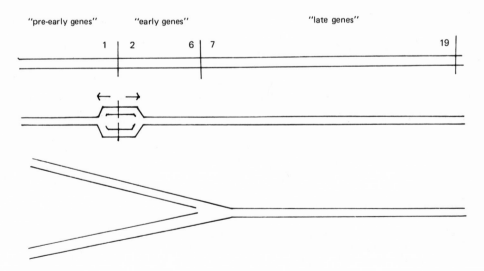

FIGURE 1. Direction of T7 DNA Replication (Dressler, Wolfson and Magazin, 1972).

polymerase (product of gene 1; Chamberlin, McGrath and Waskell, 1970) and for a virus induced ATP-dependent polynucleotide ligase (product of gene 1.3; Masamune, Frenkel and Richardson, 1971), (2) The T7 RNA polymerase transcribes the remaining 80% of the genetic map. "Early" genes contain the information for enzymes of DNA metabolism, (3) "Late" genes (genes 7 through 19) are coding for components of the viral envelope (Review: Studier, 1972). A list of the gene products involved in DNA replication is given in Table 1.

TABLE 1

Gene	Function	Molecular Weight	Phenotype
1	RNA polymerase	100,000	no expression of genes 2 through 19
1.3	Polynucleotide ligase	40,000	can be replaced by host cell ligase
2	Unknown	?	reduced DNA synthesis
3	Endonuclease	13,500	no DNA maturation
4	Unknown	67,000	no DNA synthesis
5	DNA polymerase	80,000	no DNA synthesis
6	Exonuclease	31,000	no DNA maturation

Bacteriophage T7 genes involved in DNA replication (Studier, 1972).

In addition to the genes listed above, it has been reported that genes 18 and 19 are required for DNA maturation.

T7 DNA REPLICATION IN-VIVO

It has been shown by Dressler, Wolfson and Magazin (1972) that T7 DNA replication is initiated at an internal DNA site which seems to be located near the leftmost promoter site of the T7 RNA polymerase, i.e. about 17% inward from the left end of the conventional genetic map of T7 (Figure 1). According to Dressler et al. (1972), replication proceeds bidirectionally until the left fork arrives at

FIGURE 2.   T7 Synthesis in Gene 3 Amber Mutant Infected Cells.
E. coli H 560 (pol A⁻ endoI⁻ su⁻; Vosberg   and Hoffmann-Berling,
1971) was grown in 100 ml minimal medium.   At a cell density of
2-3x10⁸/ml, the cells were pelleted and resuspended in 5 ml medium.
Mitomycin C (25 µg/ml) was added for about 20 minutes to suppress
host cell DNA synthesis.   The cells were then resuspended in 100 ml
prewarmed minimal medium and infected with gene 3 amber mutants
(multiplicity: 10 infective phages/cell).   (a) After 8 minutes at
30° C, a 50 ml sample was removed and exposed to 100 mC ³H-thymidine
for 20 seconds.   (b) The same amount of ³H-thymidine was added to
the remaining 50 ml at 16 minutes after infection for 20 seconds.
The cells were quickly cooled and pelleted in the cold.   The DNA
was extracted with Sarkosyl and Pronase and sedimented through a
neutral sucrose gradient which was formed on top of a dense CsCl-
layer (δ=1.75 g/ml).   Centrifugation was in the SW 27 rotor of the
Spinco centrifuge for 14 hours at 23 000 rpm.   ³²P-T7 DNA served as
sedimentation marker.   ³H-cpm/fraction (●).   ³²P-cpm/fraction (o).

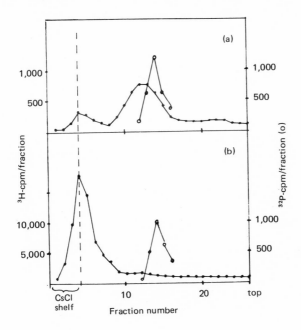

Multiplication of T7 in Temperature Sensitive E. Coli Cells.
The E. coli strains were grown at 3° C in 20 ml tryptonebroth (plus
20 mg/ml thymine).   At a cell density of 2x10⁸/ml, 10 ml samples were
removed from the cultures and placed in a 42° C waterbath.   After 15
minutes with aeration, all cultures were infected with T7 wild type
in a multiplicity of 0.2 infective particles/cell.   After lysis had
occurred few drops of chloroform were added to each culture.   Progeny
phages were plated with E. coli B as indicator strain.

the end of the parental DNA when a Y-like fork continues to move to
the right end of the DNA (Figure 1). New rounds of replication can
be initiated before the first cycle is finished.

From pulse label experiments, we concluded that this process
does not require the products of gene 3 and gene 6, i.e. the viral
endonuclease 1 and the viral exonuclease. We have described before
(Strätling, Krause and Knippers, 1973) that incorporation of [3]H-
thymide into T7 DNA can be observed in cells infected with amber
mutants in genes 3 and 6. [3] H-thymidine added early after infection
to a gene 3 amber mutant infected culture was recovered in DNA structures which sedimented slightly faster than intact T7 DNA (Figure 2,
a). The sedimentation characteristics of these DNA molecules were
those expected for the intermediates observed by Dressler et al.
(1972) (Figure 1). When ^{3}H-thymidine was added later after infection,
almost all labelled DNA sedimented much faster than unit length T7
DNA (Figure 2, b) as expected for DNA concatemeters of several times
the length of normal T7 DNA. No conversion of this fast sedimenting DNA into unit length DNA was observed. Similar data were obtained for gene 6 amber mutant infected cells. In wild type T7
infected cells, concatemers were also observed. But, in addition
to these fast sedimenting DNA forms, a large peak of unit length T7
DNA occurred in sucrose gradient analysis.

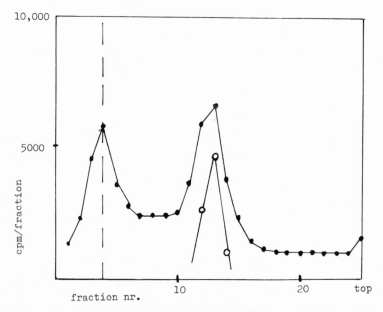

FIGURE 3. Sucrose Gradient Analysis of Pulse Labelled DNA from
          T7 Wild Type Infected Cells
Conditions: See legend to Figure 2, b. ^{3}H-cpm/fraction (●) ^{32}P-cpm/
fraction (o).

We conclude that endonuclease 1 and T7 exonuclease are involved in a DNA maturation process. It had been shown by others (Sadowski and Yerr, 1970) that both enzymes have at least one other function in the T7 infection cycle: the degradation of host cell DNA. The mononucleotides (final products of the degradation process) can be reused by the phage for the synthesis of progeny DNA. Very little ^{3}H-thymidine was incorporated into DNA when cells infected with amber mutants in genes 2, 4 and 5 (Table 1) were pulse labeled (Hausmann and LaRue, 1968; Strätling et al., 1973). These experiments showed that the products of genes 2, 4 and 5 are required for either initiation or propagation of T7 DNA replication.

T7 DNA replication does not depend on any of the known Escherichia coli genes which have to be functional for cellular DNA replication. This was concluded from the data shown in Table 2 (Knippers, Strätling and Krause, 1973). A representative strain of each complementation group of the unknown E. coli mutants temperature sensitive in DNA replication was infected with wild type bacteriophage T7 at permissive and non-permissive temperature. Table 2 shows that burst sizes measured after incubation at 30° C and at 42° C, respectively, were similar.

TABLE 2

		Average number of phages/cell produced at	
Gene locus	Host strains	30°C	42°C
A	F 117	79	49
B	E 279, 266	23	44
C	PC 1	31	36
D	PC 7	137	171
E	E 486	51	62
F	E 101	49	80
G	CR 34	140	116

Burst Size of Bacteriophage T7 Infected E. coli Cells.

T7 DNA REPLICATION IN TOLUENIZED CELLS

Bacterial cells are made permeable for nucleotides by a brief treatment with ether (Vossberg and Hoffmann-Berling, 1971) or with toluene (Moses and Richardson, 1970). Under proper conditions, these preparations corporate deoxynucleosidetriphosphates into DNA in a process which has much in common with DNA replication. One of the most interesting observations made with ether- or toluene-treated

cells was the discovery that DNA replication seems to depend on a
sufficient supply of ATP.

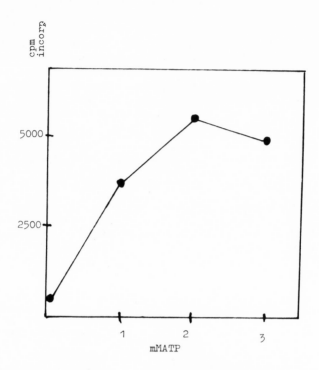

FIGURE 4. T7 DNA Synthesis in Toluenized Cells. E. coli H 560 cells
were infected with gene 3 amber mutants (multiplicity: 10 infective
phages/cell). About 15 minutes after infection (at 30° C), the infected cells were collected and toluene treated (Moses and Richardson,
1971). For incubation, 0.05 ml. of toluenized cells (total of 10^{9}
cells) were mixed with 0.05 ml. of a buffer, containing: 100 mM KCl,
50 mM trisHCl, pH 7.6; 0.1 mM dithioerythritol; 8 mM $MgSO_4$; 3 mM
spermidine-HCl and 12 mM each of dGTP and dCTP as well as 2.4 mM
^{3}H-dTTP (spec. act. 14 C/mmole). ATP was added in the indicated final
concentrations. Incubation was for 12 minutes at 30° C. Acid precipitable counts are given.

The data of Figure 4 show that ATP is also required for optimal
T7 DNA synthesis in toluenized T7 infected cells. At least a

fraction of the required ATP seems to be used for DNA chain initiation.
It is quite conceivable that another part of the ATP was used to re-
phosphorylate deoxyribonucleotides and ribonucleotides. No data on
the extent of rephosphorylation are presently available. The requirement of ATP for DNA chain initiation was suggested by an experiment where ^{14}C-labeled ATP and ^{3}H-labeled deoxynucleosidetriphosphates were added simultaneously to a preparation of toluenized T7
infected cells. After incubation for 6 minutes at 30° C, the nucleic
acids were extracted, heat denatured and centrifuged in a Cs_2SO_4
equilibrium gradient. The result, as presented in Figure 5, shows
that the majority of the ^{14}C-labeled material banded, as expected,

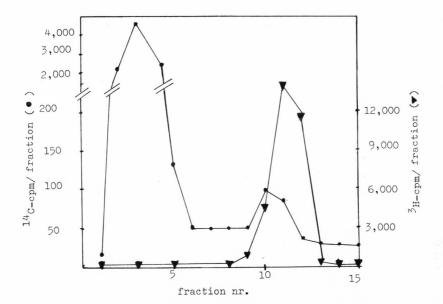

FIGURE 5. Equilibrium Centrifugation in a Cs_2SO_4 Gradient. 0.5 ml.
gene 3 amber mutant infected toluenized H 560 cells were incubated
under standard conditions (see: legend to Figure 4) with ^{14}C-labeled
ATP (final concentration: 2 mM; spec. act.: 5 C/mole) and [3]H-
deoxynucleosidetriphosphates for 6 minutes at 30°C. The nucleic
acids were extracted with 5% Sarkosyl in the presence of 2 M NaCl.
The extracted nucleic acids were denatured at 100°C for 5 minutes
and rapidly cooled. Formaldehyde (final concentration: 1% was then
added to prevent the formation of unspecific RNA aggregates. The
volume was adjusted to 2 ml before 1.4 g Cs_2SO_4 was added. Centrifugation was at 35,000 rpm for two days in the SW 56 rotor of the
Spinco centrifuge. Part of each fraction was removed, precipitated
with trichloracetic acid on Millipore filters and counted.

at a buoyant density where RNA (in the presence of formaldehyde) should band.   Between 3 and 6 percent reproducibly appeared, however, in fractions of the Cs_2SO_4 gradient where the ^{3}H-labeled single stranded DNA banded.   Recentrifugation of this material did not change the apparent buoyant density.   All ^{14}C-labeled material was alkaline labile and sensitive to RNase.   We conclude that some RNA was covalently connected to DNA.   The production of this RNA-DNA copolymer may indicate that initiation of T7 DNA replication could proceed via an RNA primer strand as has been proposed before for other systems (Wickner, Brutlag, Schekman and Kornberg, 1972; Sugino, Hirose and Okazaki, 1972).

We have also investigated the structure of the T7 DNA synthe-sized in toluenized cells by neutral sucrose gradient sedimentation.

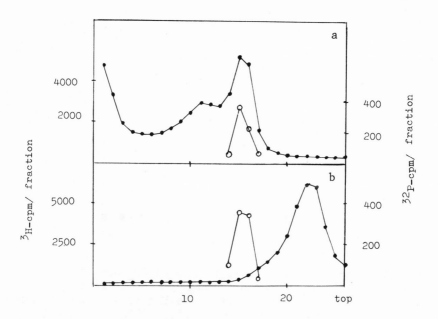

FIGURE 6.   Sucrose Gradient Analysis of T7 DNA Synthesized in Toluenized Cells.   H 560 cells, infected with gene 3 amber mutants (a) and with T7 wild type phage (b) were toluene treated.   After 15 minutes incubation under standard conditions (see Figure 4) with 2 mM ATP, the DNA was extracted with detergents at high salt and sedi-mented through a neutral sucrose gradient.   (20%-5%; 1 M NaCl in tris-EDTA; 16 ml; SW 27 rotor of the Spinco centrifuge; 13 hours at 25 000 rpm and 8°C).

Cells infected with T7 wild type and with gene 3 amber mutants were
compared. It can be seen from the sedimentation profiles of Figure
6 (which should be compared to those of Figure 2 and 3) that the
DNA synthesized in wild-type T7-infected toluenized cells seems to
be degraded while the DNA from gene 3 mutant infected cells has the
sedimentation characteristics of in-vivo pulse labeled DNA (Knippers
and Krause, 1973). This seems to indicate that the endonuclease I
of T7 (the product of gene 3) degrades most of the in-vitro synthesized DNA.

T7 DNA SYNTHESIS IN DNA MEMBRANE COMPLEXES

DNA polymerase I deficient (pol A⁻) E. coli cells were infected
with ^{32}P-T7 gene 3 amber mutants. After gentle lysis (Strätling
and Knippers, 1971) of the infected cells, a sucrose gradient analysis of the extracts was performed. Figure 7 shows that most of the
^{32}P-labeled DNA appeared in a fast sedimenting complex. Comparing
this result with our previous data (Strätling and Knippers, 1971)
we conclude that T7 DNA was attached to the cellular wall-membrane
complex. For reasons which will become apparent below we do not
think that this association is significant for T7 DNA replication.
The preparation of DNA membrane complexes is, however, a convenient
technique to separate structurally bound T7 proteins from free "cytoplasmic" proteins.

To obtain some information about the distribution of DNA synthesizing activity in the gradient of Figure 7, samples were removed
from each fraction and incubated with ATP and ^{3}H-labeled deoxynucleosidetriphosphates. As recorded in Figure 7, much DNA synthesizing activity was detected in the DNA membrane complexes. "Free"
DNA polymerase became detectable after addition of single stranded
E. coli DNA to each sample. Single stranded DNA is an excellent
primer for the isolated T7 DNA polymerase (Oey, Strätling and Knippers, 1971). The external primer DNA, however, did not further stimulate the DNA synthesizing activity in DNA membrane complexes.

In fact, essentially all of the DNA synthesized in DNA membrane
complexes hybridized with T7 DNA even when an excess of single stranded
E. coli DNA had been present in the incorporation assay. We conclude
from this experiment that the T7 DNA polymerase (and possibly other
functions required for T7 DNA replication) were firmly associated
with their authentic template.

To obtain some information on the nature of these structurally
bound proteins, a T7 infected culture was labeled with a ^{14}C-labeled
amino acid mixture during the first half of the eclipse period of
the infection cycle. DNA membrane complexes were prepared. It was
found that about 20% of all labeled proteins cosedimented with the
fast sedimenting complex. The remaining ^{14}C-labeled material stayed
at the top of the sucrose gradient.

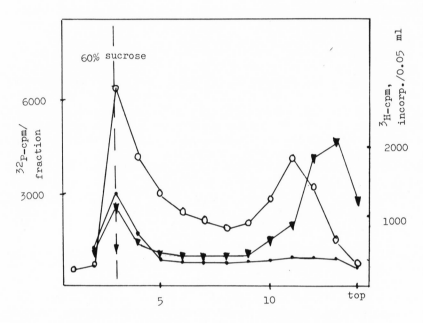

FIGURE 7.   DNA Membrane Complexes from T7 Infected Cells.
H 560 cells were infected with ^{32}P-labeled gene 3 amber mutants
in a multiplicity of 5 - 10.   About 16 minutes after infection at
30° C, the cells were rapidly cooled.   The preparation of DNA
membrane complexes had been described (Strätling and Knippers, 1972):
briefly, cells were incubated with lysozyme in a EGTA-containing
buffer with 20% sucrose, an equal volume of 2% Brij 58 was then
added.   The transparent suspension was then layered on to a 30% -
10% sucrose gradient which had been poured on top of a 60% sucrose
shelf.   The sample was centrifuged for 4 hours at 25 000 rpm (SW 27
rotor).   The distribution of ^{32}P-counts (o) was determined in 0.1
ml samples from each fraction.   Incorporation of deoxynucleotides
was determined as described (see: Figure 4) in the presence of 2 mM
ATP: without addition of single stranded primer DNA (●); with 10 mg
denatured E. coli DNA/assay (▼).

More than 95% of the structurally bound proteins can be removed from the DNA membrane complex by treatment with DNase or 1
M KCl (Strätling, in prep.). We conclude that the proteins are
associated with the DNA rather than with the membrane.

A sample from the DNA bound protein fraction was investigated
by polyacrylamide gel electrophoresis in the presence of sodium-
dodecylsulfate. The radioactive bands were detected by autoradiography and quantitated by microdensitometry. A total of about 12
protein bands were detected in the unfractionated protein sample.
(Figure 8). Six proteins appeared to be preferentially associated
with the DNA. By comparison with Studier's assignments of protein
bands to gene products (Studier, 1972), we assume that the DNA-
bound proteins include: RNA polymerase, DNA polymerase, polynucleotide ligase, a fraction of the gene 4 protein and at least two
smaller proteins; one of these may be the gene 6-coded exonuclease.
Several of the DNA bound proteins form tight aggregates with total
molecular weights of about 400,000 (Strätling, in prep.). The
significance of these aggregates is unknown.

T7 DNA SYNTHESIS IN CRUDE EXTRACT

The isolated T7 DNA polymerase prefers single stranded DNA over
double stranded DNA as template-primer (Oey et al., 1971). In crude
protein preparations from T7 infected cells, a DNA synthesizing
activity can be detected which accepts double stranded intact T7 DNA
as template (Figure 9). This activity does not require an active
viral endonuclease I or an active exonuclease. In fact, when an
active nuclease was present in the extracts the in-vitro synthesized
DNA was rapidly degraded (cf. Figure 6). Unit length (or longer)
DNA structures were only synthesized when an extract from gene 3
amber mutant infected cells had been used. The DNA synthesizing
activity depended on an optimal concentration of ribonucleoside-
triphosphates (rNTPs). ATP could replace the mixture of the four
rNTPs. Optimal activity was observed in the presence of 1 - 2 mM
spermidine. This is a concentration which does stimulate the isolated T7 RNA polymerase by a factor of 5 - 10 (Knippers, Strätling
and Ferdinand, 1973). The activity of isolated DNA polymerase was
not altered by addition of spermidine. Actinomycin inhibited the
DNA synthesizing activity in crude extracts. This drug also inhibited the activity of isolated T7 RNA polymerase but not that of
isolated T7 DNA polymerase. The rifamycin derivative AF/013 was
also inhibitory on both, the DNA synthesizing system in crude extracts and the isolated RNA polymerase, while it did not, in the
concentrations used, influence isolated DNA polymerase (Knippers
et al., 1973). We conclude from these observations that an RNA
polymerase was required for DNA polymerization in crude extracts.
This RNA polymerase is not the host cell enzyme since rifampicin
and streptolydigen did not influence the reaction. Both drugs are
potent inhibitors of the E. coli RNA polymerase. It seems to be

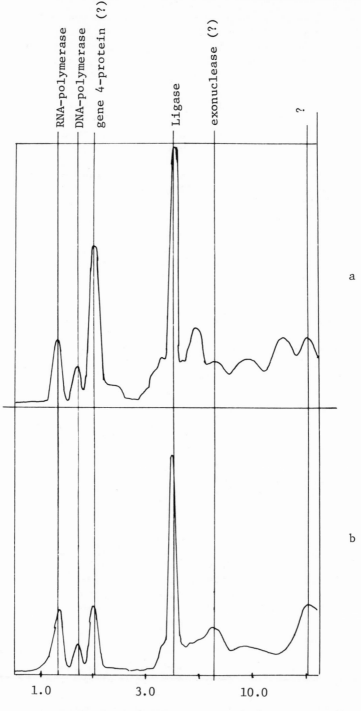

FIGURE 8

FIGURE 8. Polyacrylamide Gel Electrophoresis.
A T7 wild type infected E. coli culture was labeled from the 6th to
8th minute after infection with a mixture of ^{14}C-labeled aminoacids.
The culture had been irradiated with ultraviolet light before infection to suppress residual host cell protein synthesis. After 2 minutes in the presence of ^{14}C, an excess of unlabeled aminoacids was
added in order to reduce further incorporation of labeled aminoacids
into protein. The samples were incubated another two minutes to allow
any ^{14}C-containing protein chains to be finished. The cells were then
rapidly cooled and centrifuged. (a) A part of the cell suspension
was then lysed with lysozyme. SDS (final concentration: 1%) and
mercaptoethanol (final concentration: 1%) were added. The cell extract was then electrophoresed in 7.5% gels according to Weber and
Osborn (1968). The gel was longitudinally sliced; the central slice
was dried and exposed to an X-ray film. The band pattern was evaluated by microdensitometry (Vitatron) - redrawn. (b) Another part
of the cell pellet was gently lysed with Brij 58 and sedimented
through a sucrose gradient to prepare DNA membrane complexes (Figure
7). A sample with a similar total radioactivity as the sample taken
for the control electrophoresis (a) was treated with SDS and mercaptoethanol. Electrophoresis and autoradiography: see (a). The following arguments support the notion that the labeling time corresponded
to the early eclipse period of the infection cycle: (1) High proportion of "pre-early" proteins, note the high ligase peak; (2) absence
of typical "late" proteins, the components of the phage envelop.

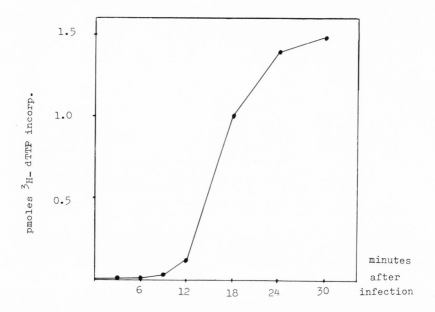

FIGURE 9.   DNA Synthesizing Activity in Crude Extracts from Gene 3
Amber Mutant Infected Pol A⁻ E. Coli Cells (from: Knippers, Strätling and Ferdinand, 1973). A 1 1-culture of H560 cells was infected
with T7 gene 3 amber mutants under the conditions described under Figure 2. After various times at 27° C, 100 ml samples were removed and
rapidly cooled. DNA free extracts were prepared. Briefly, this preparation procedure included the following steps: (1) lysis of the pelleted cells with lysozyme in a 20% sucrose buffer; (2) addition of an
equal volume of 2 M KCl to elute DNA bound proteins; (3) centrifugation at 100 000 xg for 2 hours to remove DNA membrane complexes and
ribosomes; (4) ammonium sulfate precipitation of the proteins in supernatant; (5) dialysis of precipitated proteins against 10% glycerol in
50 mM tris-HCl, pH 7.6, 0.1 mM EDTA; 0.1 mM dithioerythritol; 2 mM
spermidine-HCl. Incorporation assay: 0.050 ml buffer (100 mM KCl;
50 mM tris-HCl, pH 7.6; 8 mM MgSO₄; 2 mM spermidine HCl; 0.1 mM
dithioerythritol) plus 2 mg T7 DNA, 2 mM ATP, 24 mM each of dATP,
dGTP, dCTP as well as 0.5 mM ³H-dTTP (spec. act.: 15 C/mmole). To
this mixture was added about 15 mg protein of the preparation. Incubation was for 10 minutes at 30° C. The trichloroacetic acid precipitable radioactivity was measured. No DNA synthesis was observed
in the absence of T7 DNA.

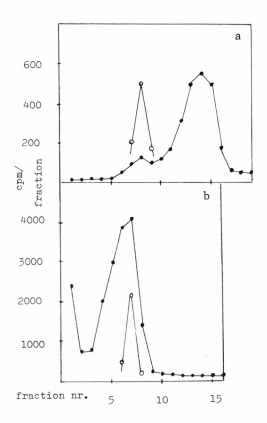

FIGURE 10. Sucrose Gradient Analysis of the DNA Structures
  Synthesized in Crude Extracts (from: Knippers, Strätling and
  Ferdinand, 1973).
About 20 mg from an extract prepared, respectively, 15 minutes after
infection (a) and 24 minutes after infection (b) was incubated in
a 0.1 ml volume under standard conditions (see Figure 9). The
reaction was stopped after 6 minutes at 30° C by EDTA and detergents.
The samples were then centrifuged through a neutral sucrose gradient
(see Figure 6). ^{32}P-T7 DNA was a sedimentation marker. ^{3}H-cpm/
fraction (●). ^{32}P-cpm/fraction (o).

apparent, therefore, that the virus induced RNA polymerase is required for T7 DNA synthesis in crude extracts.

An experiment, similar to the one shown in Figure 5, was performed with crude extracts. An RNA-DNA copolymer was detected which
was composed of 500 - 1000 deoxyribonucleotides with a chain of 50 -
100 ribonucleotides on the 5'-end. The synthesis of this copolymer
may indicate that bacteriophage T7 DNA strands are initiated via RNA
primer strands as has been proposed before (Knippers, Strätling and
Krause, 1973).

No DNA synthesis was observed in extracts from gene 5 amber
mutant infected cells. A greatly reduced DNA synthesis was detected
in extracts from gene 4 amber mutant infected cells. These observations indicate that the products of genes 4 and 5 are required for
the DNA synthesis observed. It is not clear why isolated T7 DNA
polymerase requires single stranded DNA as template primer while
the DNA synthesizing activity in crude extracts (which among others
contains the DNA polymerase) accepts double stranded intact T7 DNA.
It could be excluded that the DNA template was degraded or otherwise altered in the reaction mixture. ^{32}P-T7 DNA sedimented after
incubation through neutral and alkaline sucrose gradients with the
sedimentation characteristics of a control DNA (Knippers et al.,
1973). It may be possible that a DNA unwinding protein (Alberts and
Frey, 1970) is operative in our system but so far we were not able
to unambiguously identify such a function. A neutral sucrose gradient analysis of the in-vitro synthesized T7 DNA is shown in Figure
10. The profile is similar to the one shown in Figures 2 and 3
where in-vivo pulse labeled DNA was examined under the same conditions. After alkaline treatment, however, the in-vitro synthesized
DNA was degraded to a spectrum of small fragments with sedimentation
coefficients around 9 S. No conditions have been found under which
these fragments were converted to unit length DNA strands in spite
of the fact that the virus induced polynucleotide ligase is active
under the usual incubation conditions. We conclude that the initiation mechanism apparently worked in our system while chain elongation and/or fragment connection was unsufficient.

We finally would like to mention that the in-vitro system seems
to be specific for T7 DNA: only limited DNA synthesis was observed
when either double stranded bacteriophage or highly polymerized E.
coli or calf thymus DNA was added to the reaction mixture.

The E. coli strains used in this study were kindly provided by Dr.
M. Kohiyama. (Review: Gross, 1972) The biological functions of
the E. coli genes are not known, except those of gene E which is
coding for the bacterial DNA polymerase III (Gefter, Hirota, Kornberg, Wechsler and Barnoux, 1971) and of gene F which codes for a
ribonucleotide reductase (Fuchs, Karlström, Warney and Reichard,
1972). Bacteriophage T7 does not require this enzyme because the
breakdown products of the E. coli DNA are reused by the phage to
synthesize its own DNA.

REFERENCES

1. Alberts, B. and Frey, L. (1970) Nature 227: 1313.
2. Chamberlin, M., McGrath, J. and Waskell, L. (1970) Nature 228:
   227.
3. Dressler, D., Wolfson, J. and Magazin, M. (1972) Proc. Nat.
   Acad. Sci. USA 69: 998.
4. Fuchs, J.A., Karlström, H.O., Warney, H.A. and Reichard, P.
   (1972) Nature New Biology 238: 69.
5. Gefter, M.L., Hirota, Y., Kornberg, T., Wechsler, J.A. and
   Barnoux, C. (1971) Proc. Nat. Acad. Sci. USA 68: 3150.
6. Gross, J. (1972) Current Topics in Microbiology and Immunology
   57: 39.
7. Hausmann, R. and LaRue, K. (1969) J. Virology 3: 278.
8. Knippers, R., Strätling, W. and Krause, E. (1973) in "DNA
   Replication In-Vitro", R.B. Inman and R.D. Wells, Eds.,
   University Park Press, Baltimore.
9. Knippers, R. and Krause, E. (1973) submitted to Europ. J.
   Biochem.
10. Knippers, R., Strätling, W. and Ferdinand, F.J. (1973) sub-
    mitted to Europ. J. Biochem.
11. Masamune, Y., Frenkel, G.D. and Richardson, C.C. (1971) J.
    Biol. Chem. 246: 7874.
12. Moses, R.E. and Richardson, C.C. (1970) Proc. Nat. Acad. Sci.
    USA 67: 674.
13. Oey, J.L., Strätling, W. and Knippers, R. (1971) Europ. J.
    Biochem. 23: 497.
14. Sadowski, P.D. and Kerr, C. (1970) J. Virol. 6: 149.
15. Strätling, W. and Knippers, R. (1971) Europ. J. Biochem. 20: 330.
16. Strätling, W., Krause, E. and Knippers, R. (1973) Virology,
    in press.
17. Studier, F.W. (1972) Science 176: 367.
18. Sugino, A., Hirose, S. and Okazaki, R. (1972) Proc. Nat. Acad.
    Sci. USA 69: 1863.
19. Vossberg, H.P. and Hoffmann-Berling, H. (1971) J. Mol. Biol.
    58: 739.
20. Weber, K. and Osborn, M. (1969) J. Biol. Chem. 244: 4406.
21. Wickner, W., Brutlag, D., Schekman, R. and Kornberg, A. (1972)
    Proc. Nat. Acad. Sci. USA 69: 965.

RNA AND COLICINOGENIC FACTOR E1 REPLICATION

D. J. Sherratt,[1] D. G. Blair,[2] D. B. Clewell[3] and
D. R. Helinski

Department of Biology, University of California

San Diego, La Jolla, California, U.S.A.

The bacterial non-sex factor plasmid, colicinogenic factor E1 (ColE1), has a molecular weight of 4.2×10^6 daltons and there are normally about 20 copies present per E. coli cell (Clewell and Helinski, 1972). Unlike sex-factors and the E. coli chromosome, ColE1 continues to replicate in chloramphenicol treated cells for many generations (Bazaral and Helinski, 1970). However addition of rifampicin to chloramphenicol treated cells results in rapid cessation of ColE1 synthesis (Clewell, Evenchick and Cranston, 1972), indicating that RNA synthesis but not protein synthesis is required for ColE1 replication.

Initial examination of ColE1 DNA synthesized in chloramphenicol showed that, like that isolated from exponentially growing cells, it consists of covalently closed supercoiled duplex DNA. However further analysis of the plasmid DNA synthesized in chloramphenicol showed a proportion of it to be both alkali and RN'ase labile (Fig. 1). In contrast, uniformly labelled plasmid DNA isolated from exponentially growing cells was not significantly affected by these treatments. The supercoiled and open-circular nature of the DNA in each

---

[1]Present address: School of Biology, University of Sussex, Falmer, Brighton, England

[2]Department of Chemistry, U.C.S.D.

[3]Present address: Departments of Oral Biology and Microbiology, University of Michigan, Ann Arbor, Michigan, U.S.A.

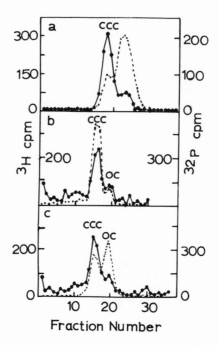

Fig. 1.   Alkali and RN'ase lability of ColEl DNA synthesized in
chloramphenicol.

a)   Alkaline CsCl equilibrium centrifugation of ColEl DNA
synthesized in the absence and presence of chloramphenicol (150
μg/ml, 18 hours at 37°C)

b) and c) Velocity sedimentation through 5-20% neutral sucrose
density gradients of ColEl DNA before (b) and after (c) pancreatic
RN'ase treatment (1 mg/ml for 10 min at 37°C)

●————● ColEl DNA synthesized in absence of chloramphenicol
●– – – ● ColEl DNA synthesized in presence of chloramphenicol
CCC = covalently closed duplex circles
OC = nicked duplex circles

Growth, labelling and assay conditions are as described in
Blair et al. (1973).

of the two peaks obtained after pancreatic RN'ase treatment (Fig. 1c), was verified by electron microscopy. RN'ase H, which is specific for the RNA portion of RNA-DNA hybrids (Miller, Gill and Riggs, 1972), also induced the rapid conversion, to the open circular form, of plasmid DNA synthesized in chloramphenicol. Ribonuclease T1 had no effect. The fraction of ColEl DNA insensitive to pancreatic RN'ase was not alkali labile.

To characterize the nature of the open circles resulting from RN'ase treatment, they were sedimented through an alkali sucrose density gradient to separate the single strand circles from the single strand rods (e.g. see Blair et al., 1972). The duplex open circles consisted of approximately equal numbers of unit length rods and circles, indicating that a single break is introduced into the RN'ase sensitive supercoils. The separated linear and circular fractions were then analysed by poly (UG)-CsCl equilibrium centrifugation (Blair, Clewell and Helinski, 1971; Szybalski et al., 1971) to determine if the break introduced by RN'ase is confined to a single specific strand. Both circular and linear strands consisted of equal numbers of each of the complementary strands, showing that the RN'ase sensitive site is distributed randomly with respect to the two strands.

The proportion of ColEl DNA sensitive to alkali and RN'ase increases approximately linearly with time of incubation in chloramphenicol reaching a maximum when about 60% of the supercoils are sensitive. When this maximum is reached ColEl synthesis stops, suggesting a close relationship between the ability to replicate ColEl in chloramphenicol and the appearance of RNA containing supercoils. Addition of rifampicin (3 µg/ml) to a chloramphenicol treated culture synthesizing ColEl DNA, results not only in the cessation of ColEl synthesis, but also stopped the accumulation of RNA-containing supercoils, indicating that their generation required de novo RNA synthesis, and again emphasizing the relationship between replication of the plasmid and the generation of RNA containing supercoils. The absence of such supercoils in uniformly labelled ColEl DNA isolated from exponentially growing cells, suggests that either they have a short half life or such structures do not occur in normally growing cells. Fig. 2 illustrates a possible model to explain the above observations. RNA acts as the primer for DNA synthesis, which eventually generates two daughter molecules, each containing a nick between the 3' terminus of the newly synthesized DNA and the 5' terminus of the RNA primer. In vitro, E. coli ligase will not seal such a structure at a significant rate (Kornberg, this meeting). In route a) the priming RNA is removed by 5' exonuclease activity and the ensuing gap filled by DNA; in vitro, DNA polymerase I will fulfill both of these functions (Kornberg, this meeting). The nick can now be sealed by E. coli ligase to give a covalently closed duplex circle containing no RNA.

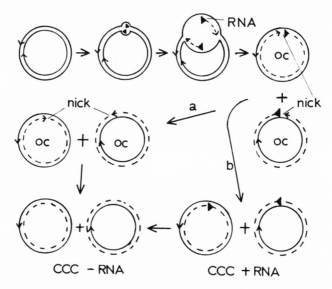

Fig. 2.   Primer model for the involvement of RNA in ColEl replication.
For simplicity, bidirectional replication only has been considered.
For unidirectional replication RNA priming could be on either of
the two strands.   For explanation see text.

      CCC = covalently closed duplex circle.
      OC = open duplex circle.

In route b) some uncharacterized sealing activity forms a ribo-deoxyribonucleotide bond to generate a covalently closed circle containing RNA. This RNA might be subsequently converted to DNA or excised and replaced by DNA.

At least the first part of route b) must operate in chloramphenicol, though we have no evidence to distinguish between the routes in exponentially growing cells. Taking into account the known properties of DNA polymerase I and its requirement for ColEl DNA replication (Kingsbury and Helinski, 1970), route a) seems to be attractive for replication in normally growing cells. Whatever the actual mechanism, the appearance of RNA containing supercoils presumably reflects the instability of an enzyme activity which is responsible for removing or converting the RNA; attempts to identify the functions necessary for this are in progress.

ACKNOWLEDGMENTS

We thank Mr. Bernard Ashcraft for his excellent technical assistance and Dr. Gordon Gill for his generous provision of the enzyme RNase H. This work was supported by U. S. Public Health Service Research Grant A1-07194 and National Science Foundation research grant GB-29492. D. G. B. was supported by a U.S. Public Health Service Predoctoral Traineeship (2-To1-6M-1045). D. R. H. is a U.S. Public Health Service Research Career Development Awardee (K04-6Mo7821).

REFERENCES

Bazaral, M. and D. R. Helinski. 1970. Biochemistry. 9:399-406.

Blair, D. G., D. B. Clewell and D. R. Helinski. 1971. Proc. Nat. Acad. Sci., U.S.A. 68:210-214.

Blair, D. G., D. J. Sherratt, D. B. Clewell and D. R. Helinski. 1972. Proc. Nat. Acad. Sci., U.S.A. 69:2518-2522.

Clewell, D. B. and D. R. Helinski. 1972. J. Bacteriol. 110:1135-1146.

Clewell, D. B., B. Evenchick and J. W. Cranston. 1972. Nature New Biology. 237:29-31.

Kingsbury, D. T. and D. R. Helinski. 1970. Biochem. Biophys. Res. Commun. 41:1538-1544.

Miller, H. I., G. N. Gill and A. D. Riggs. 1972. Fed. Proc.
   31:500 abs.

Szybalski, W., H. Kubinski, Z. Hracecna and W. C. Summers. 1971.
   In L. Grossman and K. Moldave, eds., Methods in Enzymology,
   Vol. XXI, part D., pp. 383-413.

OVERLAPPING PATHWAYS FOR REPLICATION, RECOMBINATION AND REPAIR IN

BACTERIOPHAGE LAMBDA

A. Skalka and L. W. Enquist*

Department of Cell Biology, Roche Institute of
Molecular Biology

Nutley, New Jersey, U.S.A.

INTRODUCTION

Work in our laboratory over the last few years has centered on the molecular events which occur during intracellular DNA replication of bacteriophage lambda. We have approached the problem in two ways: one series of experiments has involved analysis of replicative intermediates formed at various times after normal phage infection. A second series involved study of presumed intermediates that accumulate after the introduction of mutational blocks. Our results, together with those of others (see Kaiser, 1971, for review) have suggested that replication of lambda DNA occurs in two distinct stages.

The lambda chromosome, as it exists in the phage head, is a linear duplex DNA molecule of approximately 31×10^6 daltons. It contains short single stranded regions (12 nucleotides) at each molecular end which are complementary to each other. Because of these "sticky ends," the lambda chromosome circularizes soon after its injection into the host. Covalent closure of this circular molecule most likely occurs through the action of the host enzyme, polynucleotide ligase. During the first (early) stage of replication, lambda DNA replicates as a circular molecule from a fixed origin, often bi-directionally, to form daughter circles. In a normal infection, replication via this early mode occurs once, or perhaps twice. It is followed by a second mode of replication whose

*Present address: Laboratory of Molecular Genetics, National Institute of Child Health and Human Development, NIH, Bethesda, Maryland

products are long, linear molecules ("concatemers") containing tandem repeats of the lambda chromosome. Pulse-chase radioisotope-labeling experiments indicate that concatemers, not the circular molecules, are precursors to the DNA found in phage heads.

The existence of intracellular concatemers of lambda DNA was taken as presumptive evidence for a rolling circle mechanism in late replication (Gilbert and Dressler, 1968; Eisen et al., 1968). An alternative hypothesis, that concatemers were formed by a recombination mechanism was also recognized. The availability of mutations specifically affecting phage or host recombination functions made it possible to test this alternative directly. These test systems have led us to our current studies on the relationship between replication, recombination and repair in lambda. In the following section, we review the major features of the repair and recombination genes, mutants of which have been used in our study.

PHAGE AND HOST RECOMBINATION AND REPAIR FUNCTIONS

The recABC Genes

The primary general recombination pathway in E. coli requires the recA, B and C genes (Table 1). The recA gene specifies a protein whose function is still unknown. Mutations in the recA gene are pleiotropic in that they affect cell growth, prophage induction

TABLE 1. Main host and viral general recombination genes

E. coli: the rec ABC system	
1) rec A:	function unknown. "reckless" phenotype
2) rec B:	ss exonuclease – ATP dependent ⎫
	ds exonuclease – ATP dependent ⎬ = Exonuclease V*
	ss endonuclease– ATP stimulated
rec C:	ATPase

Lambda: the red system	
1) exo:	ds exonuclease – ATP independent ⎫ Probably function
2) bet:	function unknown ⎬ together = red
3) gam:	"associated with recombination"; maps next to red; lowers recombination frequency 3-fold

*Wright, Buttin and Hurwitz, 1971

and cell survival following UV, mitomycin or X-ray treatment. In
addition to being defective in general recombination, recA mutants
display the "reckless" phenotype characterized by breakdown of
cellular DNA. Such breakdown occurs spontaneously at a low level,
but increases markedly following irradiation with relatively small
doses of UV. This phenomenon is attributed to an uncoupling of the
recA product from the recB and C products which results in uncontrolled nuclease action.

The recB and C genes code for independent polypeptide chains
which are subunits of a complex exhibiting four separate enzymatic
activities: a double-strand specific DNA exonuclease and a single-
strand specific DNA exonuclease, both ATP-dependent; an ATP-
stimulated single-strand specific DNA endonuclease; and an ATPase.
The recBC nuclease has been extensively purified and is being
studied in several laboratories. A mutation in either the recB or
C gene abolishes all four activities. RecB and C mutants are not
as defective in general recombination as are recA mutants. Furthermore, they exhibit a "cautious" phenotype; their DNA is broken down
only slowly after UV irradiation.

Studies of separate mutations which suppress the recombination
defects of recB or C mutants have revealed other recombination pathways in E. coli (see Clark, 1971 and 1973 for general review).
Because these activities are not detectable under our conditions,
(i.e. recA cells), they are not pertinent to the present discussion.

The Red Genes
General recombination specified by bacteriophage lambda is
catalyzed by the products of two genes designated exo (or redX) and
beta (or redB)(see Radding, 1973 for review). The product of the
exo gene is an exonuclease specific for double-stranded DNA. The
enzyme acts processively and specifically attacks the 5'-phosphate-
ended strand of a DNA duplex exposing the complementary single strand.
Lambda exonuclease is not dependent on ATP. The protein product of
the beta gene has been isolated but has no demonstrable enzymatic
activity. Beta protein does, however, interact with lambda exonuclease and seems to enhance its affinity for DNA. A mutation in
either exo or beta causes a severe defect in general recombination.

The gene, gamma (gam) has recently been described by Zissler,
Signer and Shaefer (1972) who suggested that its activity was somehow "associated with recombination." This suggestion was made
because general recombination in the absence of gam was reduced to
about 30% of the normal level and also because the gam gene is
adjacent to exo and beta. (The lambda chromosome shows a strong
clustering of genes of related function). Evidence for an interaction
between the phage and host general recombination pathways comes
from the recent finding that the gam protein can inhibit the activity

of the recBC nuclease in vitro (Unger and Clark, 1972).

E. coli DNA Polymerase I

DNA Polymerase I is known to have, in addition to its polymerizing activity, two exonuclease functions (Setlow et al., 1972):
a 3' →5' exonuclease and a 5' →3' exonuclease. The polymerase
functions in UV repair by patching gaps remaining after pyrimidine
dimer excision. Its role in normal DNA replication is presumed to
involve a similar activity on gaps between "Okazaki" fragments.
DNA Polymerase I mutants (polA1) of E. coli, which lack polymerizing
and 3' →5' exonuclease activity but retain the 5' →3' exonuclease
activity (Lehman and Chien, 1973), are sensitive to UV, X-rays and
methylmethane sulfonate. An interaction between the pol and rec
recombination pathways of E. coli may exist because polA⁻rec(A⁻,
B⁻, or C⁻) mutants are inviable. The reason for this lethality is
unknown (Monk, Kinrose and Town, 1973).

RESULTS AND DISCUSSION

Origin of λ DNA Concatemers

Our initial efforts were designed to distinguish between two
possible origins of concatemers. Are they formed by replication
or by recombination? Several variations of the possible alternatives are diagrammed in Fig. 1. Model A shows formation of concatemers from a rolling circle intermediate. Possible pathways involving
enzyme catalyzed recombination are shown in Models B and C. A third
variation can also be envisioned; formation via the spontaneous end-
to-end joining of mature monomers (model D). These models are useful
because they make definite predictions which can be tested experimentally. A critical test of model B and C is to eliminate the
known recombination pathways (rec, red and int) and to examine the
concatemers (if any) that are formed. A further test of model C is
to analyze the concatemers for the presence of E. coli DNA. Model D
can be evaluated by using mutants unable to form cohesive ends. In
addition, polymers formed by this mechanism may be expected to begin
and terminate at the mature end-joins. The next section describes
the experiments which tested the four models.

Lambda Replication in the Absence of Recombination

Experimental conditions were selected to eliminate, as far as
possible, the pathways for concatemer production outlined in models
B, C and D. RecA hosts were infected with red int mutants of lambda
which contained, in addition, a mutation in genes A or E, known to
prevent the formation of mature DNA molecules (unit length chromosomes with cohesive ends) (Weissbach et al., 1968; MacKinlay and
Kaiser, 1969). Bacterial DNA was density labeled by growth in
medium containing D_2O and ^{15}N (and ^{14}C-thymidine) prior to infection.
These bacteria were infected in H_2O-^{14}N medium containing tritiated

Genes responsible	Molecular Origin	Molecular Product

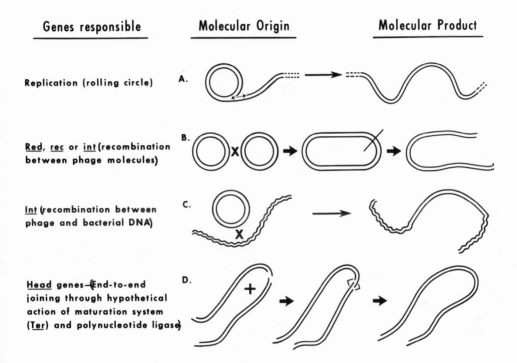

Replication (rolling circle) A.

Red, rec or int (recombination
between phage molecules) B.

Int (recombination between
phage and bacterial DNA) C.

Head genes—End-to-end
joining through hypothetical
action of maturation system
(Ter) and polynucleotide ligase) D.

Fig. 1.   Four possible modes of origin for λDNA concatemers.

thymine. Intracellular phage DNA was isolated by density-centrifu-
gation in CsCl just before cell lysis, a period when concatemers are
found in abundance. The DNA so prepared was analyzed by DNA-DNA
hybridization, sucrose gradient sedimentation and by density banding
(after shearing) in $Cs_2SO_4-Hg^+$ gradients (Skalka, 1971). Finally
after partial denaturation, the DNA was examined in the electron
microscope (Skalka, Poonian and Bartl, 1972). This last method is
particularly useful because the DNA of lambda gives an easily distinguished pattern after partial denaturation. This allowed specific identification of individual lambda DNA molecules, enabled us to
distinguish lambda DNA concatemers from host DNA molecules, and
showed the physical location of start and end points in the lambda
concatemers.

The results of these studies were clear; under conditions in
which the frequency of genetic recombination was reduced by at least
two hundred fold, the percentage of molecules present as concatemers
remained unchanged from that in the wild type infection. Approximately 20% of the molecules were scored as concatemers (average
length about 2.1 to 2.6 monomer units) with start and end points at
random locations. Furthermore, no E. coli sequences were detected.
We concluded that concatemers are normal intermediates in the late
stage of replication and that Model A best illustrates their mode of
origin.

The Fec System

At about the time our studies of λ replication in the absence
of recombination were concluding, the gene, gamma, was discovered
(Zissler et al., 1972). In their studies Zissler and co-workers
observed that red gam mutants do not form plaques on recA strains
whereas each of the single mutants plates quite well. They called
the plating defect the fec⁻ (for "feckless") phenotype. These
observations together with the results from our studies of red⁻
infection of recA cells led us to suppose that the gam function
might play an important role in replication. Further analysis of
the fec⁻ system (Zissler et al., 1972), indicated that the defect
depended on the presence of the recBC product because fec⁻ mutants
could make plaques on recAB strains. We analyzed the growth
properties of fec⁻ mutants and their component mutations, red and
gam, in rec⁺, recA, recAB and polA cells (Table 2). PolA and
ligase deficient E. coli mutants belong to a class termed feb⁻
by Zissler et al. (1971). Lambda replication in the feb⁻ mutant,
polA will be discussed in a following section.

Although fec⁻ (red⁻gam⁻) mutants make very few viable phage
per cell in strains which express the recBC nuclease (rec⁺ or recA),
they make a significant amount of λ DNA (Table 2). The amount of
λ-specific DNA synthesized amounted to about 30-40 phage equivalents
per cell or about half that of the wild-type. A comparison of phage

TABLE 2. Growth of λ mutants after infection of
various host bacterial strains*

Phage	Hosts			
	rec⁺ (W3110)	recA⁻ (204)	recA⁻B⁻ (KL254)	polA⁻ (W3110)

(Yield relative to wild type phage)

Phage	rec⁺ (W3110)	recA⁻ (204)	recA⁻B⁻ (KL254)	polA⁻ (W3110)
red⁺gam⁺(wild type)	1.0 (167)**	1.0 (60)	1.0 (90)	1.0 (140)
red⁻ gam⁺	0.34	0.40	0.38	0.04
red⁺ gam⁻	0.37	0.36	1.07	0.03
red⁻ gam⁻	0.16	0.09	0.30	0.02

*Growth was in supplemented liquid minimal medium (Joyner et al., 1966). The multiplicity of infection was 1-2 phages/bacteria and burst sizes were determined by standard one-step growth techniques. The red mutations were either the missence bet 113 or amber red 270 and the gam mutations, the missence gam 5 or amber gam 210.

**Numbers in parentheses indicate actual burst size.

equivalents of DNA versus actual plaque forming particles gave a relative packaging efficiency of 0.8 for wild type, 0.6 for red⁻ (or gam⁻ alone) but only about 0.1 for the red gam mutant. These findings suggested that red⁻gam⁻ intracellular DNA may be defective in some way and encouraged us to analyze its structure.

To eliminate potential ambiguity caused by DNA maturation and to insure optimal recovery of late replication intermediates, we included, in addition to red and gam mutations, a λA gene mutation. An example of the sucrose sedimentation analyses is presented in Fig. 3. From these results it is clear that red⁻gam⁻ intracellular DNA differs markedly from that formed during red⁺gam⁻ and red⁺gam⁺ infections. Intracellular DNA from red⁺gam⁺ sedimented much more rapidly than linear, monomeric DNA in neutral sucrose (3a). Analysis in alkaline gradients (3d), indicated that this DNA contains strands longer than unit lambda length. Such a sedimentation pattern is typical of DNA concatemers. In contrast the DNA formed by red gam mutants did not contain concatemers. In neutral sucrose, (3c) red⁻gam⁻ DNA sedimented as a mixture of two distinct species, at positions characteristic for covalently closed supercoils and "nicked" circles of lambda DNA. This result was verified by sedimentation in alkaline gradients (3f). Here an amount of radioactive material approximately equal to that at the position of the first component in the neutral gradient, sedimented into the dense pad at the bottom (indicated by the black bar). Such behavior is

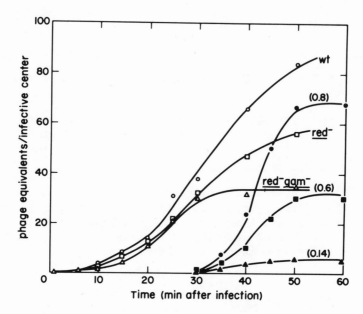

Fig. 2.   Synthesis of free phage and phage equivalents in intracellular DNA during infection of recA Bacteria.   Open symbols
indicate phage equivalents of DNA per infective center.   Closed
infected with red⁻; triangles, infected with red⁻gam⁻.

The infected cells were suspended in 5 ml minimal medium
containing 10 µCi[³H]thymine/ml (methyl labeled, 11·7 Ci/mmol; New
England Nuclear).   Unlabeled thymine was added to give a spec. act.
of 2.2 µCi/µg.   Samples (0·2 ml) were taken at the indicated
intervals and precipitated with trichloroacetic acid as described
by Enquist and Skalka (1973).   The amount of lambda-specific DNA
in the precipitate was determined by DNA hybridization as described
by Skalka (1971a) and Skalka and Hanson (1972).   The DNA content of
the phages was taken as 5.08×10^{-10} µg and the content of thymine
per lambda DNA molecule was estimated at 4.8×10^{-12} µg.   (The calculation of DNA phage equivalents was then determined as discussed
by Carter and Smith (1970).)   Plaque-forming units were determined
by standard plaque assay using E. coli C600 (suII⁺) as the indicator
bacteria.   The number of infectious centers was determined at 5 min
after infection by plating a suitable dilution on E. coli C600.
(Reproduced from Enquist and Skalka, 1973, with the permission of
Acad. Press, Ltd, London.)

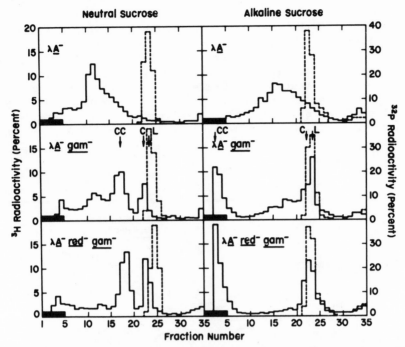

Fig. 3.   Sedimentation analysis of late intracellular phage DNA from infected <u>recA</u> bacteria.   Gradients (a) to (c) are neutral sucrose and (d) to (f) are alkaline sucrose.   Infected cells (multiplicity of infection, 2) were suspended in 5 ml minimal medium (plus 5μg thymine/ml) at 47° C for 5 min with aeration, followed by aeration and incubation at 37° C for the duration of the experiment.   At the 23rd min, enough d[^{3}H]Thd (methyl labeled, 20 Ci/mmol; New England Nuclear was added to give a spec. act. of 20 μCi/0 25 μg dThd in the presence of 5 μg unlabeled thymine/ml.   At the 25th min enough unlabeled dThd was added to give 2 mg/ml final concentration (chase).   DNA was isolated at 25, 35 and 45 min as described by Enquist and Skalka (1973).   Only the results from the 35-min chase are shown here.

Neutral and alkaline sucrose gradients (5 ml) were prepared and the samples applied as described by Skalka <u>et al</u>. (1972).   All gradients contained as a marker ^{32}P-labeled linear monomer λDNA extracted from phage particles (broken lines).   In neutral gradients, arrows indicate the position of the linear marker (L), relaxed circular molecules (C) and covalently close circular molecules (CC). In alkaline gradients, arrows mark the position of the single-stranded linear marker (L), single-stranded circles (C) and covalently closed circles (CC).

All gradients contain dense pads consisting of 0.2 ml 70% sucrose and 0.2 ml 80% sodium iothalmate (Angio-Conray, Mallinckrodt Pharmaceuticals) which had been placed on the bottom of the tubes.

Legend to Fig. 3 (cont'd).
Fractions containing this material are indicated by black rectangles.
Sucrose gradients were centrifuged in the SW50·1 rotor at 28,000
revs/min for 210 min at 11° C (neutral gradients) and 45,000 revs/
min for 120 min at 11° C (alkaline gradients). Sedimentation is
from right to left. Neutral gradients contained about 0·4 μg
bacterial DNA (unlabeled) and about 0·02 to 0·06 μg phage DNA (spec.
act. from 1 to 5x10⁵ cts/min/ μg). Alkaline gradients contained
about 0.8 μg bacterial DNA and 0·04 to 0·12 μg phage DNA. (Reproduced from Enquist and Skalka, 1973, with the permission of Acad.
Press, Ltd, London)

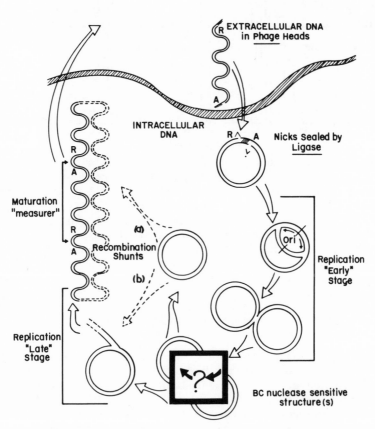

Fig. 4. A schematic representation of the infectious cycle of bacteriophage lambda. The model emphasizes a replication program
consisting of distinct early and late stages, demonstrates a possible
role for recombination, function in phage development and stresses
the essential nature of concatemers in phage morphogenesis. (Reproduced from Enquist and Skalka, 1973, with the permission of Acad.
Press, Ltd, London).

characteristic of covalently closed circles. The remaining material
sedimented slightly ahead of the single-strand mature phage DNA
marker. Independent analysis in which alkaline gradients were run
for a longer time, showed that DNA at this position could be resolved
into two components of approximately equal amounts. One sedimented
like single-stranded circular λ DNA while the other sedimented like
single-stranded linear phage DNA. Thus, we concluded that the
"nicked" circles identified in the neutral gradients (Fig. 3c) contained, on the average, one interruption per molecule. These
"nicked" circles are of some interest and will be discussed later.
In both the neutral and alkaline sucrose sedimentation experiments,
gam⁻ DNA sedimented like a mixture of the two other DNA preparations;
we observed both concatemers and circular DNA molecules. Since the
only difference between 3b and 3c and 3e and 3f is the presence of
an active Red system, we conclude that the concatemers in the gam⁻
infection are formed by Red function.

These results present an apparent paradox. As summarized in
the previous section, we could show that phage and host recombination systems were not required for concatemer formation. In the
fec⁻ system, it seemed that recombination functions were required.
Studies of replication of red, gam, and red gam mutants in rec⁺,
and recA B cells provided some clues which helped to resolve this
paradox. The results of several types of data are summarized in
Table 3. They show that when gam mutants replicated in hosts containing an active recBC nuclease (either rec⁺ or recA bacteria), the
primary product of replication at all times was circular DNA molecules.
In contrast, these mutants produced seemingly normal replicative
intermediates (i.e. concatemers at late times) in recA B hosts. Since
circular molecules are typical of the "early" mode of replication
we supposed that in the rec⁺ and recA hosts, gam mutants might be
somehow locked in the "early" stage. A mechanism which accounts for
this and proposes a role for the BC nuclease is incorporated into
our model which is depicted in Fig. 4. The model emphasizes 1) a
replication program consisting of distinct early (circle producing)
and late (concatemer producing) stages, 2) suggests a possible role
for recombination function in phage development and 3) stresses the
essential nature of concatemers in phage morphogenesis. A brief
explanation of the model follows: Lambda DNA enters the cell,
circularizes and begins to replicate as a circular molecule. At
some point after one or two rounds of this "early" replication,
possibly signaled by the act of termination (an event which takes
place in our "black box"), the late mode of replication begins. We
propose (1) that the recBC nuclease can interfere with this early to
late transition by attacking a required (transient?) replication
intermediate. The known properties of the BC nuclease suggest that
such an intermediate might contain a free end or a single-stranded
region longer than 5 nucleotides long (Karu and Linn, personal communication). Structures not attacked by this nuclease include

TABLE 3. Summary of results from analyses on the rate of synthesis and structure of late phage DNA from various combinations of mutant infections

Bacteria	rec+		recA-		recA-B-	
	Late rate	Late structures	Late rate	Late structures	Late rate	Late structures
Phages:						
Wild type	N	N	N	N	N	N
red-	D	N'	D	N'	D	N'
gam-	D	Circles (and some contatemers)	D	Circles (and some contatemers)	N	N
red-gam-	D	Circles (and some contatemers)	D	Circles (no contatemers)	D	N'

Abbreviations used: N, normal; late structures are typically linear monomers and concatemers; N', mostly normal structures, but some quantitative differences when compared with wild type; D, defective; circles, circular monomers, both covalently closed and "nicked or gapped." (Reproduced from Enquist and Skalka, 1973, with the permission of Acad. Press, Ltd, London.)

covalently closed and "nicked" circular molecules (Wright, Buttin
and Hurwitz, 1971), the very structures which seem to accumulate
in the fec⁻ infected cells. We further propose (2) that the gam
protein circumvents this potential interference by the BC nuclease
by direct inhibition of its activity (Sakaki and Echols, personal
communication).   Such activity has been demonstrated for the gam
protein in vitro and our independent in vivo experiments support
this interpretation. Thus, the transition from early to late
replication depends on negative control of the BC nuclease by gam.

Our previous results support the view that concatemers are
formed via a late replication mode. More recent results from the
fec⁻ system showed that although recombination functions are not
essential for concatemer formation--some concatemers can, and
probably are, normally formed this way. Indeed, in the absence of
normal late replication, it is the only route. We propose at least
three ways in which recombination functions could promote concatemer formation. The first, analogous to model B in Fig. 1, is
shown diagrammatically in Fig. 4 as recombination shunt (a).
Recombination functions (either Red or Rec) act to join single
monomeric circles forming multimeric molecules with no extended
single-stranded regions and no free ends (requirements for co-
existence with rec BC nuclease). A second way (shunt b) suggests
that in the absence of late replication, recombination functions
would generate concatemers from parental circular templates through
formation of new replication forks (Boon and Zinder, 1969).   Such
a mechanism could help to resolve an interesting but puzzling
finding, summarized in Table 3, which indicated that in all hosts
(rec+, recA and recAB) the absence of Red functions resulted in an
approximately two to four fold lowering in the rate of phage DNA
synthesis. Thus, formation of additional replication forks by Red
function may stimulate DNA synthesis even when late replication
occurs normally. Finally, recombination functions could act by
repairing some of the recBC nuclease damaged replication intermediates. These "repaired" intermediates might then re-enter the
normal pathway of late replication.

The last feature of the model on which we want to comment deals
with a possible relationship between DNA structure and λ morphogenesis. Our analysis of DNA synthesis in recA and rec+ cells shows a
strong correlation between appearance of plaque-forming particles
and the production of DNA concatemers. We find that monomeric
circular DNA molecules cannot be efficiently packaged. Our most
recent studies (results not shown) indicate that other late
functions, i.e. production of late mRNA and structural proteins
are normal in all the cases cited. Clearly, infectious particle
morphogenesis depends on structural features of the intracellular
DNA. Perhaps the concatemer structure is required because (as
first suggested by Szpirer and Brachet, 1970) the maturation and

packaging system of λ requires the presence of two cohesive-
end sequences in the same molecule. In such a case, the multimeric
circles presumably formed by recombination mechanisms are formally
equivalent to replication-produced concatemers.

The conclusion that the production of concatemers is an essential step in phage morphogenesis is important for several reasons.
In a teleological sense, it helps to explain why, if λ replicates
initially by a circular mode, it should adopt a second mode of
replication. Apparently the requirements of DNA maturation and
morphogenesis demand that these initially circular molecules find
an efficient way of producing multimers. Recombination functions,
in addition to late replication, are of some importance in this
regard. They can, in fact, produce multimeric DNA in an infection
which is defective in late replication, and do enhance the rate of
late DNA replication under normal conditions.

Our findings emphasize several questions concerning possible
mechanisms of control of the transition from early to late replication. What determines the number of early rounds? Are there
specific host or phage gene products involved? Is the initiation
of the late replication pathway somehow triggered by the termination
of an early round? Does late replication originate at the same
site as early replication? Answers to these questions are far from
complete, but it seems that some insight can be gained from further
study of the Fec system.

First, let us consider what determines the number of early
rounds. Recall that normally λ replicates only once or twice in
the early mode. In cells with an uninhibited recBC nuclease, λ
replicates in an early mode continuously. It follows from these
facts that under "normal" conditions, the actual number of early
rounds may depend on the rapidity with which the recBC nuclease can
be neutralized by the λgam protein. If this idea proves to be true,
then one would predict that in recBC⁻ cells, there should be, at
most, only one round of early replication.

Second, let us examine what can be learned with respect to the
origin of late replication using the fec system. In our model, the
structure formed at the late origin is sensitive to the recBC nuclease. If this is correct and if the interrupted circles which accumulate in fec⁻ infection are end-products of recBC nuclease action,
then analysis of their structure may provide clues as to the site
of origin of late replication. Such an analysis was initiated by
annealing single-stranded linear or circular DNA purified from
alkaline sucrose gradients of fec⁻ DNA with separated strands of λ
DNA. The results of these experiments indicate that each component
hybridized with equal efficiency to both of the separated strands,
suggesting that they contained a mixture of both strands. We con-

clude that the interruption can occur with equal frequency in both strands. More recent biochemical studies on purified interrupted circles indicate that the break in these molecules is not a simple nick or gap because it cannot be repaired by T4 ligase or T4 ligase plus T4 DNA polymerase or E. coli DNA polymerase I. The circles can, however, be repaired after pre-treatment with exonuclease III (a double-strand specific exonuclease attacking both 3'-OH and 3'-phosphate ended chains). Further study suggests that the structure at the interruption may contain a 3'- as well as a 5'-phosphate. Such a structure would not be sealed by T4 ligase even in the presence of T4 DNA polymerase or E. coli DNA polymerase I unless the 3'-phosphate was removed. At the present time, we are unable to determine the significance of this structure. However, limited digestion with exonuclease III will provide us with a method to identify the site of the interruption. The circles can be digested with this enzyme, the resulting gap filled with radioactive-labeled nucleotides, and the ends sealed with ligase. Sealed circles can be purified and the location of the original gap determined by hybridization techniques (Skalka, 1971). This and other approaches involving electron microscopy are now in progress.

The Feb System

What other gene products of λ or its host may be involved in the transition from the early to late replication pathways? One clue was provided by observation of Murray and later by Zissler (Zissler, Signer and Shaefer, 1972) that single mutants in λ red or gam failed to plate on polA strains of E. coli. Later experiments showed that these mutants behaved similarly on ligase defective (lig⁻) E. coli strains (Gellert, personal communication). Zissler et al. (1972) called bacterial mutants unable to plate gam phage, feb⁻. To gain more information concerning the physiological role of the gam and red genes, we analyzed the DNA synthesized by these mutant phages during infection of the feb⁻ host, polA.

The results which we obtained in comparisons of DNA synthesis and phage production (Fig. 5) with the single red or gam mutant infections in polA were strikingly similar to those obtained with the double mutant, red gam, in recA bacteria. In the polA host, either single, as well as the double mutant phage, produces about 40-55 phage equivalents of DNA, but form only about 5 plaque-forming units per cell (see also data in Table 1). This indicates a packaging efficiency of about 0.1, the same value observed in the fec⁻ infection. Similarly, sedimentation analyses (data not shown) indicated that in polA cells greater than 60% of each of the three mutant phage DNAs was in the form of circular monomers, but in these cases, most of the circles contained interruptions in one or the other of their strands. In the one case analyzed (gam⁻ infection), approximately one-third of the interrupted circles were re-

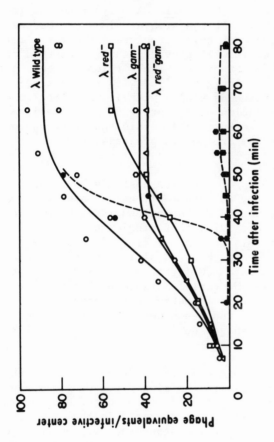

Fig. 5. Synthesis of free phage and phage equivalents in intracellular DNA during infection of pol A bacteria. Open symbols
and continuous lines indicate phage equivalents of DNA per infective center. Closed symbols and broken lines are plaques forming
units per infective center, circles, infected with wild type,
squares with red⁻, hexagons, with gam⁻ and triangles with red⁻gam⁻.
Conditions were as described in Fig. 2.

pairable by T4 ligase and polymerase. Using radioactively labeled nucleotides to fill in the gaps, we could calculate (from the specific activity of the substrate) that the gaps averaged about 9 nucleotides long. Hybridization data suggested that these gaps occurred at random location on the λ chromosome. After enzyme treatment, the ratio of covalently closed circles (those initially present plus those enzymatically produced) to interrupted circles was about 50:50, very much like the DNA produced in the fec⁻ system. We postulate that the repairable gaps result from random damage to the intracellular λ DNA not readily repaired in the polA host, and that the replication block in this case may well be similar to that found in the fec⁻ system.

Table 4 presents a summary of the results from our sedimentation analysis of the lambda DNA produced in feb⁻ infections. Our interpretation of these results in the context of our model is as follows: Infection #1: gam⁻ phage in the polA host produce only monomeric circular DNA molecules--even at late times. This is puzzling. Although our model predicts a defect in late replication for gam⁻ phage, some concatemers might be expected to arise from the red or rec-mediated recombination shunts. The absence of DNA concatemers suggests that DNA polymerase I might be required in both the red and rec shunts. Several schemes for a molecular mechanism for general recombination do include a role for DNA polymerase I (Radding, 1973). An alternative interpretation is that recombination intermediates do not survive in polA1 cells.

The host cells in the red⁻ infection #2 are phenotypically rec⁻ because the gam protein can be expected to inhibit the recBC product (Unger et al., 1972). Although general recombination is

TABLE 4. Summary of results from sedimentation analyses of structure of late phage DNA from various feb⁻ infections

Infection	Phage	Host pol	rec	Late DNA Structures*
1	gam⁻ red⁺	−	+	circles
2	gam⁺ red⁻⁻	−	(−)**	circles
3	gam⁻ red⁻	−	+	circles
4	gam⁺ red⁻⁻	+	(−)	N'
5	gam⁺ red⁺	−	(−)	N

*Definitions as in Table 3.

**Symbols in parentheses refer to phenotypic state of cell--due to presence of gam protein.

defective, our model predicts that in this infection the switch to
late replication should occur as in the wild type infection.
Because we observe only monomeric circles and no concatemers at late
times, we must conclude that either pol or red (perhaps both) are
required for the switch to (or continuation of) late replication,
or that unbalanced metabolic affects of the mutations (resulting
perhaps from partly operating pathways for enzymes) destroy late
replication intermediates. As expected, results with the double
mutant red gam in the polA host (infection #3) are the same as with
either single mutant alone.

    Some further information concerning the possible roles of red
and pol in late replication is obtained from comparison of results
from infections #4 and #5. Because red⁻gam⁺ phage replicate well
(produce concatemers) in wild type cells (#4) and because red⁺gam⁺
(wild type) phage replicates normally in the polA⁻ hosts (#5), it
seems possible that pol and red may have some complementary
activities.

    Other more complicated interpretations can be invoked, and it
is not yet possible to propose any biochemical or molecular model
based on the feb⁻ data because several variables have yet to be
examined. In considering various possibilities it is important to
recall that the polA1 mutant, while lacking polymerizing activity,
still retains normal levels of the 5'→3' exonuclease. Another mutant
(resA), presumed to lack both activities, seems to plate λ red (but
not λ gam) mutants well (Glickman et al., 1973). Analysis of DNA
synthesis in this as well as other pol 1 and lig feb⁻ mutants
should enable us to distinguish between a direct or indirect role
for the polymerase and ligase enzymes in lambda DNA replication
and recombination.

                              SUMMARY

    Our initial attempts to understand the significance of λ DNA
concatemers have led us to investigations involving the three related
aspects of DNA metabolism: replication, recombination and repair.
Our results have revealed that not only are there two distinct modes
of replication in λ, but also that the pathways for repair and recombination may exert both direct and indirect influences on phage
development--some of which not heretofore suspected. In some cases,
for example, that involving the interaction between the λ gam
protein and the host's recBC nuclease, the relationship between
phage and host function and their significance to phage growth seem
quite clear. In other cases, for example in our attempts to understand the role of DNA polymerase I in mutant and wild type infection,
we are still far from having a complete picture. However, continued
analysis of components in each pathway and the study of their
corresponding mutational blocks should provide us with deeper

understanding of the molecular mechanisms involved and should help
us to clarify the extent and significance of overlap in each of
these pathways.

REFERENCES

Boon, T. and N. D. Zinder. 1969. Proc. Nat. Acad. Sci., U.S.A.
64:573.

Carter, B. J. and M. G. Smith. 1970. J. Mol. Biol. 50:713.

Clark, A. J. 1971. Ann. Rev. Microbiol. 25:438.

Clark, A. J. 1973. Ann. Rev. Genetics, in press.

Eisen, H., L. H. Pereira da Silva and R. Jacob. 1968. Cold Spr.
Harb. Symp. Quant. Biol. 33:755.

Enquist, L. W. and A. Skalka. 1973. J. Mol. Biol. 75:185.

Gilbert, W. and D. Dressler. 1968. Cold Spr. Harb. Symp. Quant.
Biol. 33:473.

Glickman, B. W., C. A. Van Sluis, G. Van der Maas and A. Rorsch.
1973. J. Bacteriol. 114:951.

Joyner, A., L. N. Isaacs, H. Echols and W. Sly. 1966. J. Mol.
Biol. 19:174.

Kaiser, D. 1971. In A. D. Hershey, ed., The Bacteriophage Lambda,
p. 417, Cold Spr. Harb., N. Y.

Lehman, I. R. and J. R. Chien. 1973. Fed. Proc. 32:45, Abs.

MacKinlay, A. G. and A. D. Kaiser. 1969. J. Mol. Biol. 39:679.

Monk, M., J. Kinrose and C. Town. 1973. J. Bact. 114:1014.

Radding, C. 1973. Ann. Rev. Genetics, in press.

Setlow, P., D. Brutlag and A. Kornberg. 1973. J. Biol. Chem. 247:
224.

Skalka, A. 1971a. In A. D. Hershey, ed., The Bacteriophage Lambda,
p. 535. Cold Spr. Harb. Labs, Cold Spr. Harb., N. Y.

Skalka, A. 1971b. In Methods in Enzymology (Nucleic Acids), Part
D, Vol. 21, p. 341.

Skalka, A. and P. Hanson.   1972.   J. Virol. 9:583.

Skalka, A., M. Poonian and P. Bartl.   1972.   J. Mol. Biol. 64:541.

Szpirer, J. and P. Brachet.   1970.   Mol. Gen. Genetics. 108:78.

Unger, R. C. and A. J. Clark.   1972.   J. Mol. Biol. 70:531.

Unger, R. C., A. J. Clark and H. Echols.   1972.   J. Mol. Biol. 70:539.

Weissbach, A., P. Bartl and L. A. Salzman.   1968.   Cold Spr. Symp. Quant. Biol. 33:525.

Wright, M., G. Buttin and J. Hurwitz.   1971.   J. Biol. Chem. 246: 6543.

Zissler, J., E. R. Signer and F. Shaefer.   1972.   In A. D. Hershey, ed., The Bacteriophage Lambda, p. 455.   Cold Spr. Harb. Labs, Cold Spr. Harb., N. Y.

STUDIES ON THE FOLDED CHROMOSOME OF E. COLI

A. Worcel

Department of Biochemical Sciences, Princeton University

Princeton, New Jersey, U.S.A.

There are a few obvious questions about the compact tridimensional structure of the DNA inside the bacterial cell. The first question is: How does the E. coli chromosome, which is one millimeter long, fold up inside the nuclear body of only one micron diameter? A second, related question is: How do the two daughter chromosomes separate from each other at the end of the replication cycle; how do they segregate into two daughter cells? We thought that the structure of the chromosome must be quite organized (not random), particularly after we were aware of one peculiar type of chromosome segregation which we could induce after producing numerous chromosome initiations in E. coli (multifork replication). It is possible to trigger the initiation of chromosome replication in E. coli by blocking DNA synthesis. If the block in DNA synthesis is short enough, after releasing the block the replicating fork will continue and a new fork will start at the chromosome origin. We found a few years ago (Worcel, 1970) that we could block and release using the temperature sensitive strain (DNA B) of E. coli which stopped DNA synthesis upon a shift to non-permissive temperature. Coming back to the permissive temperature we found that a new chromosome initiation was induced. We could repeat that process, shifting from 30 to 42 degrees and each time a new fork was initiated. We were able to induce 5 to 10 replication forks per chromosome and the cell didn't seem to be incumbered by them. Upon return to the permissive temperature, the daughter chromosomes were eventually disentangled and became part of viable daughter cells.

Fig. 1 shows the temperature sensitive strain grown at the permissive temperature. It is essentially like the wild-type.

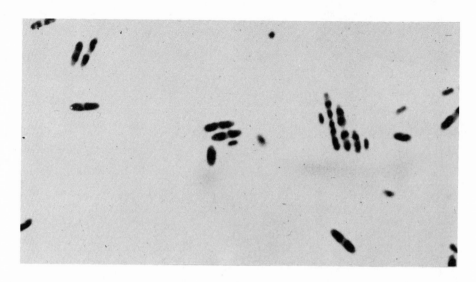

Fig. 1. Nuclear staining (Piéchaud, 1959) of exponentially growing DNA B CRT 266.

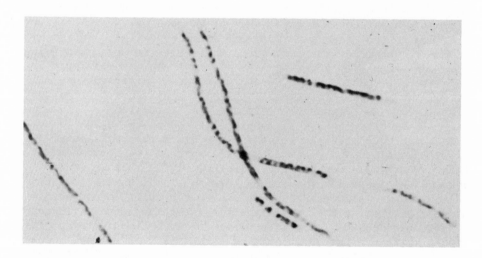

Fig. 2. Same cells after five 30°-40° shifts and subsequent growth at 30° for 30 minutes.

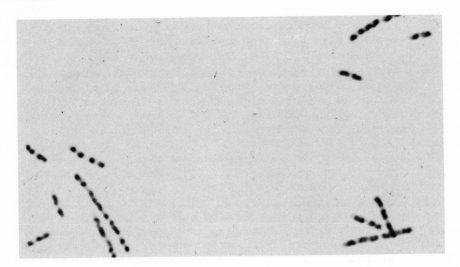

Fig. 3.   Same cells two hours later.

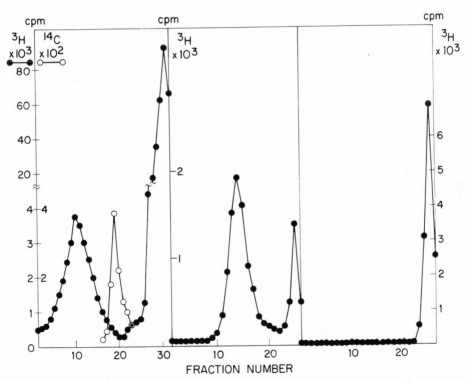

Fig. 4.   Sucrose gradient profiles of folded DNA of <u>E. coli</u>.
          a.   whole lysate
          b.   rerun of peak fraction (N° 10) of a.
          c.   same fraction after shearing.

The cells are growing at 30 degrees and one can see the nuclear bodies, the typical structure of the replicating DNA. Fig. 2 shows the cells after induction of 5 new chromosome re-initiations (made by shifting back and forth between 40 and 30 degrees 5 times, followed by growth at 30 degrees). When DNA synthesis is blocked, cell separation is immediately blocked and as the cells continue to grow, they elongate making long snakes. The DNA is more or less distributed throughout the cell with the exception of the tips of the cell. Fig. 3 shows the cells 2 hours later. At this time condensation of the DNA begins and it is possible to count nuclei. Some cells are back to normal although there are still some cells which are very long and in which septation did not yet take place. Eventually all of the cells are normal. It is in this case difficult to understand how 6 chromosomes separate without entangling the DNA. We thought then that it was an interesting problem to study the whole chromosome in its folded conformation in order to answer some meaningful questions about the segregation of the bacterial chromosome.

We started working with a system for lysing the cell described by Pettijohn and his collaborators. Stonington and Pettijohn (1971) described a new form of DNA that can be obtained by lysing the cells very gently with a high concentration of sodium chloride. DNA is a very long and highly charged molecule and inside a cell all of these charges are neutralized by counter ions: polyamines, basic proteins, etc., and by small ions. When the cells are lysed all of those counter ions are lost. In the DNA the charges will repel each other, and the DNA will unfold and will lose its folded conformation. To prevent this, Pettijohn used a high concentration of a counter ion in the lysis procedure. Any counter ion will do. The method he used, which we are using now, is simply 1.0 M NaCl in the lysis media. If one is careful enough all of the DNA is extracted into a particle having a sedimentation coefficient of about 1500 to 2000S. Associated with this DNA are all of the nascent RNA chains, but no ribosomes or most other cell proteins; the lysis is done in desoxycholate, Brij,1.0M NaCl and low magnesium and most of the proteins are removed from the folded chromosomes. The only proteins associated with the DNA appear to be the subunits of the core RNA polymerase. Pettijohn showed that transcription of the nascent RNA can be completed if the appropriate ribotriphosphates are added to this folded chromosome. Less than 1% of the cellular protein is associated with the folded chromosome, most of this being RNA polymerase. The particle by weight is about 70% DNA, 20% nascent RNA, and 10% protein.

Fig. 4 shows the results of a sucrose gradient sedimentation of a cell extract obtained under those conditions. The DNA is labeled with tritiated thymidine. In this case we didn't precipitate the gradient fractions. Centrifugation is to the left, 30 minutes at

17,000 RPM with internal T4 phage marker (sedimentation 1000S).
The radioactivity at the top of the gradient is due to free thymidine. All of the DNA sediments very fast with a broad Peak at
around 1700S. If we take a fraction from this first gradient and
rerun it we can still see that it again sediments as a folded
molecule (Fig. 4b). The viscosity of the lysate is about the same
as the viscosity of the buffer used. If on the other hand this
fraction is ejected just once from the tip of an Eppendorf pipette
it now behaves as usually extracted unfolded viscous DNA. It remains at the top of the gradient during sedimentation. The folded
complexes tend to aggregate with each other and precipitate. The
best results are obtained when the cells are lysed at room temperature. No membrane is associated with the chromosome. Less than
1% of labelled leucine is associated with this fast sedimenting
form. The cellular proteins remain at the top of the gradient.

Fig. 5 shows that the complexes in the broad Peak are heterogeneous, composed of individual fractions which have different
sedimentation coeffecients. These S values are reproducible and
meaningful. On the left is a typical preparative sucrose gradient
of a lysate. The other frames show reruns of individual fractions
of this first profile. Fraction number 9, the heaviest fraction,
runs the fastest with a sedimentation value of 2200S. The other
fractions run slower. Finally fraction 14, the lightest fraction
and the first fraction significantly above background, runs just 3
fractions ahead of T4, and has an S value of 1300. If we combine
all of the fractions 9, 10, 11, 12, 13, 14, we reconstruct the
original broad distribution profile. An increase in mass in the
complex from 1 to 2 would cause a 66% increase in S value. The
increase we observe is 70%, 2200S as compared to 1300S value, and
therefore supports our notion that we are seeing a spectrum of
replicating molecules. To prove this point we decided to look at
bacteria which are known to be at the start of replication. When
auxothophs are starved of amino acids, they complete one round of
replication gradually and then wait for new initiation (Maaløe and
Hanawalt, 1961).

Fig. 6 shows the results of sucrose gradient sedimentation of
lysates of cells starved for amino acids for various times. The
first is after 15 minutes amino acid starvation (on the left). We
can see that 15 minutes after amino acid starvation there is a
bimodal distribution (there are two peaks). One is the usual 1700S
peak we obtain in exponentially growing cells. The second peak is
1300S, 3 fractions faster than T4. 30 minutes later this 1300S
species increases at the expense of the faster sedimenting form.
We always observe a small amount of fast sedimenting forms, even
after 2 or 3 hours of amino acid starvation. Perhaps this is due
to some chromosomes which did not finish their round of replication
under amino acid starvation. We conclude from these results that we

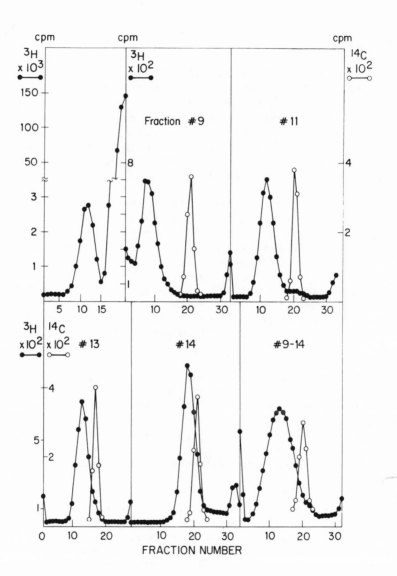

Fig. 5.   Heterogeneity in the sedimentation velocities of the folded chromosomes from exponentially growing cells.

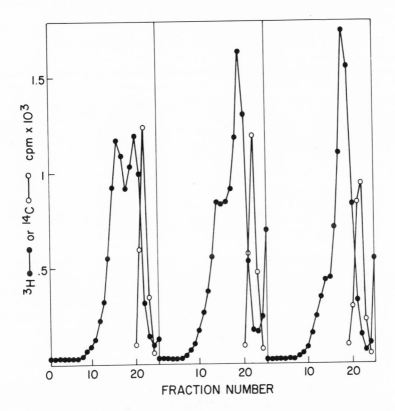

Fig. 6. Sucrose gradient profiles of folded chromosomes released from E. coli cells starved for required amino acids for 15 minutes (a), 30 minutes (b) and 60 minutes (c).

are dealing with the intact folded bacterial chromosome. We could now ask a few interesting questions about its structure: If it was intact closed circular, and if it had many nicks. We asked these questions by probing into the structure of the chromosome with inter-calating dyes, in this case ethidium bromide. Ethidium bromide can intercalate between the base-pairs of the DNA, and when there are no nicks in a closed circular molecule this intercalation causes super twisting (Waring, 1965).

Fig. 7 shows what happens when we centrifuge our folded chromo-some in ethidium bromide. We perform sucrose gradient sedimentation velocity centrifugation under exactly the same conditions as in our first experiments, but with ethidium bromide in the gradients. The amount of radioactive DNA is well below one gamma per ml, so the amount of ethidium bromide added is practically the same as the concentration of free ethidium bromide, and the values are meaningful for the estimation of the super helical concentration of the DNA. Fig. 7a is another typical preparative run of the lysate of exponen-tially growing E. coli cells and again you see there is a peak at about 1700S. When we run extra aliquots of this lysate in a gradient containing 2 gamma of ethidium bromide per ml they sedi-ment slower, closer to T4. When there is 20 gamma per ml of ethi-dium bromide all of the folded chromosomes of the exponentially growing cell lysate now run faster. The changes are very repro-ducible. The gradients are collected with the same number of frac-tions and internal T4 marker is in exactly the same fraction number. Therefore it is really the folded chromosome that is moving (and not the T4 DNA) under the influence of ethidium bromide. In Fig. 8 we plot the S values of the peak fraction, and of the first fractions above background on both sides for each concentration of ethidium bromide. The curves all have a dip within experimental error at 2 micrograms of ethidium bromide per ml. This was striking to us, particularly because that is exactly the concentration of free ethidium bromide necesssary to produce the same results with smaller closed circular DNA like SV40, indicating that the folded chromosome must have the same super helical concentration. All of the chromosomes, no matter at what stage of replication they are in, have the same super helical concentration. To really be sure that the changes in S values are due to changes in super helical content of the folded chromosome we nicked the chromosomes, introducing a swivel. We are now able to intercalate ethidium bromide without changing drastically the sedimentation coefficient of the chromosome. We treated purified fractions from exponentially growing cell lysates of folded chromosome at 0°C for a short time with pancreatic DNAse under conditions which are known to produce single strand breaks in the chromosome and no double stranded cuts (with magnesium and no calcium). After 10 minutes, at 0°C the reaction was stopped by adding a large excess of EDTA and the folded chromosomes were recentrifuged on sucrose gradients containing various concentrations

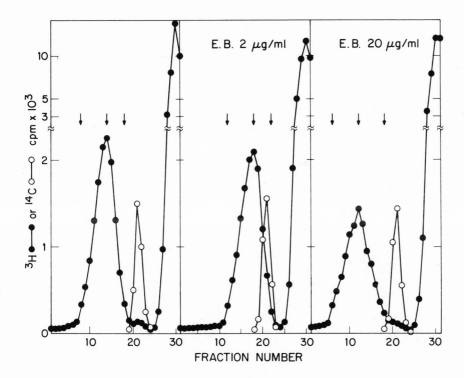

Fig. 7.   Effect of ethidium bromide on the sedimentation velocity
of the folded chromosomes.

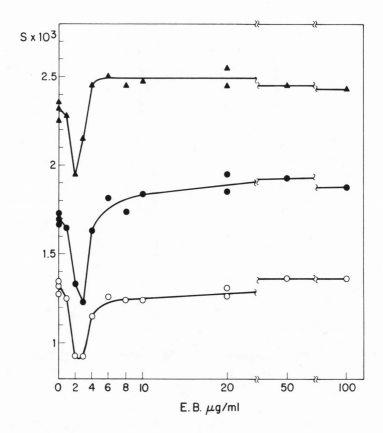

Fig. 8.   Plot of the S values (lowest, highest and peak fraction)
against ethidium bromide concentration.

of ethidium bromide, or without ethidium bromide.

Fig. 9 shows what happens with these chromosomes as we introduce single-stranded nicks with DNAse. The control was incubated at
0°C without DNAse for 10 minutes. With 1μg and 2μg per ml of DNAse
the S value has decreased. With 10 μg/ml the material has a sedimentation coefficient slightly below T4(900S) and with 20 μg to
30 μg/ml DNAse the sedimentation doesn't decrease further. If we
use a large excess of DNAse the chromosome fragments and unfolds
and stays at the top of the gradient. As the 900S particle appeared
to be a definite structure, and not just an intermediary stage in
the gradual unfolding of the chromosome, we decided to look at super
helices in this 900S particle. Fig. 10 shows the results. All of
these experiments are now done with purified fractions. We chose a
1500S complex from an exponentially growing cell lysate and we recentrifuged it. In the upper part is the native complex, non-DNAse
treated, sedimented without ethidium bromide; with 2 μg/ml ethidium
bromide the S value decreases from 1500S to about 800S, and with
6 μg/ml ethidium bromide it runs faster now again, about 1800S. If
the same complex is incubated with 10 μg/ml of pancreatic DNAse
to obtain the relaxed particle (lower frames) without ethidium
bromide the S value drops to 900S. This is what we call the relaxed
particle. Now we sediment the relaxed chromosome with 2 μg/ml
ethidium bromide and it runs slower; exactly the same as the native
chromosome. But at 6 μg/ml ethidium bromide when the native chromosome was sedimenting faster, being twisted by ethidium bromide, the
nicked form sediments even slower. It cannot super twist at high
ethidium bromide concentration; the molecule must be relaxed without super helices.

Fig. 11 shows the plots of the S values against concentration of
free ethidium bromide. We can see the same basic changes which have
been observed with smaller closed circular DNA molecules. First,
the S value of the super-coiled closed circular DNA decreases and
then increases with increasing ethidium bromide concentration,
while the nicked DNA does not. The second similarity is that at
the equivalence region, the S value of the native closed circular
DNA is the same as the S value for the nicked DNA. The third
similarity is that the concentration of ethidium bromide required
for the equivalence point is the same as the concentration of
ethidium bromide required at the equivalence region of smaller
closed circular DNAs, such as SV40. Therefore the concentration of
superhelices must be identical, namely about one negative super
helical twist per 400 base pairs (Bauer and Vinograd, 1968; Wang,
1969). However, there are minor and major differences. Minor
differences are that the S values are between 50 and 100 fold greater
than the S values obtained with a small closed circular DNA like
SV40, and lambda. The second minor difference is the relative
change that we observe in sedimentation rate as the molecule is

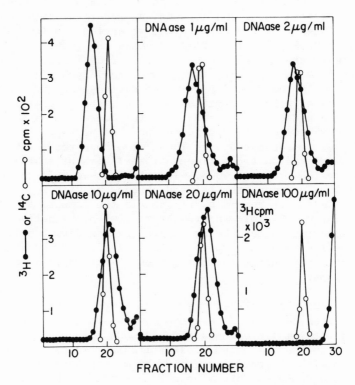

Fig. 9.   Effect of nicking with DNase on the sedimentation velocity of the complexes.

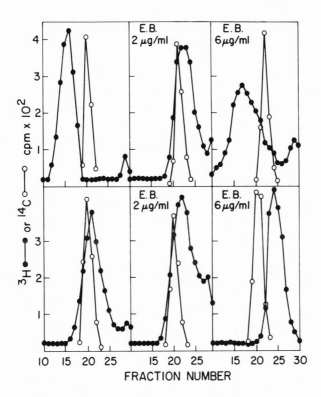

Fig. 10. Effect of ethidium bromide on the sedimentation velocity of the native folded chromosomes (upper frames) and on the nicked chromosomes (lower frames).

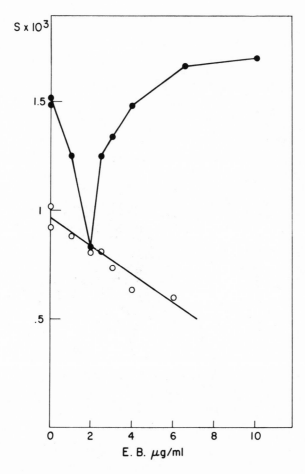

Fig. 11.   Plot of S values against ethidium bromide concentration.
Closed circles: native folded chromosomes.   Open circles: DNase
nicked relaxed chromosomes.

untwisted and super twisted. This is much greater than the relative
change obtained with a smaller molecule. Those minor differences
can be expected because of the much larger size of the E. coli DNA.
The really major differences are the following: first, the S value
at the equivalence region is still about 5 times higher than the S
value of the unfolded closed-circular DNA molecule. Estimates of
the sedimentation coefficient for intact E. coli or subtilis DNA
are about 170S (Kavenoff, 1971). The second major difference is
that in order to obtain the relaxed DNA we need definitely more
than one single-stranded nick per chromosome. This we know because
of the rather large amounts of DNAse required to relax the chromosome and because of the existence of intermediary forms between the
highly super coiled and the relaxed form. We decided to estimate
as accurately as possible the number of single strand nicks required
to eliminate all of the super helices in the E. coli chromosome.
To do this we simply analyzed the single-strand DNA fragments,
obtained from the folded chromosome before and after nicking with
DNAse, on alkaline gradients.

   Fig. 12 shows the alkaline sucrose gradients. The internal
marker in this case is denatured DNA of T4 (single-stranded DNA).
Fig. 12a is a profile of the denatured DNA obtained from the native
complex. The DNA fragments have very high sedimentation coefficients;
all run faster than T4 in a broad distribution. After nicking with
DNAse, the profile changed completely. After 1μg/ml DNAse the
molecules sediment slower, and the profile is markedly non-gaussian.
When we reach 10 γ/ml DNAse, the single stranded DNA fragments of
the relaxed particles run slower than T4. At 20 μg/ml DNAse the
DNA sediments even slower.

   Table 1 shows the estimated molecular weights of the single-
strand DNA fragments from these experiments, using Studier's
(1965) extrapolations. The fragments we are interested in are small,
and the Studier relationship must be quite accurate for their
molecular weight. We take the molecular weight of the single strand
DNA of E. coli to be exactly one half of the molecular weight of the
E. coli DNA, (M.W. 25×10^8 Daltons, Cairns, 1963). Therefore we can
calculate the number of breaks that we introduced into this chromosome. When the relaxed molecules contain no more super-helices
there are about 40 breaks per DNA strand. This is an upper estimate
for the number of breaks required to completely eliminate the super
helices in the E. coli chromosome. If we multiply that number by
2 (there are two strands per E. coli chromosome) we have an upper
estimate of about 80 single strand breaks per E. coli chromosome
required to remove all of the super helices. The lower estimate
is obtained from the partially relaxed intermediate form. Some of
the molecules have only 6 breaks per DNA strand whereas some others
have about 30. If we take the lower number, and multiply it by 2,
12 would be the real lower estimate for the number of breaks required

Molecular weight estimates of the DNA single strands fragments obtained from the E. coli folded chromosome before and after DNAase treatment

DNAase (μg/ml)	Type of DNA Complex	Molecular Weight Classes (x10^8)		Breaks/strand of E. coli DNA
0	Compact ("native")	I	15.1	intact
		II	10.7	1
		III	5.07	2
		IV	2.46	4
1	Intermediate	I	6.12	2
		II	1.91	~ 8
		III	0.45	~ 30
2	Intermediate	I	1.171	~ 9
		II	0.42	~ 30
10	Relaxed	I	0.29	~ 50
20	Relaxed	I	0.21	~ 70

The values were obtained by assuming that the relationship between molecular weight (M) and sedimentation velocity (S) obeys the empirical Burgi-Hershey relationship for denatured DNA (   ): $\frac{S_1}{S_2} = \left(\frac{M_1}{M_2}\right)^{.42}$ and that the M.W. of the T4 single strands marker is 65 x 10^6 (   ). The roman symbols refer to the fractions in the alkaline-sucrose gradients shown in Figure 6.

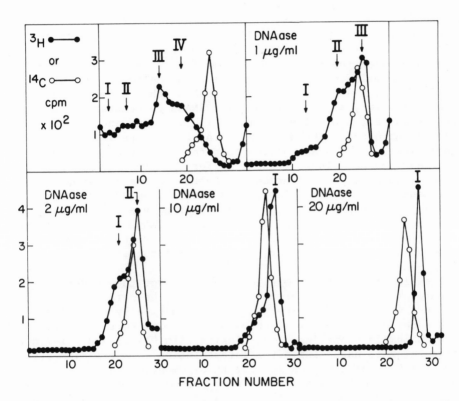

Fig. 12. Alkaline sucrose gradient profiles of DNA single strand fragments obtained from folded chromosomes before and after nicking with DNase.

to relax the chromosome.   The actual number should be between 12 and 80 cuts per chromosome.

Pettijohn and his collaborators have shown that this structure can be unfolded by various manipulations. We have shown that it can also be done by shearing. It can also be unfolded by SDS, heat (about 5 minutes at 56 degrees) and most interesting and intriguing, by RNAse. Very short incubation with RNAse completely unfolds the chromosome. We repeated this observation. We decided to investigate the kinetics of the unfolding as a function of the concentration of RNAse required for unfolding. Fig. 13 shows the results of RNAse treatment. Incubations are done in vitro, and in this case we are using the folded chromosome from a lysate of amino-acid starved cells, lined up at initiation. After incubation with 150 µg/ml RNAse for 10 minutes at 0°C the chromosome unfolds and sediments very slowly. With lower RNAse concentration, 10 µg/ml, 2 species appear, one which still is highly folded and the other completely unfolded. The S value of the folded species may be slightly lower after RNAse action than the S value of the non-RNAse treated chromosome. These results differ from the results we obtained with DNAse, which gradually decreased the S values. We do not usually see intermediary forms with the amino-acid starved complexes. RNAse appears to have an all or none effect. One interpretation for the all or none effect of RNAse treatment is that there are very few RNA molecules holding the structure together and few or perhaps one hit can unfold the molecule.

Fig. 14 illustrates a preliminary model for the folded structure of the E. coli chromosome based on the data which I have presented. On the left corner we show schematically the folded native super-coiled bacterial chromosome. Going to the right, what we think is happening after nicking with DNAse, and going down, what we think is happening after intercalation with ethidium bromide. The simplest interpretation for the requirement of many nicks per chromosome for relaxation is that when the bacterial chromosome is in its folded conformation it is not a continuum, but is divided topologically into a limited number of loops. When nicked with DNAse, each nick would relax one of those loops but the swivel introduced would not be able to swivel the whole bacterial chromo-some. If we nick extensively we obtain what we call the relaxed particle. The S value drops from 1500S to 850S. The arrows show experimentally observed transitions.   The broken arrows represent transitions which also take place although we did not study them in detail. The relaxed particle has some tertiary structure in it because its S value is about 5 times greater than the S value of the unfolded DNA. We show the structure to be composed of loops although the loops are all relaxed. The number of loops should equal the number of single strand nicks per chromosome which are required to relax the molecule completely. That number should be somewhere

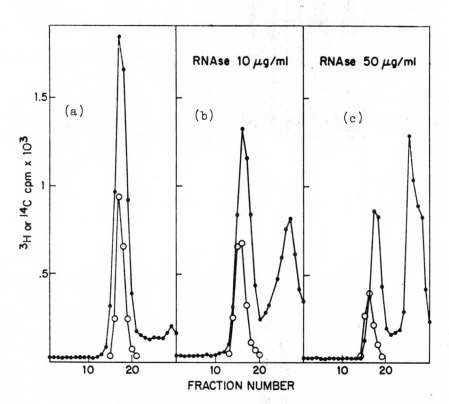

Fig. 13.   Effect of RNase on the sedimentation velocity of folded chromosomes isolated from amino acid starved cells.

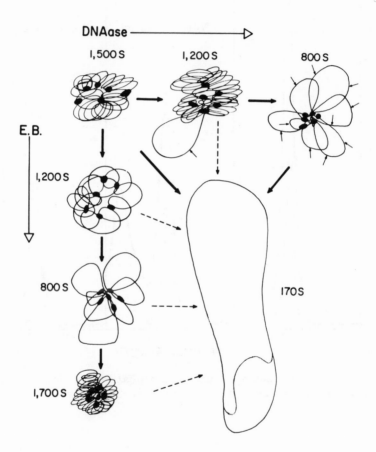

Fig. 14. Preliminary model for the folded E. coli chromosome. C.B.: ethidium bromide.

between 12 and 80. When ethidium bromide intercalates it relaxes
the chromosome gradually throughout the folded chromosome and we
obtain slower sedimenting forms, which eventually have the same
sedimentation coefficient as the DNAse-relaxed chromosome. This
is again our relaxed particle with no super helices but still
folded up in loops. The difference being that in this case if the
ethidium bromide concentration is raised and the chromosome is not
nicked, it super twists in the opposite positive sense. That
obviously does not happen with DNAse relaxed chromosomes. Knowing
that RNA is important for this folding we put it as the black dots
in the center. Somehow, we imagine that the RNA-DNA interactions
may help to fold the chromosome and at the same time may be responsible for the limitation in the rotation of the DNA strands.
If we treat with RNAse we see the transition to the unfolded
chromosome which has lost all of its tertiary structure.

The loops are large, on the average about 20 microns long,
and can still not be responsible for folding the chromosome into
the 1 micron diameter nuclear body. Therefore the loops must have
some kind of tertiary structure. A model of this is shown in
Fig. 15. If we coil up the chromosome into a broad helix as shown
in a, every 360 degree turn of this helix will create one super
helix. Since we know that the super helical concentration of this
DNA is one negative super twist per 400 base pairs, the helical
coiling here has to be 400 base pairs per 360 degree turn. Both
super helices in a and b could be shown as right-handed or left-
handed. The Watson-Crick double stranded DNA is right-handed and
therefore the tertiary coiling must be left-handed in order to
cause a topological deficiency in total helical turns (negative
super helices). If we stretch the negatively coiled helix we
obtain right-handed super helices as shown in b and those are the
super helices one sees in vitro.

I would like to show now some pictures from Hajo Delius' work
with these complexes. The electron micrographs show that the
chromosome is intact and all super twisted at high ethidium bromide
concentration. The model as shown in Fig. 14 is preliminary and
can not account for the replicating chromosome. We suggested
(Worcel and Burgi, 1972) that a plausible model for the folding of
the replicating chromosome would be one which would fold the
replicating chromosome into 2 or 3 nucleation areas. The unreplicated parental chromosome could be folded up in such a way that
this part may just unwind and each of the partially replicated
daughter chromosomes would make another particle. We should have
3 nucleation areas per chromosome. As the replication proceeds the
old folded chromosome is unrolled while the other 2 new ones are
rolled in.

The electron micrographs tell us something about the loops.

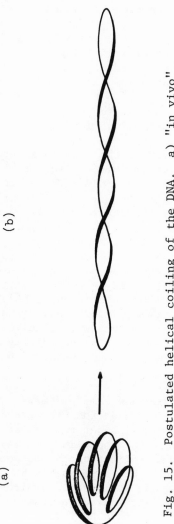

Fig. 15. Postulated helical coiling of the DNA. a) "in vivo" folding. b) "in vitro" extended conformation.

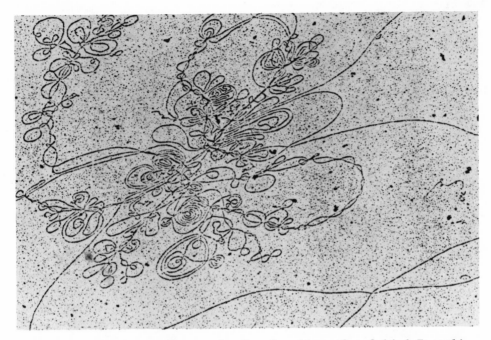

Fig. 16.   Electron micrograph of a fraction of a folded E. coli
chromosome.

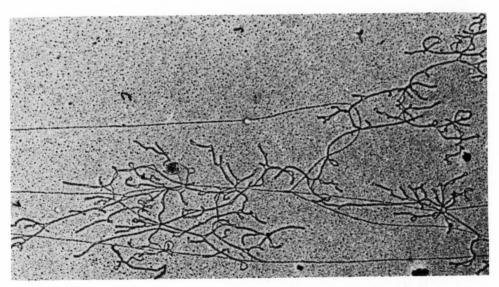

Fig. 17.   Same material as in Fig. 16, but after spreading in
100 µg/ml ethidium bromide.

At a high ethidium bromide concentration the molecule is completely super twisted. Dr. Delius has looked at partially relaxed chromosomes in which some of the DNA is nicked. Fig. 16 shows a partially relaxed chromosome which retains some of the folded super coiled loops and also some relaxed DNA fibers. It is very gently coiled somehow. This molecule is shown here after Kleinschmidt spreading. This same DNA at high ethidium bromide concentration shows highly twisted loops (Fig. 17). There is something sitting at the base of the loop which prevents the rotation of the DNA strands. We believe that something is RNA because after complete unfolding following RNAse treatment this is not seen.

I would like to summarize our findings. First, the folded replicating chromosome can be isolated intact. There are no double strand breaks and few or no single strand nicks. There are 20 to 80 DNA loops per chromosome and they are super coiled. Each nick with DNAse will relax the super coils in one loop but will not affect the super helical content of the rest of the chromosome.

I would like to finish this presentation by stressing that after 3 billion years of trial and error and natural selection it should come as no surprise if the tertiary structure of the DNA inside the cell is not random but may be programed into the DNA itself.

REFERENCES

Bauer, W. R. and J. Vinograd. 1968. J. Mol. Biol. 33:141.

Cairns, J. 1963. J. Mol. Biol. 6:208.

Maaløe, O. and P. Hanawalt. 1961. J. Mol. Biol. 3:144.

Piéchaud, M. 1959. Ann. Inst. Pasteur. 86:787.

Stonington, O. G. and D. E. Pettijohn. 1971. Proc. Nat. Acad. Sci., U.S.A. 68:6-9.

Studier, F. W. 1965. J. Mol. Biol. 11:373.

Wang, J. C. 1969. J. Mol. Biol. 43:263.

Waring, M. J. 1965. J. Mol. Biol. 13:269.

Worcel, A. 1970. J. Mol. Biol. 52:371.

Worcel, A. and E. Burgi. 1972. J. Mol. Biol. 71:127.

PATTERN OF REPLICATION OF VARIOUS GENES IN EXPONENTIAL AND SYNCHRONIZED CULTURES OF <u>ESCHERICHIA</u> <u>COLI</u>

Robert E. Bird, Jacqueline Louarn* and Lucien Caro

Département de Biologie Moléculaire, Université de Genève

Geneva, Switzerland

INTRODUCTION

Early experiments to locate the origin of replication on the <u>E. coli</u> chromosome used a variety of techniques: pulse mutagenesis with nitrosoguanidine, enzyme induction or transduction with bacteriophage P1. They provided the general conclusion that there was a fixed origin of replication, in the lower left quadrant of the <u>E. coli</u> genetic map, and that replication proceeded uni-directionally in a clockwise manner. Some of the data obtained with P1 transduction could be viewed however as indicating bi-directional replication (Caro and Berg, 1968). Recently, we have described a more precise method for determining origin and direction of chromosome replication in <u>E. coli</u> (Bird <u>et al</u>., 1972). The origin and the direction of replication were defined in two ways: the gradient of marker frequency in exponential cultures and the sequence of marker replication after partial synchronization. Gene frequency and sequence of replication were assayed by DNA-DNA hybridization using only two markers. The first, the prophage lambda, is fixed and the second, the prophage Mu-1, is moveable. We have constructed an isogenic series of strains, each lysogenic for prophage lambda and for Mu-1 integrated into a different chromosomal site. Mu-1 often integrates within a gene, thus making the identification of its location quite simple. Both the gradient of marker frequency and the sequence of marker replication after synchronization indicate an origin near <u>ilv</u> with replication proceeding simultaneously

---

*Present address: Centre de Recherche de Biochimie et de Génétique cellulaire du C N R S, Toulouse, France.

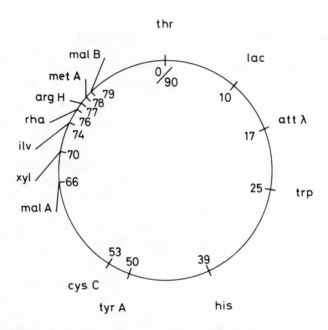

Fig. 1.   Location of phage Mu-1 induced mutations.   The map locations (Taylor, 1970) of the phage Mu-1 induced mutations discussed in the text are shown.

in both directions on the circular E. coli chromosome to end at a
point near trp (see Fig. 1). These results are summarized in this
report.

    In order to locate the origin more precisely we have also
looked at the replication of a marker near the origin after complete
synchronization. A strain carrying Mu-1 in the ilv gene and lysogenic for λ ind⁻ was synchronized by a period of amino acid starvation followed by a period of thymine starvation in the presence of
the required amino acids; synchronous replication was then initiated
by the readdition of thymine (Pritchard and Lark, 1964). Aliquots
of the culture thus synchronized were exposed at various times to
pulses of ^{3}H- thymidine, DNA was extracted and hybridized to filters
containing either lambda or Mu-1 DNA.

    The results indicate that the origin of replication is located
at a map position very close to the marker ilv (74 min.).

MATERIALS AND METHODS

    The parental strain of E. coli W1485 used was CB 0129 (Caro
and Berg, 1968). It requires B1, thymine (low requirer) and leucine. From it a series of strains lysogenic for λ⁺ and for Mu-1
were constructed, with Mu-1 integrated in various sites (Fig. 1).

    The strain used for the experiment of Fig. 3 (Strain Mx213)
carried Mu-1 in the ilv gene, was cured from λ⁺ by infection with
λimm^{21}b2 and was subsequently lysogenized for λind⁻.

    The construction of the strains and the hybridization methods
have been described (Bird et al., 1972).

    The DNA-DNA hybridization performed in Fig. 3 was made in
2 x ssc (1 x ssc = 0,15 M NaCl, 0,015 M Sodium citrate) and 50%
formamid at 42° for 36 hrs (Kourilsky, Leidner and Tremblay, 1971).

RESULTS

    Exponential cultures of the series of Mu-1 and lambda lysogens
were grown on minimal medium supplemented with casamino acids. DNA
was extracted, denatured and immobilized on membrane filters.

    Gene frequency was measured by hybridization with a mixture of
^{3}H-Mu-1 DNA and ^{14}C-λ DNA. The ratio of the two isotopes binding
to the filter gives a measure of the number of copies of Mu-1 present relative to λ. This measure is proportional to the relative
frequency of the gene into which Mu-1 is integrated. A plot of

this number versus map position will show the gradient of gene frequency.

The results are seen in Fig. 2B. The maximum is at <u>ilv</u> with a bi-directional gradient proceeding in both directions to a minimum near <u>trp</u>. This result is consistent with bi-directional replication beginning at <u>ilv</u> and terminating near <u>trp</u>. It is not consistent with half the population replicating in one direction and half in the other; in that case, the resulting gradient would approach a straight line (Bird <u>et al</u>., 1972).

The steepness of the gradient of marker frequency can be increased by shifting cultures from high thymine concentrations (20 µg/ml) to low thymine concentration (0,2 µg/ml). Such a shift causes the replication fork velocity to decrease, while cell growth and initiations of chromosome replication occur at the normal frequency (Zaritsky and Pritchard, 1971). The results of such an experiment for a shift of two generation periods (80 min) is shown in Fig. 2C. The gradient of marker frequency is considerably steeper, showing an increase in the number of replication forks per chromosome.

To demonstrate that starvation for required amino acids allowed termination of replication, cultures were grown in either minimal medium plus requirements or supplemented with casamino acids, then transferred by filtration to medium lacking amino acids. After a period of incubation such that replication had ceased the cultures were harvested and assayed for frequency of Mu-1 and λ. The results are shown in Fig. 2A. The frequency of Mu-1 and λ is always close to one. This indicates that most chromosomes are completed in the absence of amino acid starvation.

After reinitiation of replication by addition of required amino acids we have found two sequences of marker replication (Bird <u>et al</u>., 1972). The first is clockwise: <u>ilv</u>, <u>thr</u>, λ, <u>trp</u> and the second is counterclockwise: <u>ilv</u>, <u>mal</u> A, <u>tyr</u> A, <u>his</u>, <u>trp</u>. These results are only consistent with an initiation near <u>ilv</u> and replication proceeding simultaneously in both directions.

In order to define more precisely the origin of replication we have looked at the time of replication for a marker close to the origin after partial synchronization of replication cycles. The strain, lysogenic for λ ind⁻ and Mu-1 (in <u>ilv</u>), was synchronized by the following procedure: 1) exponential growth in minimal medium; 2) aminoacid starvation (removal of isoleucine, valine and leucine) for 120 min; 3) thymine starvation in the presence of required amino acids for 50 min; 4) readdition of thymine to reinitiate replication. Aliquots of the culture were pulse labeled for 30 seconds with ^{3}H⁻ thymidine at 0, 2, 4, 6, 8, 10 and 12

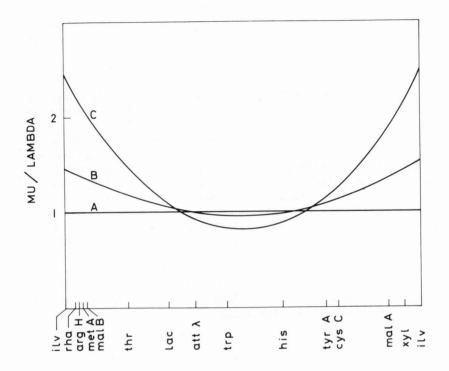

Fig. 2A. Mu-1 / λ ratio after amino acid starvation.

Fig. 2B. Gradient of marker frequency (Mu-1 / λ ratio) in exponential cultures grown in minimal medium supplemented with Casamino acids.

Fig. 2C. Gradient of marker frequency in cultures grown 80 minutes in 0-2 μg thymine/ml. The data for this figure can be found in Bird et al. (1972).

minutes after the addition of thymine. Incorporation was stopped by
addition of pyridine, the DNA was extracted, purified, denatured and
hybridized against filters loaded with λ or Mu-1 DNA. Controls
showed no increase in the titer of either phage 60 min after initiation.

Fig. 3 shows the specificity of the DNA thus labeled. The λ
filters hybridized little radioactivity and were undistinguishable
from blank filters. There are, therefore no or almost no replication forks operating in the λ region of the chromosome nor is there
any significant level of repair replication. Mu-1, by contrast, is
replicated at a maximal rate during the first 30 seconds following
initiation. The extent of Mu-1 replication falls off rapidly to
reach a basal level 6 minutes after initiation. The thymidine pool
in the cell, following the treatment used for synchronization, is
very low (Caro, unpublished observations; McKenna and Masters, 1972).
If the velocity of the replication is estimated from the time of
40 minutes taken to replicate one half chromosome (Cooper and
Helmstetter, 1968; Bird et al., 1972) the data shown here indicate
that the origin of replication is located at 74 min $\pm$ 1 on the
standard E. coli map. They also show that, using the procedure
outlined, a synchronized wave of chromosome replication is initiated
at the normal origin while little or no replication takes place
elsewhere on the chromosome. Similar experiments, to be published
elsewhere, show that the region thr (position 0 on the map) is
replicated at a maximal rate at 20 min, approximately, after initiation while att λ (position 17 on the map) reaches its maximum replication rate 30-35 min after initiation.

DISCUSSION

From the experiments summarized here the following general
conclusions can be drawn:

1)   The origin of replication of the E. coli chromosome is
      located within 1 minute of map of the marker ilv (at 74
      min on the standard map). This gene is replicated with-
      in 30 seconds after initiation.

2)   The replication is bi-directional.

3)   The velocity of the two forks is, on average, the same in
      both directions. The average velocity of a fork is,
      approximately, 15 μ per minute but is subject to important
      fluctuations.

4)   The average velocity of the growing forks depends, for a
      thy⁻ cell, on the thymine concentration in the medium. It

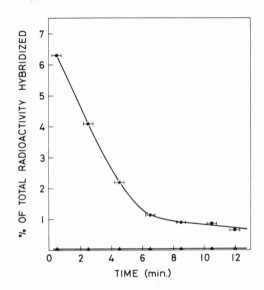

Fig. 3.   Mu-1 (————●————) and λ (————◆————) specific DNA
labeled during short pulses of thymidine-³H at various times after
initiation in synchronized cultures of the strain MX lysogenic for
Mu-1 (in the <u>ilv</u> gene) and for λ ind⁻.

is slowed down considerably below 1 µg of thymine/ml.

5) The two growing forks meet at a point diametrically opposed to the origin, near the gene _trp_.

6) After a period of amino acid starvation, the growing forks have all, or nearly all, moved to that terminal point.

7) After restitution of amino-acids new growing forks are created at the normal origin. If the amino acid starvation has been followed by a period of thymine starvation the initiation is synchronous within a few minutes. No, or nearly no, other growing forks are present elsewhere on the chromosome.

A number of these conclusions have been reached independently by others and no effort has been made, in this summary, to review the literature. Our assignment of origin is in agreement with the DNA-DNA hybridization data of Yahara (1971, 1972) and the pulse mutagenesis experiments of Hohlfeld and Vielmetter (1973). Bi-directional initiation has been supported, in E. coli, by data obtained by P1 transduction (Caro and Berg, 1968; Masters and Broda, 1971), by autoradiography (Prescott and Kuempel, 1972; Rodriguez, Dalbey and Davern, 1973), 5-bromuracil labeling (McKenna and Masters, 1972) as well as by the experiments of Yahara and of Hohlfeld and Vielmetter already mentioned. Bi-directional initiation has also been demonstrated in B. subtilis (Gyurasits and Wake, 1973).

REFERENCES

Caro, L. G. and C. M. Berg. 1968. Cold Spring Harbor Symp. Quant. Biol. 33:559-573.

Cooper, S. and C. E. Helmstetter. 1968. J. Mol. Biol. 31:519-540.

Gyurasits, E. B. and R. G. Wake. 1973. J. Mol. Biol. 73:55-63.

Hohlfeld, R. and W. Vielmetter. 1973, in press.

Kourilsky, Ph., J. Leidner and G. Y. Tremblay. 1971. Biochimie. 53:1111-1114.

Masters, M. and P. Broda. 1971. Nature New Biol. 232:137-140.

Prescott, D. M. and P. Kuempel. 1972. Proc. Nat. Acad. Sci., U.S.A. 69:2842-2845.

Pritchard, R. H. and K. G. Lark. 1964. J. Mol. Biol. 9:288-307.

Rodriguez, R. L., M. S. Dalbey and C. I. Davern. 1973. J. Mol. Biol. 74:599–604.

Taylor, Austin L. 1970. Bacteriol. Rev. 34:155–175.

Yahara, I. 1971. J. Mol. Biol. 57:373–376.

Yahara, I. 1972. Japan. J. Genetics. 47:33–44.

Zaritsky, A. and R. H. Pritchard. 1971. J. Mol. Biol. 60:65–74.

BIDIRECTIONAL REPLICATION OF E. COLI DNA: A PHYSICAL DEMONSTRATION

Millicent Masters and W. G. McKenna

MRC Molecular Genetics Unit, University of Edinburgh

Edinburgh, Scotland

A number of genetic studies (1-3), some of which have been discussed at this meeting, make it clear that the chromosome of Escherichia coli is replicated bidirectionally. Such studies, however, cannot eliminate the possibility that the single linkage group of E. coli is not in fact composed of two separate structures, each of which is replicated unidirectionally. (One would need to presume, for this to be the case, that these structures associate during Hfr transfer in such a way so as to behave as a single structure.)

We would like to report here an experiment which, as well as providing evidence of a biochemical nature that replication of the E. coli chromosome is bidirectional, also renders it untenable that there is a discontinuity in the DNA double helix at the origin of bidirectional replication. In addition, the results of this experiment are not consistent with replication according to rolling circle models of replication (4,5), since daughter DNA is found as a separate structure not linked to parental DNA. The autoradiograms of bidirectionally replicating DNA obtained by Prescott and Kuempel (6) also support these conclusions.

We adapted to E. coli the technique employed by Weintraub (7) to demonstrate that DNA replication in developing chick erythroblasts is bidirectional. Cells of a thymine requiring mutant of E. coli were allowed to initiate DNA replication synchronously with bromouracil present in place of thymine. After 15 sec the BU is replaced by H^3-thymine, replication allowed to continue for a further 100 sec and the cells then killed. (The ^{3}H-thymine is incorporated linearly from the moment it is added [data not shown] thus the BU pulse is not longer than the 15 sec exposure.) The

235

DNA is then extracted, taking care to avoid degradation, and divided into two aliquots. One of these is irradiated with UV light to degrade the sections of DNA containing BU residues (8). The newly synthesized radioactive DNA strand segments are separated from the parental DNA strands and analysed with respect to size in an alkaline sucrose gradient.

If replication were unidirectional (Fig. 1 A) BU would be

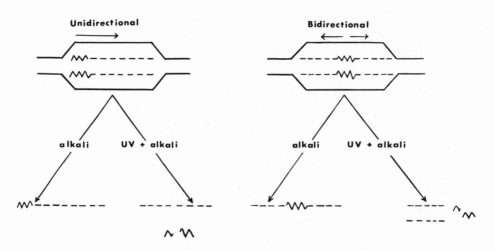

Fig. 1: Diagrammatic representation of the predicted behaviour of unidirectionally (A) as opposed to bidirectionally (B) replicating origin DNA after the following treatment. Synchronized cells are pulsed for 15 sec with 5-bromouracil (∿ BU-DNA) and then with ^{3}H-thymine for 100 sec (---^{3}H-DNA, —— unlabeled DNA). Lysates of cultures are either irradiated with UV light, or left unirradiated and centrifuged on alkaline sucrose gradients.

found at only one end of each single stranded daughter segment.
Were this BU to be removed by irradiation the molecular weight
of the segments would be reduced by only a small amount. Both
irradiated and unirradiated samples would band in the same region
of the sucrose gradient. If replication were bidirectional (Fig.
1 B) the BU would be located in the middle of each segment and UV
treatment would halve their molecular weight: irradiated and unirradiated DNA would band at different places on the sucrose
gradient.

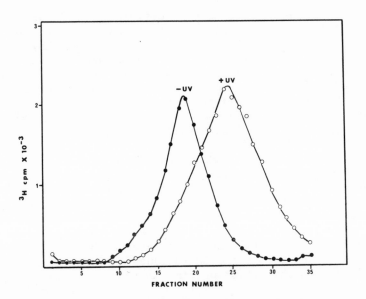

Fig. 2: Centrifugation of irradiated and unirradiated DNA in which
initiation was allowed to take place in bromouracil.

50 ml cultures at densities of 5 x 10^8 cells/ml were synchronized with regard to DNA replication by 100 min of amino acid
starvation followed by thymine starvation for one mass doubling
time (21). The cells were collected on millipore 47 mm HA membrane filters and washed with prewarmed buffer. The cells were
pulsed for 15 sec on the filter with 5-bromouracil (10 μg/ml) and
washed with prewarmed buffer. Finally the cells were suspended
in 10 ml M9-glucose + aa + thymine (1 μg/ml) + ^{3}H-thymine (20 μci/
μg). The cells were vigorously aerated for 100 sec. DNA replication was stopped and the cells lysed (see legend to Fig. 3). In
this case 0.2 ml samples of the crude lysates were exposed to UV
if required (dose 3 x 10^5 ergs) and layered on 5 ml linear 5-20%
alkaline sucrose gradients (0.1 M NaOH, 0.9 M NaCl) for centrifugation in a Beckman L2-65B centrifuge using a SW50.1 rotor at
20 Krpm for 12 hr. 10 drop fractions were collected on Whatman
3 MM paper filters, washed in 5% ice cold TCA followed by 80%
ethanol and counted – see legend to Fig. 2. Open circles,
irradiated DNA; closed circles, unirradiated DNA.

The results of this experiment (Fig. 2) are consistent with bidirectional replication. Irradiation has approximately halved the molecular weight of the single stranded segments (M.W. untreated/M.W. treated = 2.6 ± 0.3; 2.3 is predicted assuming total loss of BU residues). We conclude that the BU incorporated in this experiment is incorporated into the middle of a DNA fragment. This fragment might either be a daughter strand of DNA which has replicated bidirectionally (Fig. 1B), or might have arisen in one of the ways set out below.

If the fragments are indeed bidirectionally replicating daughter DNA the BU they contain will be flanked on both sides by ^{3}H-thymine containing daughter DNA sequences. If they arose in another way the ^{3}H-thymine will be found to one side of the BU only: the DNA on the other side would have been synthesized before the addition of label and would hence contain unlabeled thymine. Such fragments could arise as follows:

    I.   If parental DNA were used to prime daughter DNA replication, BU could be incorporated in continuation of nicked parental DNA.

   II. If termination of rounds of chromosome replication did not occur during the synchronization procedure but only after BU was returned to the medium, both the BU and the ^{3}H-thymine could have been used to terminate rather than initiate replication.

  III. If initiation had occurred before the addition of BU due to retention of thymine in an internal pool or to leakiness of the mutant the BU residues would be at some distance from the actual origin.

If the DNA were then to fragment (I, II) to yield pieces homogeneous in size and with BU in the middle, or if the synthesis occurring before BU addition were equal in quantity to that allowed afterward (III) pieces could be generated which would halve in size on exposure to UV.

We excluded these possibilities by showing that BU used to initiate replication is not added to a previously synthesized piece of DNA. Bacteria, with DNA uniformly ^{14}C-thymine labeled were synchronized and allowed to initiate and continue replication for 100 sec in ^{3}H-BU. The DNA of the cells was isolated and the strands separated and analysed with respect to density and label on CsCl gradients.

The parental ^{14}C containing DNA is of light density: daughter DNA which is fully substituted with ^{3}H-BU is of heavy density (Fig. 3). The absence of any DNA either containing both labels (I, II, possibly III) or intermediate in density (I, II and III) eliminates the possibility that the results set out in Fig. 1 result from BU having been incorporated in continuation of a previously synthesized piece of DNA.

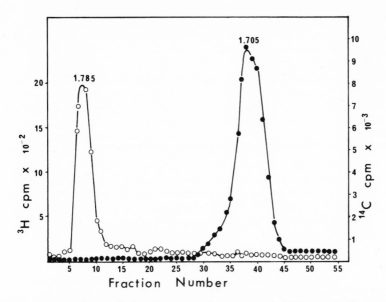

Fig. 3: Separation of daughter and parental DNA on CsCl gradients.
E. coli B/r (thy-cys-leu-his-) was grown in M9 medium + 0.2% glucose
+ 10 µg/ml thymine, 20 µg/ml each of leucine, histidine and cysteine
(aa) with rotary shaking at 37° C.   The cells were labeled with
^{14}C-thymine (10^{-2} µci/µg) for 3 generations and synchronized by
100 min aa starvation followed by thymine starvation for 1 mass
doubling time.   The cells were washed with prewarmed buffer and
resuspended in M9-glucose + aa + 10 µg/ml bromouracil + 2 µci/µg
^3H-bromouracil for 100 sec and vigorously aerated.   DNA replication
was stopped in 50 ml aliquots (5×10^8 cells/ml) by pouring the
cells with crushed ice into buffer.   After washing twice in 1 x SSC
the cells were suspended in 2 ml 0.1 x SSC + 10^{-3} M EDTA and lysed
with 1% Na dodecyl sulfate (SDS) + pronase 1 mg/ml.   The DNA lysate
was extracted twice with an excess volume of 88% phenol - 12% meta-
cresol - 0.1% 8-hydroxyquinoline followed by gentle shaking in
24:1 mixture of chloroform:octanol to remove the SDS.   The DNA
was denatured and centrifuged according to the procedure of Stein
and Hanawalt (10) except that we used BDH Analar grade CsCl and
diluted the DNA in 10^{-3} M EDTA.   10 drop fractions were collected
into ice-cold 5% TCA and counted on Whatman GF/C filters in
PPO-POPOP scintillant in a Packard Tricarb liquid scintillation
counter.

We therefore conclude that BU used to initiate replication is incorporated into what is to become the middle of a segment of DNA afterward extended in both directions. Since this segment can be isolated intact it is clear that there is no discontinuity in the integrity of the double helix at the origin of replication, and that elongation takes place in both directions from the origin to yield a continuous daughter strand.

Our experiments cannot show if any primer, whether DNA or RNA is involved in the act of initiation. A very short DNA primer could be included between BU residues and not be detected by our experiments. A longer one however, would have to be detached within 100 sec of initiation. Rolling circle models, which postulate a long term connection between parental and daughter DNAs are therefore excluded. Any priming RNA would have to be removed within 100 sec as any RNA residues present in the DNA would result in breakage of the daughter strands on exposure to alkali.

The experiments reported here are published in more detail elsewhere (9).

REFERENCES

1. Masters, M. and Broda, P. Nature (New Biol.) 232: 137, 1971.

2. Bird, R. E., Lovarn, J., Martuscelli, J. and Caro, L. G. J. Mol. Biol. 70: 549, 1972.

3. Hohlfield, R. and Vielmetter, W. (in press).

4. Gilbert, W. and Dressler, D. Cold Spring Harbor Symp. Quant. Biol. 33: 473, 1968.

5. Watson, J. D. In Molecular Biology of the Gene, p. 291, W. A. Benjamin, Inc. New York, 1970.

6. Prescott, D. M. and Kuempel, P. L. Proc. Nat. Acad. Sci. USA 69: 2842, 1972.

7. Weintraub, H. Nature (New Biol.) 236: 145, 1972.

8. Hotz, G. and Walser, R. Photochem. Photobiol. 12, 207, 1970.

9. McKenna, W. G. and Masters, M. Nature 240, 539, 1972.

10. Stein, G. H. and Hanawalt, P. C. J. Mol. Biol. 64: 393, 1972.

CHROMOSOME REPLICATION IN BACILLUS SUBTILIS

Tatsuo Matsushita, Aideen O'Sullivan, Kalpana White
and Noboru Sueoka

Department of Biological Sciences
Princeton University
Princeton, New Jersey

AN INITIATION MUTANT (DNA-1)

A number of initiation mutants in bacteria have been reported which are essential for the study on the mechanism of initiation. In B. subtilis initiation and termination of the chromosome replication can be analyzed rigorously by using genetic transformation.

Kalpana White in our laboratory has recently isolated a number of temperature sensitive dna mutants (White and Sueoka, 1973). One of the initiation mutants (dna-1) has been characterized by combining density transfer experiments with transformation. The results in Figure 1 show the evidence for termination of chromosome replication upon a raise of temperature of an exponentially growing culture from 30° to 45°C with a concomitant transfer to BU medium. The evidence for the cessation of initiation by temperature raise is excellent. The theoretical prediction for the transfer of markers to hybrid peak as a function of their position on the chromosome is borne out by the result (Figure 2).

The dna-1 culture at 45°C can resume initiation synchronously upon a lowering of temperature to 30°C (Figure 3). As is evident from the above results, the replication of ade 16 marker can give a convenient measure of initiation.

MEMBRANE ATTACHMENT OF CHROMOSOME

The idea of chromosomes being attached to the membrane was proposed first by Jacob, Brenner and Cuzin (1963). Subsequently.

Ryter and Jacob (1963) reported electron microscopic evidence of
the attachment in B. subtilis. In 1965, attachment of the replication fork to the membrane was observed by Ganesan and Lederberg
(1965) in B. subtilis and by Smith and Hanawalt (1965) in E. coli.
We obtained evidence in B. subtilis that the chromosome may always
be attached to membrane at the replication origin and possibly
also at the terminus (Sueoka and Quinn, 1968). It was found that
genetic markers near the origin, and also, though to lesser extent,

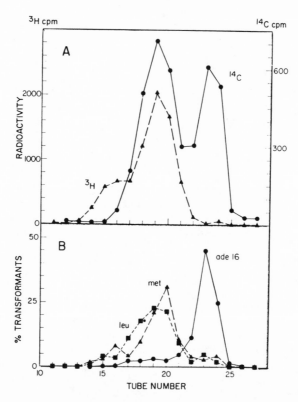

Fig. 1. Replication in mutant dna-1 at 45°C (White and Sueoka,
1973). The mutant culture was uniformly labelled with 14C-thymine
at 30°C. The growth was monitored by Klett-Summerson colorimeter
(66 Filter). In logarithmic growth between Klett units 30 to 40
cells were quick-filtered, washed and shifted to 45°C in a BrU-C+
medium (25 μg/ml bromouracil and 2.5 μg/ml thymine) containing
3H-thymine (2 μc/ml). At the end of 80 min at 45°C crude lysates
were prepared and DNA was subjected to CsCl density gradient
centrifugation. Fractions were analyzed for TCA precipitable
radioactivity and transforming activity. A. Radioactivity profile;
B. Transforming profile. The marker, ade-16 is located close to
the origin, leu in the middle and met towards the terminus.

those close to the terminus, were enriched in the membrane fraction and that the radioactive label with ^{3}H-thymidine introduced in the region close to the origin stayed in the membrane fraction in spite of a long chase with cold thymidine. The enrichment of markers close to the origin and the terminus has been further substantiated (Figure 4; Sueoka and Bell, unpublished results). In the origin labelling experiment, in order to improve the synchrony, the initiation during the spore germination was held temporarily by the addition of nalidixic acid with subsequent release by its removal. Since the effect of nalidixic acid is not completely understood, the result is not devoid of possible artifact. Recently, similar experiments without using nalidixic acid have been performed (O'Sullivan and Sueoka, 1972). The replication of the chromosomes of germinating spores was held temporarily by thymine starvation without using nalidixic acid. Then the replication was released by pulsing with ^{3}H-BUdR and the pulse label was chased with non-radioactive BUdR.

As seen in Figure 5, the "origin pulse" stayed in the membrane fraction even after the chase. In this experiment, the possibility of the pulse label ending up in repair synthesis is excluded from the density transfer pattern of the pulse label (Figure 6). The

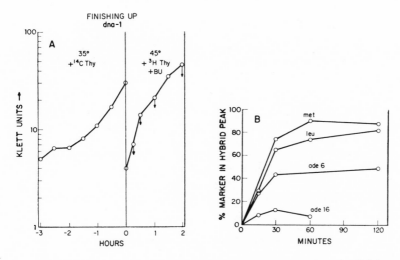

Fig. 2. The kinetics of finishing up in mutant dna-1 (White and Sueoka, 1973). The experimental procedure is similar to that described in Figure 1 except that the cultures were grown at 35°C and were washed and resuspended in five times the amount of medium BrU-C⁺. Samples were withdrawn at the indicated times. A. Protocol of the experiment; B. Kinetics of transfer of genetic markers to the hybrid DNA.

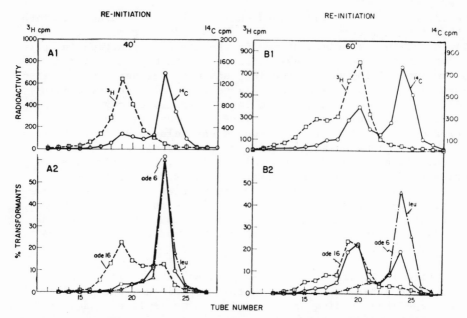

Fig. 3.   Reinitiation in mutant dna-1 (White and Sueoka, 1973).
dna-1 was uniformly labelled with ^{14}C thymine at 35°C.   At Klett
30, cultures were shifted to 45°C.   After 2 hr at 45°C cells were
harvested by suction on a Millipore filter, washed and resuspended
at 35°C in BrU-C$^+$medium (25 µg/ml bromouracil and 2.5 µg thymine
and ^{3}H-thymine, 3 µc/ml).   Samples were withdrawn at 40 and 60
min, cell lysates made and DNA analyzed by CsCl centrifugation.
A1, B1.   Radioactivity profiles; A2, B2.   Transforming activity
profiles.

fact that the pulse label was indeed incorporated into the origin
area is evident from the comparison of the pulse with the profile
of the ade 16 marker (Figure 6).   The fact that the pulse label is
transferred faster than the ade 16 marker shows that the label is
incorporated in the area closer to the origin than the ade 16
marker.   This also eliminates the possibility that the labelling is
by repair synthesis rather than replication.

UNIDIRECTIONAL VERSUS BIDIRECTIONAL REPLICATION

Our ways of mapping markers by using marker frequency analysis
and by synchro-transfer experiments give a temporal replication
order of the markers (see Figure 7).   In 1964, dichotomous replication was discovered (Yoshikawa, O'Sullivan and Sueoka, 1964; Oishi,
Yoshikawa and Sueoka, 1964) and later its symmetric feature was
proven (Quinn and Sueoka, 1970).   The dichotomous replication did,
however, raise a configurational problem: how to connect four (or

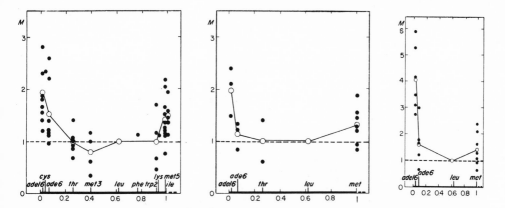

Fig. 4. Cell-growth conditions and membrane enrichment of genetic markers of B. subtilis (from Sueoka and Bell, unpublished results). The membrane-enrichment index (M) is defined as: $M = (X_m/K_f)/(S_m/S_f) = (X_m/S_m/(X_f/S_f)$ where X_m and X_f are transformants of marker X in membrane-bound and membrane-free DNA while S_m and S_f are transformants of marker S (standard, leucine marker in the present analysis) in membrane and free fractions, respectively (Sueoka and Quinn, 1968). The large open circle is the average of values for a marker. It is noted that variability of the index for the leucine is necessarily added to variability of the other markers.

Left: GM11 medium, cell-generation time 60 min. Same as reported in Sueoka and Quinn (1968), except that more data are added.

Middle: C+ medium (a minimal medium plus 0.05% casein hydrolysate, Difco; Yoshikawa and Sueoka, 1963), cell-generation time 40 min.

Right: Penassay medium (1.2% antibiotic medium 3, Difco; Yoshikawa, O'Sullivan and Sueoka, 1964), cell-generation time 20 min.

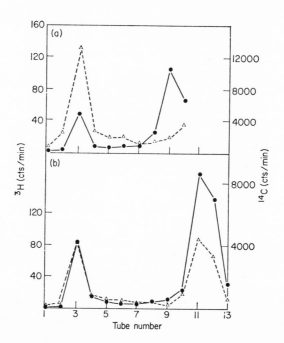

Fig. 5. Sucrose gradient profiles of pulse-labelled DNA at the origin (from O'Sullivan and Sueoka, 1972). DNA was uniformly labelled with ^{14}C-thymine, pulse-labelled to release DNA replication at the origin with ^{3}H-bromouracil and chased to dichotomy in medium containing bromouracil. (a) 15-sec pulse; (b) 45-sec pulse. Cells of 168 <u>thy⁻</u> <u>ind⁻</u> <u>dna-1</u> were grown for several generations in C^{+} medium supplemented with ^{14}C-thymine (0.2 μCi/ml, 55·75 μCi/ μmole). The cells were diluted 5-fold into 40 ml of the same medium at 45°C and incubated for 60 min. The cells were washed with minimal medium (80 ml) at 45°C and resuspended in C^{+} medium without thymine at 35°C. After 20 min the cells were placed on a Millipore filter apparatus, prewarmed to 35°C. Suction was applied, ^{3}H-bromouracil in 1 ml C^{+} medium was added (80 μCi, 0·0079 mg/ mCi), the cells washed with minimal medium (80 ml) and resuspended in C^{+} medium (40 ml) supplemented with 5-bromouracil (40 μg/ml) and thymine (3 μCi/ml). Incubation at 35°C was continued for 30 min. A gentle lysate was prepared and centrifuged in sucrose steps (20%, 62%). ^{14}C, —o——o—; ^{3}H, —△——△—.

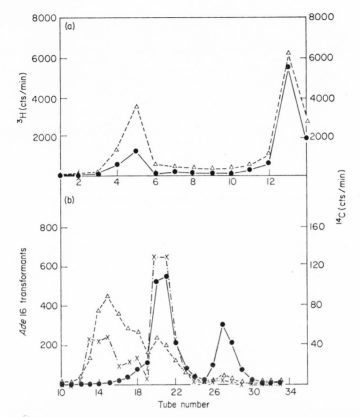

Fig. 6. (a) Sucrose gradient profile of DNA pulse-labeled and chased (from O'Sullivan and Sueoka, 1967). The experiment is the same as that described in Figure 5 except that the pulse was performed with ^{3}H-bromodeoxyuridine (2 μCi/ml, 0·0326 mg/mCi) for 10 sec. (b) CsCl gradient profile of the membrane fraction, shown in (a), which was centrifuged in a CsCl gradient. [14]C, ——●——●——; ——△——△——, ade-16 transformants, ·—X—·—X—·.

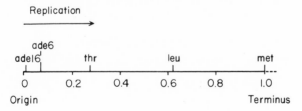

Fig. 7. Map locations of markers used in the present study on the B. subtilis chromosome. The map is based on the data by Yoshikawa and Sueoka (1963) and O'Sullivan and Sueoka (1967).

more) origins to one terminus. The maximum number of double strands
which can be connected to the terminus by covalent bonds is only
two out of four and with this it is hard to replicate while maintaining circularity of the chromosome. Alberts (1971) proposed a
model to connect one end to four beginnings based on an experiment
and the model of Yoshikawa (1970). Essentially Alberts' point was
that the chromosome initiates replication at the origin and replicates bidirectionally. In dichotomous replication then, before
the chromosome is completed, the origin is reinitiated (Figure 8).

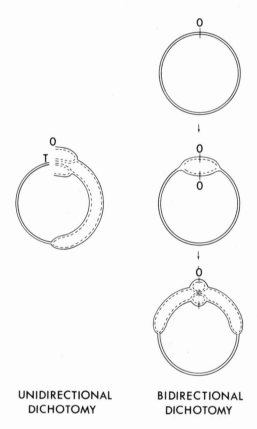

UNIDIRECTIONAL   BIDIRECTIONAL
DICHOTOMY   DICHOTOMY

Fig. 8. Unidirectional and bidirectional dichotomous replication.
The symmetric feature of dichotomous replication of normal chro-
mosome replication has firmly been established in B. subtilis
(Quinn and Sueoka, 1970) and in E. coli (Fritch and Worcel, 1971).
The topological problem of connecting one terminus to the four (or
more) origins has been solved by bidirectional replication which
was raised by Alberts (1970) and demonstrated by Wake (1972).
In this figure, the nick-less initiation at the origin (Cairns-
Inman type is shown for bidirectional dichotomy.

In this way, a "4 to 1" connection which is encountered in the unidirectional dichotomy can be avoided. We took the model (bidirectional dichotomy) seriously along with the evidence that all four origins of this chromosome are attached on the membrane (O'Sullivan and Sueoka, 1972). When we were discussing this, we received the beautiful result of autoradiographic studies from R.G. Wake in Sydney, Australia. Wake's result demonstrates, beyond reasonable doubt, that the basic configuration of bidirectional dichotomous replication shown in Figure 7 is right. His result, which was first reported at Cold Spring Harbor in the summer of 1971 has been published (Wake, 1972).[1]

I would like to mention here that there is one dilemma in the B. subtilis system. I do not doubt that at the beginning replication is bidirectional. But there is a question about complete or partial bidirectionality. When the linkage groups are constructed by transduction, the largest linkage group contains approximately 75% of the markers so far analyzed (see Dubnau, 1971; Young and Wilson, 1972). Within this group the replication order is unidirectional. The logical conclusion to be drawn from the map concerning the replication behavior of the subtilis chromosome would be that at least 75% of the chromosome may cover one arm because threonine, leucine and methionine is the exact order of replication in time. Even if we include all of the remaining markers the other arm should be shorter than 25%. It could be argued that bidirectionality may be symmetric as in coli, but there is for some reason a big silent region of the chromosome in which no markers have been isolated, or that there is some uncertainty in the linkage map of subtilis. If these two possibilities be proven wrong, then B. subtilis chromosome replication should be partially bidirectional. In any case, the origin must be fixed at around the ade-16 marker. The next obvious question concerns the existence or absence of the genetically fixed terminus for replication. In the case of perfect bidirectionality (i.e. both arms have an equal length), this question is hard to answer. However, in a partial bidirectional case, there should be a fixed terminus.

---

1. Signs of bidirectionality in E. coli were seen in earlier works of Caro and Berg (1968), Yahara (1971) and Masters and Broda (1971). More direct evidence has recently been reported by Bird et al. (1972) and Prescott and Kuempel (1972). In B. subtilis, in retrospect, the evidence was there in the discrepancy between the transduction mapping data of ade-16 and suc markers (see Lepesant et al., 1969) and temporal replication order of these markers by synchro-density transfer analysis (O'Sullivan and Sueoka, 1967). Direct proof came from the autoradiographic analysis of Wake (1972) and Gyurasits and Wake (1973).

An interesting clue for bidirectionality existed for quite some time at least around the origin. The gene for levan-sucrose (sucA) replicates after ade-16 by synchro-transfer experiment (O'Sullivan and Sueoka, 1967; Kennett and Sueoka, 1971). Lepesant et al. (1969) reported that the map position analyzed by transduction is to the left of ade-16. This discrepancy is resolved if replication is bidirectional, and if the ade-16 is closer to the origin than sucA.

DNA REPLICATION, IN VIVO AND IN VITRO

Our in vitro DNA replication study emphasizes a comparison between in vivo and in vitro replications. Our recent results have been reported at the Second Annual Steenbock Symposium on "DNA Synthesis In Vitro" which was held in July of 1972 at the University of Wisconsin and will be published (Sueoka et al., 1973). Therefore, only a summary is given below.

The present status of our studies on B. subtilis DNA replication in vitro is:

(1) ATP-stimulated DNA synthesis in toluenized cells satisfies two criteria of normal DNA replication: semi-conservative replication and sequential elongation.

(2) So far, the maximum replication in toluenized cells is about 10% of the chromosome length per replication fork in 60 min at 37°C.

(3) There is no new initiation in toluenized cells.

(4) Effective and rigorous tests for elongation and initiation have been established.

(5) A simple test for locating the site of synthesis has been formulated.

(6) At least one PCMB sensitive DNA polymerase has been found.

(7) ATP-stimulated DNase has been characterized.

REFERENCES

Alberts, B.M. (1970) Fed. Proc. 29, 1154.
Bird, R.E., J. Louarn, J., J. Martuscelli and L. Caro (1972) J. Mol. Biol. 70, 549.
Caro, L.G. and C.M. Berg (1968) Cold Spring Harb. Symp. Quant. Biol. 33, 559.
Dubnau, D. (1970) In "Handbook of Biochemistry", 2nd ed. (H.A. Sober and R.A. Harte, eds), p. I-39, Chemical Rubber Co., Ohio.
Fritsch, A. and A. Worcel (1971) J. Mol. Biol. 59, 207.

Ganesan, A.T. and J. Lederberg (1965) Biochem. Biophys. Res.
    Commun. 18, 824.
Gyurasits, E.B. and R.G. Wake. J. Mol. Biol., in press.
Jacob, F., S. Brenner and F. Cuzin (1963) Cold Spring Harb.
    Symp. Quant. Biol. 28, 329.
Kennett, R.H. and N. Sueoka (1971) J. Mol. Biol. 60, 31.
Lepesant, J.A., F. Funst, A. Carayon and R. Dedonder (1969)
    L.R.H. Acad. Sci., Paris 269, 1712.
Masters, M. and P. Broda (1971) Nature New Biol. 232, 137.
Oishi, M., H. Yoshikawa and N. Sueoka (1967) Nature 204, 1069.
O'Sullivan, M.A. and N. Sueoka (1967) J. Mol. Biol. 27, 349.
O'Sullivan, M.A. and N. Sueoka (1972) J. Mol. Biol. 69, 237.
Prescott, D.M. and P.L. Kuempel (1972) Proc. Nat. Acad. Sci. 69
    2842.
Quinn, W.G. and N. Sueoka (1970) Proc. Nat. Acad. Sci. Wash.·
    67, 717.
Ryter, A. and F. Jacob (1964) Ann. Inst. Pasteur 107, 384.
Smith, D.W. and P.C. Hanawalt (1967) Biochim. Biophys. Acta
    149, 519.
Sueoka, N. and W.G. Quinn (1968) Cold Spring Harb. Symp. Quant.
    Biol. 33, 695.
Sueoka, N., T. Matsushita, S. Ohi, M.A. O'Sullivan and K. White
    (1973) In "DNA Synthesis In Vitro", University Park Press,
    Baltimore, in press.
Wake, R.G. (1972) J. Mol. Biol. 68, 501.
White, K. and N. Sueoka (1973) Genetics 73, 185.
Yahara, I. (1971) J. Mol. Biol. 57, 373.
Yoshikawa, H. (1970) J. Mol. Biol. 47, 403.
Yoshikawa, H. and N. Sueoka (1963) Proc. Nat. Acad. Sci. 49, 559.
Yoshikawa, H., M.A. O'Sullivan, and N. Sueoka (1964) Proc. Nat.
    Acad. Sci. U.S.A. 52, 973.
Young, F. and G.A. Wilson (1972) In "Spores V" (H.O. Halvorson,
    R. Hanson and L.L. Campbell, eds.), Amer. Soc. for Microbiol-
    ogy, p. 77-106.

EUKARYOTIC DNA POLYMERASES

F. J. Bollum and Lucy M. S. Chang

Department of Biochemistry, University of Kentucky

Medical Center, Lexington, Kentucky 40506

INTRODUCTION

Biochemical processes are often studied in prokaryotic systems
due to a lower degree of complexity in the organization of structural and enzymatic components in these simple organisms. In
prokaryotic cells biochemical processes and the components associated with these processes are often compartmentalized in cell
organelles such as the nucleus, mitochrondria, microsomes and etc.
In eukaryotic cells, cellular DNA and DNA synthesis are localized
in the nucleus. Although one might assume that all enzymes involved in DNA synthesis and maintenance are localized in the nucleus,
experimental findings contradict this assumption. The number of DNA
polymerases and their intracellular locations in an eukaryotic cells
are probably related to their in vivo functions, and to the complex
control mechanisms in the replication and maintenance of DNA.

INTRACELLULAR LOCATIONS OF DNA POLYMERASES

All eukaryotic cells contain several forms of DNA polymerases.
For example, DNA polymerase activity has been isolated from the
mitochondrial fraction of mammalian tissues (1), and this activity
exhibits different properties from the nuclear and cytoplasmic DNA
polymerases. Whether the mitochondrial DNA polymerase is related
to non-mitochondrial enzymes is not known, but it is probably safe
to say that the total DNA polymerase activity present in the mitochondrial fraction represents a very small percentage of the total
DNA polymerase activity present in the cell. The rest of this
discussion will be restricted to non-mitochondrial DNA polymerases.

Since DNA polymerases found in mammalian cells differ in molecular

253

weight as well as reaction properties, the simplest demonstration
of the presence of multiple enzymes, and classification of these
enzymes, is by the difference in their sedimentation rates analyzed
by sucrose gradient centrifugation (2). Figure 1 shows the molecular species of DNA polymerases present in the nuclear and cytoplasmic soluble fractions of rabbit bone marrow on linear 5 to 20%
sucrose gradients. The nuclear fraction analyzed was an extract
made from carefully purified rabbit bone marrow nuclei, and the
cytoplasmic fraction analyzed had all of the particulates removed.
DNA polymerase activities were analyzed at pH 7.0 and pH 8.6. The
results of the analyses showed the presence of multiple molecular

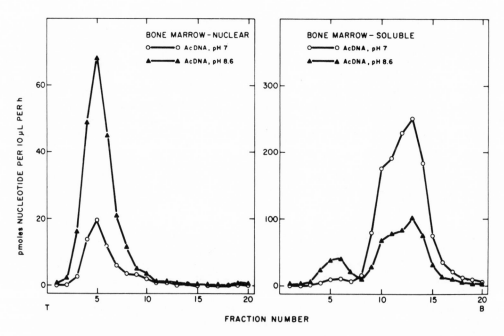

FIGURE 1. Sucrose gradient analysis of the nuclear and soluble
cytoplasmic fractions from rabbit bone marrow. The direction
of sedimentation is from left to right. The open circles represent DNA polymerase activities assayed at pH 7.0, and the closed
triangle represent DNA polymerase activities assayed at pH 8.6.

species of DNA polymerases in the cell. Only one species of DNA polymerase was found in the nuclear fraction. This enzyme had an optimum response under alkaline conditions and had a sedimentation constant of about 3.4 S. Three molecular species of DNA polymerases, including the 3.4 S species found in the nuclear fraction, were present in the cytoplasmic soluble fraction. The two high molecular weight species with sedimentation constants of 6 and 8 S appeared to be present exclusively in the cytoplasmic soluble fraction and they had optimum response under neutral condition. The two high molecular weight DNA polymerases will be discussed here as the 6 to 8 S enzyme, not because they are the same enzyme but because our unpublished work suggests the existence of some common subunits. In the absence of the knowledge of the precise relationships between these two species of DNA polymerases and our inability to separate the two species during purification, we continue to refer to the high molecular weight species of DNA polymerases as the 6 to 8 S DNA polymerase.

ALTERATION OF THE LEVELS OF VARIOUS DNA POLYMERASES IN MAMMALIAN CELLS

In contrast to genetic techniques available for the correlation of enzyme with its in vivo function in the prokaryotic systems, no genetic techniques are available to relate enzyme function with physiological role in eukaryotic cells. Manipulation of the physiological conditions of the cells can induce changes in the rate of certain biochemical processes in the cells. A corresponding change in the level of an enzyme with the change in the rate of a biochemical process is good circumstantial evidence for the involvement of that enzyme in the biochemical process. Cells with different growth rate, and therefore different rate of DNA synthesis, can be obtained from tissues from animals at different stages of development. Alternatively one can induce adult mammalian tissues to proliferate by surgical or chemical manipulations such as induction of liver regeneration by partial hepatectomy, induction of erythropoesis in mouse spleen by injection of phenylhydrazine, DMBA induction of leukemic spleens in rats, and stimulation of blast formation of human lymphocytes by phytohemagglutanin. The growth rate of tissue culture cells can be controlled and the level of DNA polymerases in these cells can be correlated with the levels of DNA synthesis.

Studies were carried out using the above systems to correlate the level of each of the DNA polymerases with growth rate and DNA synthesis. The results from all these studies are similar. In general, high molecular weight DNA polymerases, present only in the cytoplasmic fraction, were found to be the predominant activity in rapidly dividing cells. The 3.4 S DNA polymerase, present in both cytoplasm and nucleus, accounted for less than 10% of the total DNA polymerase activity in rapidly growing cells, while it accounted for up to 50% of the total activity in some adult mammalian tissues.

When the levels of various DNA polymerases were examined in
extracts of regenerating rat liver at 0, 24 and 48 hours after
partial hepatectomy, the results obtained are representative of all
other biological systems mentioned above (3). Figure 2 shows the
level of 3.4 S DNA polymerase in the nuclear fraction of regenerating rat liver at various time after partial hepatectomy. No
significant change in the level of DNA polymerase in the nucleus
occurs during liver regeneration. Figure 3 shows the sucrose
gradient profiles of DNA polymerase activities in the soluble
cytoplasmic fraction of regenerating rat liver at various time
after hepatectomy. The level of the 3.4 S enzyme in the cytoplasm

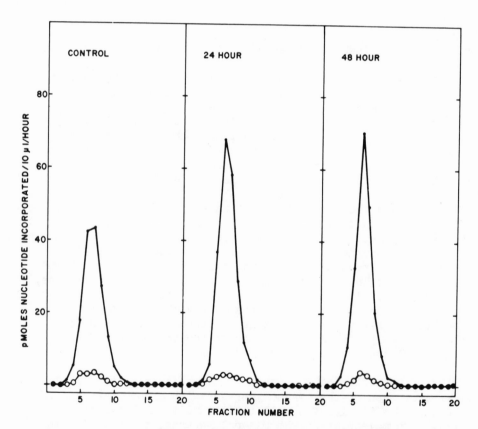

FIGURE 2.   Sucrose gradient analysis of the DNA polymerase in the
nuclear fraction of regenerating rat liver. The filled circles
represent activities assayed at pH 8.6 and the opened circles
represent DNA polymerase assayed at pH 7.0.

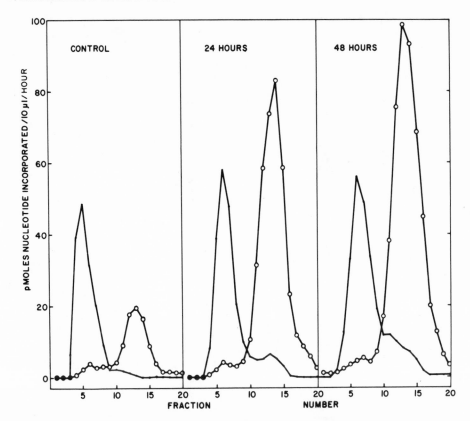

FIGURE 3. Sucrose gradient analysis of DNA polymerase activities in the cytoplasmic fraction of regenerating rat liver. The filled circles represent DNA polymerase activities assayed at pH 8.6 and the opened circles represent DNA polymerase activities assayed at pH 7.0.

remained relatively constant during the hyperplastic response, while the level of the 6 to 8 S polymerases increased remarkably (varying from 3 to 12 fold increases in different experiments).

Another system well suited for studies on the correlation of levels of DNA polymerases with DNA synthesis and cell growth is found in tissue culture cells grown under various conditions (4). Figure 4 shows the growth curve and corresponding rates of DNA synthesis of mouse L-cells in suspension. As the cells reached stationary phase, DNA synthesis as measured by ^{3}H-thymidine

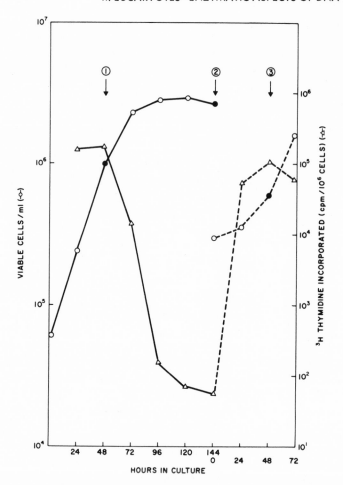

FIGURE 4. DNA synthesis and cell growth of mouse L-cells in suspension culture. At times indicated by the cross-hatched points,
numbered 1, 2, and 3, samples of the culture were removed for the
studies on levels of DNA polymerases.

incorporation decreased to very low levels. When the stationary
cells were diluted into fresh medium, DNA synthesis turned to the
normal level and cell growth continued. DNA polymerase levels
were examined in the cytoplasmic soluble and nuclear fractions of
cells growing in log phase, in stationary phase and after release
from stationary phase (indicated by numbers 1, 2, and 3 in Figure
4), and the results of these analyses are shown in Figure 5 and
Figure 6. Sucrose gradient profiles of DNA polymerase activities
in the cytoplasmic soluble fractions of cells grown at various
rates (Figure 6) showed that three species of DNA polymerases are
present. As the cells reached stationary phase, the level of the

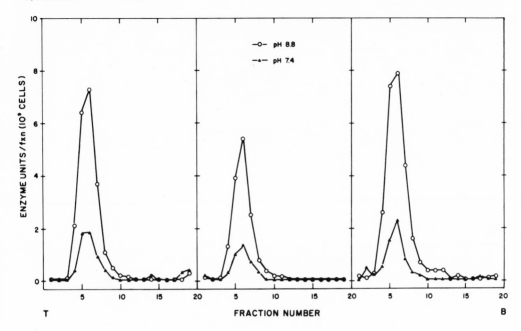

FIGURE 5. DNA polymerase activities in the nucleus of mouse L-cells during the growth cycle.

3.4 S enzyme remained relatively unchanged while the level of the 6 to 8 S enzymes decreased remarkable (3 to 12 fold decreases in different experiments). Sucrose gradient profiles of the nuclear extracts (Figure 5) showed that only one species of DNA polymerase (3.4 S species) was present, and the level of the nuclear 3.4 S enzyme remained relatively constant with respect to different rates of DNA synthesis and cell growth. Releasing the cells from stationary phase resulted in a reappearance of the 6 to 8 S enzymes in the cytoplasm (Figure 6) and relatively no change in the level of 3.4 S enzyme in the nucleus and the cytoplasm (Figures 5 and 6).

The results from studies using phytohemagglutanin-treated human lymphocytes, erythropoeitic mouse spleen or DMBA-induced

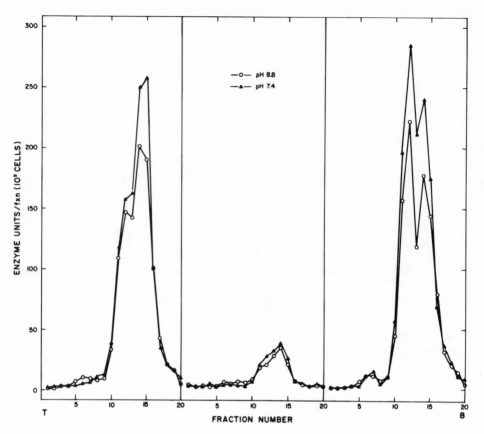

FIGURE 6.   DNA Polymerase activities in the cytoplasm of mouse L-
cells during the growth cycle.   The analyses of DNA polymerase
activities at various time of the growth cycle were carried out
on sucrose gradients, and the enzyme activities were assayed at
two pH levels.

mouse leukemic spleen were the same as shown for the regenerating
rat liver and the mouse L-cells studies; that is, only the high
molecular weight DNA polymerases in the soluble cytoplasm showed
significant responses with increases in growth rate and DNA synthesis.
No significant alteration in the level of the low molecular weight
DNA polymerase in the soluble cytoplasmic fraction or the nuclear
fraction has been shown in any of the biological systems studied
thus far.   The results from these studies clearly suggest the
involvement of the high molecular weight DNA polymerases in DNA
replication.   The precise role of each of these enzymes in vivo
still remains indefinite.

PROPERTIES OF MAMMALIAN DNA POLYMERASES

Both molecular species of mammalian DNA polymerases have similar reaction requirements, requiring initiator-template systems,
complementary deoxynucleoside triphosphates (dNTPs), and a divalent
cation.

The divalent cation requirements of mammalian DNA polymerases
are quite similar. When nuclease treated DNA ("activated" DNA)
was used as the template, the preferred divalent cation was Mg++
and the optimum concentration was about 6 to 12 mM. When initiated-
polydeoxynucleotide was used as the template, the preferred divalent
cation was Mn++ and the optimum concentration was about 0.2 to
0.6 mM.

One reaction property that discriminates between various mammalian DNA polymerases is the preferred pH of the reaction. The
pH optimum of the calf thymus 3.4 S enzyme for both DNA templates
and homopolymer templates was found to be 8.6 in Ammediol:HCl buffer, while the pH optimum of the calf thymus 6 to 8 S DNA polymerases was found to be about 7.0 in potassium phosphate buffer.
The pH optimum of the 6 S and the 8 S enzymes are not identical.
The pH optimum of the 6 S enzyme was somewhat higher than the 8 S
enzyme and this observation is reflected in the ratios of enzyme
activities assayed at pH 7.0 and pH 8.6 in the sucrose gradient
profile of rabbit bone marrow cytoplasmic DNA polymerases (Figure 1).

All mammalian DNA polymerases require the presence of all complementary dNTPs for maximum activity. The stringency of dNTP requirement is dependent upon the template used, and is therefore a property of the template rather than the enzyme. Table 1 shows the
requirement for all four dNTPs of rabbit bone marrow 3.4 S DNA
polymerase when DNAs treated with nucleases were tested for template
activity. When pancreatic DNase I-treated calf thymus DNA was
used as the template, large amounts of dNTPs were converted into
products in the presence of only one to three dNTPs. DNase I
treated DNA was probably composed of only very small regions (about
3% of total DNA) available for DNA polymerase action, and these
regions may be rich in short homopolymer runs. When sonicated
calf thymus DNA was treated with E. coli exonuclease III larger
template regions were exposed, and the requirements for all four
dNTPs became more stringent when this DNA was used as the template.
Using exonuclease III treated lamda DNA as the template less than
5% of the total activity could be detected when one to three dNTPs
were omitted from the reaction.

None of the mammalian DNA polymerase can initiate new chains,
therefore these enzymes cannot carry our de novo synthesis nor can
they replicate single stranded DNA, single stranded polydeoxy-

TABLE I

Deoxynucleoside Triphosphate Requirements for Rabbit Bone Marrow DNA Polymerase Reaction

Reaction Mixture	Template		
	DNase I treated calf thymus DNA (pmoles/15 min)	DNase I nicked exonuclease III treated calf thymus DNA (pmoles/10 min)	Exonuclease III treated lambda DNA (pmoles/15 min)
Complete	419 (100%)	119 (100%)	20.0 (100%)
-dCTP	199 (45%)	48 (40%)	
-dGTP	176 (42%)	55 (46%)	
-dATP	204 (49%)	49 (41%)	1.0 (5%)
-dCTP and dGTP	149 (36%)	33 (28%)	
-dATP and dCTP	144 (34%)	28 (24%)	
-dATP and dGTP	162 (39%)	36 (30%)	0.35 (1.8%)
-dATP, dCTP and dGTP	151 (36%)	21 (18%)	0.28 (1.4%)

All reaction mixtures contain 18 μg of DNase I treated calf thymus DNA or DNase I nicked, exo-nuclease III treated calf thymus DNA, or 9 μg of exonuclease III-treated lambda DNA, 50 mM Ammediol buffer at pH 8.6, 8 mM MgCl₂, 1 mM mercaptoethanol, and 14 μg of enzyme protein in a final volume of 0.25 ml. The complete reaction mixture contains [³H]dTTP, dGTP, dCTP and dATP, each at 0.1 mM.

nucleotides, or native DNA. DNA treated with nuclease or homopolymer initiated with complementary oligonucleotides are suitable
templates for mammalian DNA polymerases (5). Figure 7 shows the
minimum chain length of oligodeoxynucleotides required for initiation of poly dA or poly dT replication catalyzed by calf thymus
6 to 8 S DNA polymerases. The calf thymus 3.4 S enzyme was found
to have an identical minimum chain length requirement for initiation of poly dA replication. Oligodeoxynucleotide initiated
polyribonucleotides are not suitable templates for mammalian DNA
polymerases described here with the exception that polyribo-
adenylate initiated with oligothymidylate was found to be an equivalent or better template than its deoxyribose counterpart for the

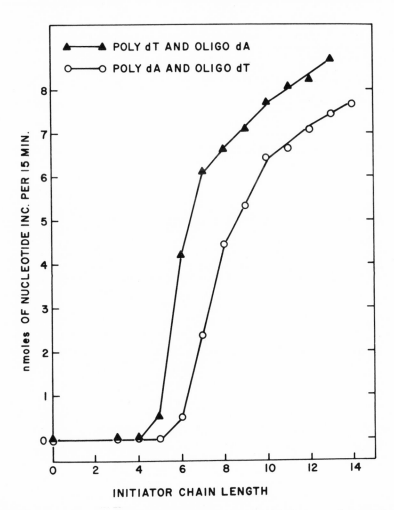

FIGURE 7. Effect of chain length of initiator on homopolymer
replication.

3.4 S DNA polymerase (2). Oligoriboadenylates of chain length larger than 5 are good initiators for calf thymus 6 to 8 S DNA polymerase, but they are relatively poor initiators for the calf thymus 3.4 S DNA polymerase in poly dT replication. Oligouridylates are poor initiators for calf thymus 6 to 8 S enzymes and 3.4 S enzyme in poly dA replication (6).

PRODUCTS OF MAMMALIAN DNA POLYMERASE REACTIONS

Due to the lack of precise knowledge of physical and chemical properties of DNA templates, analyses of the products of DNA polymerase reactions is rather difficult when DNA was used as the template. Polydeoxynucleotides initiated with complementary oligodeoxynucleotides, complementary oligoribonucleotides, or terminal deoxynucleotidyl transferance can be used as model templates to study the mechanism of DNA polymerase actions (8).

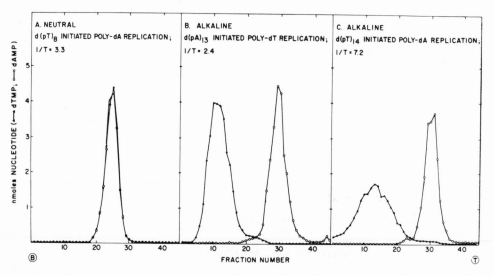

FIGURE 8. CsCl gradient analysis of the products of homopolymer replication.

When polydeoxynucleotides were initiated with complementary oligodeoxynucleotides, the product chains were not covalently linked to the template chains. Figure 8 shows neutral and alkaline CsCl gradient analyses of the products of poly dA replication catalyzed by calf thyms 6 to 8 S DNA polymerase at various input initiator to template ratios. The product chains were not linked to the template chains, as demonstrated in the alkaline gradients, but the product chains are hydrogen-bonded to the template chains as demonstrated in the neutral gradient. Figure 9 shows the size of the template and products chains in neutral and alkaline sucrose gradients when initiator to template ratios were varied. On neutral sucrose gradients, the products of poly dA replication appeared as linear duplexes with the nucleotide ratio (template to product nucleotide) equal to about one. The alkaline sucrose gradient analyses showed that increasing initiator to template ratio resulted in decreasing in the chain length of the product chains. The

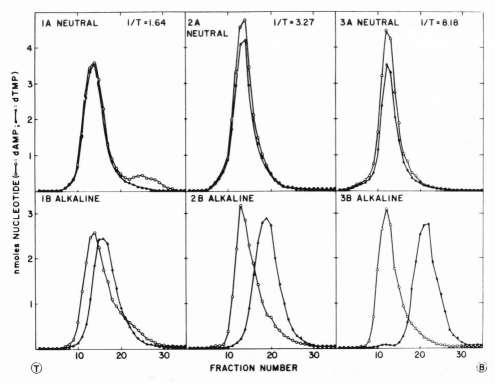

FIGURE 9. Neutral and alkaline sucrose gradient analyses of the products of poly dA replication.

initiator molecules added into the reaction were found to be covalently linked to the product chains. When radioactive initiator is used in a reaction with dNTP labeled with a different
isotope, chain length of the product chains can be calculated directly in each fraction of the sucrose gradient (Figure 10).

An alternative method for initiating a polydeoxynucleotide
templates in DNA polymerase reactions is through the action of
terminal deoxynucleotidyl transferance. Reaction of a polydeoxynucleotide with its complementary dNTP catalyzed by terminal transferance proceeds only until a stable double-stranded region is
formed at the 3'-terminus of the polydeoxynucleotide, resulting in
a "hook" polymer. The advantages of using a hook polymer synthesized with terminal transferase as template for DNA polymerase

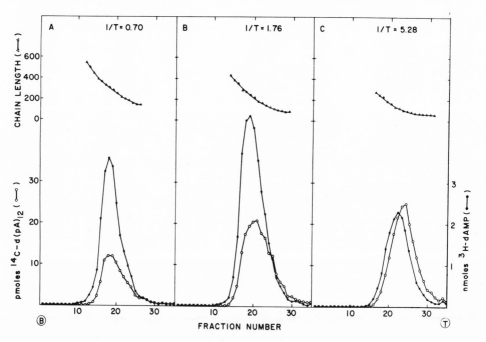

FIGURE 10. Number avarage molecular weight of the products of
multiply-initiated poly dT replication analyzed on alkaline sucrose
gradients.

reactions are that each template contains only one initiated
region and all template chains are initiated. Replication of the
hook polymers with mammalian DNA polymerases gives rise to products
in which the template chains are covalently linked to the product
chains. Figure 11 shows the product of poly dA replication catalyzed by calf thymus 6 to 8 S DNA polymerase in the presence of
of terminal transferase in alkaline CsCl gradient. The result
demonstrates the hybrid density of the product of this reaction
and establishes covalent linkage between the product chain and
the template chain.

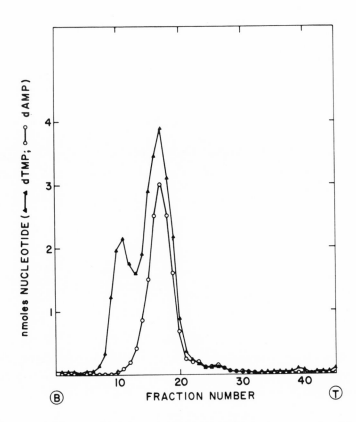

FIGURE 11. Alkaline CsCl gradient analysis of terminal deoxynucleo-
tidyltransferase-initiated poly dT replication.

ASSOCIATED ENZYME ACTIVITIES IN MAMMALIAN DNA POLYMERASES

Detailed studies on the enzyme activities associated with mammalian DNA polymerases have not been carried out because of the lack of highly purified enzyme preparations. The 3.4 S DNA polymerase has now been purified to homogeneity from calf thymus chromatin (7). The 6 to 8 S DNA polymerases preparations we have are not homogeneous at the present time, but the enzymes had been purified 2000 to 5000 fold from calf thymus soluble extract. Since the associated activities of E. coli polymerase I have been extensively studied (8,9), polymerase I serves as a model enzyme for studies on DNA polymerases from other sources. We carried out a comparison of the activities associated with mammalian DNA polymerases with E. coli polymerase I (10). The activities examined are pyrophosphate exchange, pyrophosphorolysis, 3' to 5' exonuclease, template dependent deoxynucleoside triphosphate degradation, and 5' to 3' exonuclease.

The ability of E. coli polymerase I, calf thymus 6 to 8 S DNA polymerase and calf thymus 3.4 S DNA polymerase to catalyze pyrophosphate exchange is compared in Table II. Experiment I showed

TABLE II

Pyrophosphate Exchange

Experiment I

Enzyme	Pyrophosphate (mM)	Exchange (nmole/h)	Polymerization exchange
E. coli polymerase I	1	0.69	1.2
Calf thymus 6-8 S	2	0.90	13.6
	5	1.23	6.2
Calf thymus 3.4 S	1	<0.01	>100
	2	<0.01	>100
	5	<0.01	>100

Experiment II

Enzyme	Pyrophosphate (mM)	Exchange (nmole/h)	Polymerization exchange
E. coli polymerase I	1.2	29.8	3.34
Calf thymus 3.4 S	1.2	<0.001	$>10^5$

easily demonstrable pyrophosphate exchange reaction with polymerase
I and calf thymus 6 to 8 S polymerase, confirming published information concerning these two enzymes (9,11), but no measurable exchange was found for the calf thymus 3.4 S enzyme. The second experiment recorded the results obtained when the enzyme level was
increased about 30 fold and the specific activity of $[^{32}P]$ pyrophosphate was increased about 100 fold. No significant exchange
can be demonstrated for the 3.4 S enzyme.

Considerable experimentation has been done to verify and eval-
uate the significance of the negative results obtained with the 3.4
S polymerase. Studies at various pHs; in different buffers, and
with synthetic templates in Mm^{++} and Mg^{++} were carried out. All
attempts to observe an exchange reaction using the 3.4 S enzyme
were negative. The 3.4 S enzyme was tested directly for pyro-
phosphatase activity and none was found. We conclude that the 3.4
S DNA polymerase does not catalyze a pyrophosphate exchange.

The ability of E. coli polymerase I, calf thymus 6 to 8 S DNA
polymerases and calf thymus 3.4 S polymerase to catalyze pyrophosphorolytic degradation of DNA in the absence of dNTPs was compared
in Table III. Using template:initiator systems containing 3'-labeled
termini, the nature of the products of degradation in the presence
of pyrophosphate was examined. With a matched template, polymerase
I produced dTTP, dTMP and some dTDP. With a mismatched template,
polymerase I produced a large amount of dCMP and no dCTP. These
findings are in agreement with those reported by Brutlag and Kornberg (9). Calf thymus 6 to 8 S polymerase produced only dTTP with
the matched template. It produced only trace amounts of dCMP with
the mismatched template. The 3.4 S polymerase produced no hydrolytic or pyrophosphorolytic products.

All three DNA polymerases were tested for 3' to 5' exonuclease
activity. The results are shown in Figure 12. The 3' to 5' exonuclease activity described as the proofreading exonuclease in E.
coli polymerase I proceeds from the 3'-terminus toward the 5'-end.
The template systems used were the same as in the pyrophosphorolysis
experiment. Results in Frame A and Frame B showed that the proofreading exonuclease can be easily demonstrated for polymerase I.
The 6 to 8 S enzyme produced no hydrolytic product with the matched
template (Frame C) and only trace amounts of hydrolytic product
with the mismatched template (Frame D). The 3.4 S calf thymus DNA
polymerase had no proofreading exonuclease (Frames E and F).

Both E. coli polymerase I and T_4 DNA polymerase carry out the
conversion dNTP to dNMP in a template dependent reaction (12, 13).
This reaction is assumed to be a combination of hydrolytic and
polymerization reactions. As might be expected for enzyme not
having 3' to 5' exonuclease activity, neither calf thymus 6 to 8 S
nor the calf thymus 3.4 S DNA polymerase degrades dNTP in the presence

TABLE III

Pyrophosphorolysis and Hydrolysis with Matched and Mismatched Templates

Enzyme	Templates	Products			
		Polymer	dNTP	dNDP	dNMP
E. coli polymerase I	$d(pA)_{600} \cdot d(pT)_{46} \, d(pT^*)_{1.14}$	4884	1008	84	1020
	$d(pA)_{600} \cdot d(pT)_{46} \, d(pC^*)_{0.86}$	231	0	0	6462
Calf thymus 6–8 S	$d(pA)_{600} \cdot d(pT)_{46} \, d(pT^*)_{1.14}$	6642	1956	0	0
	$d(pA)_{600} \cdot d(pT)_{46} \, d(pC^*)_{0.86}$	5984	0	0	54
Calf thymus 3.4 S	$d(pA)_{600} \cdot d(pT)_{46} \, d(pT^*)_{1.14}$	8342	0	0	0
	$d(pA)_{600} \cdot d(pT)_{46} \, d(pC^*)_{0.86}$	6325	0	0	0

* Radioactive nucleotide

Numbers in the body of the table are actual counts per minute above background.

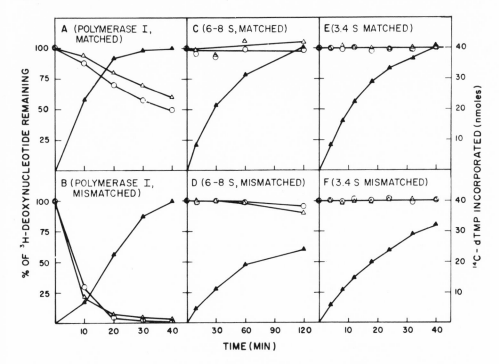

FIGURE 12. Proofreading exonuclease (3' to 5' exonuclease) associated with E. coli polymerase I (Frames A and B), calf thymus 6
to 8 S DNA polymerase (Frames C and D), calf thymus 3.4 S DNA polymerase (Frame E and F).

of template. Figure 13 shows chromatographic analyses of the monomer in calf thymus 3.4 S DNA polymerase catalyzed reaction. The
small amount of dTMP present initially in the reaction did not increase during the polymerization reaction. Similar results were
obtained for the calf thymus 6 to 8 S enzymes.

Using polydeoxynucleotides labeled at the 5'-termini, the 5'
to 3' exonuclease can be easily demonstrated in E. coli polymerase
I. The calf thymus DNA polymerases do not have any detectable
5' to 3' exonuclease activity.

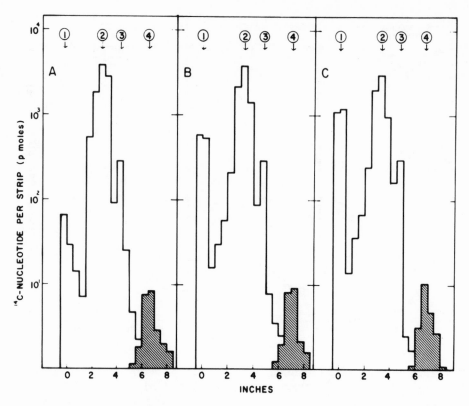

FIGURE 13. Degradation of deoxynucleoside triphosphate by calf thymus 3.4 S polymerase. Frame A shows the distribution of products after 1 min. of reaction. Frame B is after 40 min. of incubation and Frame C is after 160 min. of incubation. The numbers 1, 2, 3, and 4 designate the positions of polymer product, dTTP, dTDP and dTMP respectively.

A summary of the results of the comparison of associated enzyme activities in mammalian DNA polymerases and E. coli polymerase I is shown in Table IV.

In conclusion, several molecular species of DNA polymerase have been found in mammalian cells. The precise role of each of these enzymes in vivo still remains unknown although physiological experiments suggest the involvement of 6 to 8 S enzymes found in the soluble cytoplasm of the cell in DNA replication. Some generalized statements have been made throughout this discussion concerning

TABLE IV

Summary of DNA Polymerase Associated Activities

Enzyme activity	Enzyme		
	E. coli polymerase I	Calf Thymus 6-8 S	Calf Thymus 3.4 S
Polymerization	+	+	+
Pyrophosphate exchange	+	+	-
Pyrophosphorolysis	+	+	-
3' to 5' exonuclease	+	±	-
dNTP degradation	+	-	-
5' to 3' exonuclease	+	-	-

the properties of all mammalian DNA polymerases although detailed studies have been carried out for various DNA polymerases isolated from a few sources. We have examined the number and the distribution of DNA polymerases in a large number of mammalian cells and our results indicate similarities in the number and level of DNA polymerases in these tissues. Furthermore, immunological cross-reactivity exists between various molecular species of mammalian DNA polymerase, and between DNA polymerases from different species of animals (14). No immunological cross-reactivity can be detected between E. coli polymerase I and polymerase II with the mammalian enzymes. Mammalian DNA polymerases are clearly different from prokaryotic DNA polymerases in both the levels and the kinds of associated enzyme activities. These differences strongly suggest that the mechanism of actions of prokaryotic and eukaryotic DNA polymerases are quite different. Any generalized statements concerning DNA polymerases from prokaryotic and eukaryotic sources may be quite misleading in our understanding of eukaryotic DNA polymerases.

Acknowledgement: The research reported here was supported by a grant (No. CA-08487) from the National Cancer Institute.

REFERENCES

1. Meyer, R.R. & Simpson, M.V. (1970) J. Biol. Chem. 245: 3426.
2. Chang, L.M.S. & Bollum, F.J. (1972) Biochemistry 11: 1264.
3. Chang, L.M.S. & Bollum, F.J. (1972) J. Biol. Chem. 247: 7948.
4. Chang, L.M.S., Brown, McKay & Bollum, F.J. (1973) J. Mol. Biol. 74: 1.
5. Chang, L.M.S., Cassani, G.R. & Bollum, F.J. (1972) J. Biol. Chem. 247: 7718.
6. Chang, L.M.S. & Bollum, F.J. (1972) Biochem. Biophys. Res. Comm. 46: 1354.
7. Chang, L.M.S. (1973) J. Biol. Chem. 248:
8. Kornberg, A. (1969) Science 163: 1410.
9. Brutlag, D. & Kornberg, A. (1972) J. Biol. Chem. 247: 241.
10. Chang, L.M.S. & Bollum, F.J. (1973) J. Biol. Chem. 248:
11. Bollum, F.J. (1960) J. Biol. Chem. 235: 2399.
12. Deutscher, M.P. & Kornberg, A. (1969) J. Biol. Chem. 244: 3019.
13. Hershfield, M.S. & Nossal, N.G. (1972) J. Biol. Chem. 247: 3393.
14. Chang, L.M.S. & Bollum, F.J. (1972) Science 175: 1116.

DNA POLYMERASES AND A POSSIBLE MULTI-ENZYME COMPLEX FOR DNA

BIOSYNTHESIS IN EUKARYOTES

Earl Baril, Betty Baril

Worcester Foundation for Experimental Biology

Shrewsbury, Massachusetts, U.S.A.

Howard Elford, Ronald B. Luftig

Duke University Medical Center

Durham, North Carolina, U.S.A.

During the past decade considerable progress has been made toward understanding the mechanism and control of DNA replication. The greatest advances have come in studies of prokaryotic DNA replication and these results have facilitated studies of eukaryotic DNA replication. The points emerging are that DNA replication in prokaryotes and eukaryotes involves the concerted action of several enzymes, non-enzymic factors and an integrated intracellular arrangement.

The discovery of the pol Al mutation in Escherichia coli (DeLucia and Cairns, 1969) led to a revision of our concept of the enzymatic basis of DNA biosynthesis and to the demonstration that E. coli cells contain at least two DNA polymerases, designated II and III, in addition to Kornberg's DNA polymerase I (Moses and Richardson, 1970; Knippers, 1970; Kornberg and Gefter, 1971; Kornberg, 1969). Multiple DNA-dependent DNA polymerases have recently been demonstrated in mammalian cells (Baril et al., 1971; Weissbach et al., 1971; Chang and Bollum, 1971; Sedwick, Wang and Korn, 1972; Smith and Gallo, 1972). In rat liver there are two DNA polymerases that are distinguishable from the mitochondrial DNA polymerase (Baril et al., 1971). The activity of one of the enzymes, designated DNA polymerase II, correlates positively with the proliferative rate in regenerating rat liver and hepatomas (Baril and

275

Laszlo, 1971; Baril et al., in press). A similar enzyme is reported
to increase during the S phase of the cell cycle of mouse L cells
in culture (Chang and Bollum, this symposium).

In regenerating rat liver and hepatomas DNA polymerase II is
found in the nucleus and with membrane fragments that are isolated
from the post-microsomal supernatant solution (Baril and Laszlo,
1971; Baril et al., in press). The membrane fragments have associated not only DNA polymerase II (DP_{II}), but also ribonucleotide
reductase (RR), thymidylate synthetase (TS) and thymidine kinase
(TK) activities. These enzymes continue to co-fractionate after
the combined treatment of the membrane fragments with 2.5% n-butanol
and 5% Triton X-100. The enzymes are present with particulate
material after this treatment, and electron microscopic analysis
indicates that the particles are 8–12 nm in diameter.

The membrane fragments or the 8–12 nm particles derived from
them can by sequential reactions in vitro produce (^{3}H)-dTTP from
(^{3}H)-thymidine and subsequently incorporate the (^{3}H)-dTTP into DNA.
It is possible that DNA polymerase II, thymidine kinase, thymidylate
synthetase and ribonucleotide reductase in regenerating rat liver
and hepatomas are associated with a multi-enzyme complex for DNA
biosynthesis.

DNA POLYMERASES IN RAT LIVER AND HEPATOMAS

Dr. Bollum has already discussed his rather convincing evidence
for the existence of two non-mitochondrial forms of DNA polymerase in
mammalian cells. Simultaneously, independent studies by Baril and
coworkers (1971) and Weissbach et al. (1971) also demonstrated the
presence of these enzymes in normal and abnormal mammalian cells.
Two non-mitochondrial forms of DNA polymerase are present in normal
and regenerating rat liver and transplantable hepatomas (Baril et al.,
in press). The sub-cellular distributions of these enzymes were
investigated in these tissues and the enzymes have been partially
purified and characterized (Baril et al., in press). The DNA polymerases have different physical and enzymatic properties, as also
shown by Chang and Bollum (1971 and this symposium).

We designate the non-mitochondrial DNA polymerases in rat liver
and hepatomas as DNA polymerases I and II (Baril et al., in press).
This is based on the chronological order of their discovery (Baril
et al., 1971; Howk and Wang, 1969). DNA polymerase I is a low
molecular weight protein, about 30,000 daltons, that does not bind
to DEAE-cellulose at pH 7.5 but does bind firmly to phosphocellulose
at pH 7.0. DNA polymerase II has a molecular weight of about
250,000 daltons, as estimated by gel filtration, and binds to DEAE-
cellulose at pH 7.5 but is weakly bound to phosphocellulose at

pH 7.0. The activity of DNA polymerase II is completely inhibited
by thiol-blocking agents at a concentration of 0.6 mM, while the
activity of DNA polymerase I is not inhibited. Both of the purified
enzymes lack deoxyribonuclease activity and require activated DNA
for maximal activity. Both enzymes show a strict requirement for
a template and an initiating strand (Baril et al., 1971). These
enzymes are probably counterparts to the 6-8 S and 3.3 S DNA polymerases that have been described by Chang and Bollum (1971). However, the enzymatic properties of the DNA polymerases from rat
liver and hepatomas have not yet been as thoroughly characterized
as the DNA polymerases discussed by Dr. Bollum.

   We find that DNA polymerase I is the predominant activity in
normal rat liver and is probably the same enzyme that was described
earlier in rat liver by Howk and Wang (1969). DNA polymerase I is
present in the nuclei and with cytoplasmic ribosomes from normal
and regenerating rat liver and hepatomas (Baril et al., in press).
The activity of DNA polymerase I is not increased significantly in
the proliferating tissues. In contrast, the activity of DNA polymerase II is low in normal rat liver and is detectable only with
membrane fragments that are isolated from the post-microsomal
supernatant solution (Baril et al., 1971). In regenerating rat
liver DNA polymerase II is present in the nucleus, as well as the
post-microsomal membrane fragments. A comparison of the phosphocellulose elution profiles of DNA polymerase from the nucleus,
ribosomes and post-microsomal membrane (smooth) fragments from
24 hour regenerating rat liver is shown in Fig. 1. DNA polymerase
II is eluted from phosphocellulose by 0.20 M potassium phosphate at
pH 7.2 and DNA polymerase I is eluted by 0.50 M potassium phosphate. The nucleus of 24 hour regenerating rat liver contains DNA
polymerases I and II, whereas nuclei from normal and early regenerating rat liver, before 18 hours post-hepatectomy, contain only
DNA polymerase I. The ribosomes and the post-microsomal membrane
fragments from normal or regenerating rat liver have associated
only DNA polymerase I and DNA polymerase II respectively. The
activity of DNA polymerase II in the nucleus and with the post-
microsomal membrane fragments increases between 18 and 30 hours
post-hepatectomy and then declines (Baril et al., in press). In
contrast, the activity of DNA polymerase I in the nucleus and with
ribosomes does not change significantly during liver regeneration
(Baril et al., in press).

   A comparison of the phosphocellulose elution profiles of DNA
polymerase in nuclear extracts from normal rat liver and Morris
hepatomas of different growth rates is shown in Fig. 2. All of the
DNA polymerase in nuclear extracts of normal liver has the chromatographic properties of DNA polymerase I; it is eluted by 0.5 M
potassium phosphate. Most of the DNA polymerase in the nuclear
extract from Morris hepatomas 7800, 7794A and 7777 has the chromato-

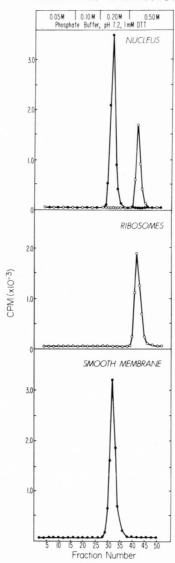

Fig. 1. Phosphocellulose chromatography of DNA polymerases from
nuclei, ribosomes and post-microsomal membranes of 24 hour regenerating rat liver. Composite elution profiles are from phosphocellulose
chromatography of DEAE-cellulose fractions that contain DNA polymerase activity from the respective sub-cellular fractions (Baril
et al., 1971). DEAE-cellulose fractions with the highest DNA polymerase activity were pooled and dialyzed overnight against 4 liters
of 0.05 M potassium phosphate, pH 6.8, 0.001 M dithiothreitol and
20% ethylene glycol. The dialyzed fractions were chromatographed
separately on 0.9 x 10 cm columns of phosphocellulose equilibrated
with the same buffer. Elution was stepwise with increasing

Legend to Fig. 1 (cont'd).

concentrations of potassium phosphate from 0.05 to 0.5 M.   The
collected fractions were dialyzed against TKMD and assayed for DNA
polymerase activity using "activated" DNA, as described previously
(Baril et al., 1971).   Open circles, DNA polymerase from DEAE-
cellulose column wash; closed circles, DNA polymerase from 0.25 M
KCl fraction from DEAE-cellulose chromatography.

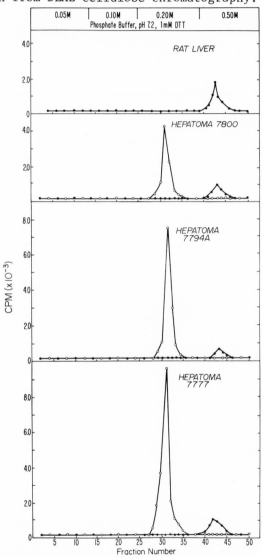

Fig. 2.   Phosphocellulose chromatography of rat liver and hepatoma
nuclear DNA polymerases.   The experimental procedures are described
in the legend to Fig. 1.   Closed circles, DNA polymerase from DEAE-
cellulose column wash; open circles, DNA polymerase from 0.25 M KCl
fraction from DEAE-cellulose chromatography.

graphic properties of DNA polymerase II; it is eluted by 0.20 M
potassium phosphate. A minor fraction of DNA polymerase, having the
chromatographic properties of DNA polymerase I, is also present in
the nuclear extracts from the hepatomas. DNA polymerase II is also
present with the post-microsomal membrane fragments from the hepatomas and its activity with this fraction is increased in hepatomas
relative to rat liver (Baril et al., in press). The enzymes from
regenerating rat liver and hepatomas have been characterized and
were shown to have the physical and enzymatic properties ascribed
to DNA polymerases I and II (Baril et al., 1971; Baril et al., in
press).

DNA polymerases I and II probably have different physiological
functions since only DNA polymerase II responds to stimuli for cell
proliferation. It is tempting to implicate DNA polymerase II in
DNA replication and DNA polymerase I in a constitutive process such
as DNA repair. However, the evidence for their physiological functions is necessarily indirect at this time.

PRESENCE OF MULTIPLE ENZYMES FOR DNA BIOSYNTHESIS WITH THE MEMBRANE FRAGMENTS

It was shown previously that all of the cytoplasmic DNA polymerase II is sedimentable and is localized with membrane fragments
that sediment with the post-microsomal pellet, P-4 (Baril et al.,
1971). The post-microsomal membrane fragments have the following
characteristics: a phospholipid to protein ratio of 0.15;
^{14}C-palmitate is incorporated in vivo; a low RNA to protein ratio of
0.06; no detectable DNA; no cytoplasmic membrane markers such as
monoamine oxidase, succinate dehydrogenase, 5'-nucleotidase, glucose-
6-phosphatase and cytochrome P_{450}. The origin of these membrane
fragments is not yet known.

The demonstration of DNA polymerase II with the post-microsomal
membrane fragments led us to question whether other enzymes for DNA
biosynthesis were also present with this sub-cellular fraction.
Table 1 shows the sub-cellular distributions of DNA polymerase,
thymidine kinase, thymidylate synthetase and ribonucleotide reductase
in Novikoff tumor. Most of the DNA polymerase activity is present
with the nuclear, microsomal and post-microsomal (P-4) fractions.
DNA polymerase II occurs only in the nucleus and the P-4 fraction
from proliferating tissues. The activity with the microsomes is
that of DNA polymerase I (see Fig. 1) (Baril et al., 1971).

Ribonucleotide reductase activity is found predominantly with
the P-4 fraction but a small amount of activity appears in the
soluble fraction (S-4). Thymidine kinase and thymidylate synthetase
activities are distributed between the P-4 and supernatant (S-4)

TABLE 1. Sub-cellular distribution of DNA polymerase
TK, TS and RR activities in Novikoff tumor

Fraction	DNA Polymerase Unit/mg	TK Unit/mg	RR Unit/mg	TS Unit/mg
Nuclei	2.6	<0.1	<0.2	<0.1
Mitochondria	0.1	<0.3	Not detectable	Not detectable
Microsomes	5.8	<0.3	Not detectable	Not detectable
Post-microsomal Pellet (P-4)	6.5	18.6	91.1	15.1
Soluble (S-4)	0.4	14.1	5.2	16.0

The procedures for sub-cellular fractionation and assay of the enzymes were as described previously (Baril et al., 1971; Elford et al., 1970). The fractions were resuspended in TKMD and dialyzed for 12 hours against 4 liters of this buffer. One unit of activity for the respective enzymes is as follows: DNA polymerase, 1 nanomole of (^{3}H)-dTMP incorporated per hour at 37°; TK, 1,000 CPM of (^{3}H)-dTMP formed in 18 minutes at 37°; RR, 1,000 CPM of (^{3}H)-dCDP formed in 40 minutes at 30°; TS, 1,000 CPM of (^{3}H)-dTMP formed in 20 minutes at 37°.

fractions. Similar sub-cellular distributions of these enzymes were observed with fetal and regenerating rat liver and Morris hepatomas (Baril, Baril and Elford, 1972).

The activity of each of the enzymes with P-4 shows a good correlation with the proliferative rate of the tissue. As shown in Table 2, only DNA polymerase II (DP_{II}) is detectable with the P-4 fraction isolated from normal liver and regenerating liver before 24 hours post-hepatectomy. At about 24 hours post-hepatectomy the activity of DP_{II} with this fraction begins to increase and thymidine kinase (TK) and ribonucleotide reductase (RR) activities become detectable. The activities of these enzymes continue to increase during liver regeneration. These enzymes are also present with the post-microsomal pellets from fetal liver, Novikoff tumor, Morris hepatomas and rat spleen.

The association of these enzymes with the post-microsomal pellet does not appear to be due to electrostatic absorption or aggregation at low salt concentrations. As shown in Table 3, most of the activity of each enzyme remains sedimentable even after treatment of the P-4 fraction with 0.5 M KCl. Some loss of thymidine kinase, thymidylate synthetase and ribonucleotide reductase activities occurs during this treatment. We attribute this loss to partial

TABLE 2. Comparison of DP II, TK and RR activities
in post-microsomal pellet of tissues

Tissue	DP II Unit/mg	TK Unit/mg	RR Unit/mg
Normal Rat Liver	1.02	Not detected	Not detected
Regenerating Liver			
5 Hrs. Post-Hepx	1.08	Not detected	Not detected
18 Hrs. Post-Hepx	1.53	<0.3	Not detected
24 Hrs. Post-Hepx	2.78	10.1	12.1
40 Hrs. Post-Hepx	3.44	16.4	31.9
18 Day Fetal Rat Liver	4.30	74.0	178.2
Novikoff Tumor	6.50	60.0	115.1

The Post-microsomal pellet (P-4) was obtained by centrifugation
of the post-microsomal supernatant from the respective tissues at
78,000 x $g_{(ave)}$ for 15-20 hours (Baril et al., 1971). The units of
enzyme activity are indicated in Table 1. TS was not assayed in
these particular experiments but has been assayed and increases
similarly to TK and RR.

TABLE 3. DP II, TK, RR and TS activities after washing Novikoff P-4

Fractions and washing conditions	DP II Total units	TK Total units	RR Total units	TS Total units
Novikoff P-4	24.7	326.4	5,140.0	269.3
Control-TKMD wash of Novikoff P-4				
Pellet	12.3	161.9	2,570.0	130.5
Supernatant	ND	ND	ND	1.8
0.5 M KCl, TKMD wash of Novikoff P-4				
Pellet	11.5	120.6	1,790.0	82.0
Supernatant	ND	ND	6.3	3.4

The post-microsomal pellet (P-4) from Novikoff tumor was resuspended in TKMD buffer and divided into 2 aliquots. KCl was added
to one aliquot to 0.5 M and both aliquots were centrifuged at
105,000 x $g_{(ave)}$ for 20 hours. The supernatants were decanted, the
pellets resuspended in TKMD and all fractions were dialyzed overnight against TKMD. Units of activity as indicated in Table 1.
ND - not detectable.

inactivation of the enzymes by the high salt concentration since the activities are not recovered in the supernatant.

It was shown previously that the post-microsomal pellet (P-4) can be further resolved by centrifugation in a discontinuous gradient of 0.8 - 2.0 M sucrose for 20-24 hours (Baril et al., 1971; Baril et al., 1970). The DP_{II} of the post-microsomal pellet bands in the 1.3 M sucrose layer of the gradient. Most of the TS, TK and RR activities with the post-microsomal pellet from proliferating tissues, such as Novikoff tumor, also band in the 1.3 M sucrose layer of the gradient (Table 4). Electron microscopy and chemical analysis of material in the 1.3 M sucrose layer indicated that it is composed of membrane fragments (Baril et al., 1970).

ISOLATION OF THE MULTI-ENZYME COMPLEX

The enzymes (DP_{II}, TK, TS and RR) can be isolated as an apparent aggregate by combined treatment of the post-microsomal membrane fragments with 2.5% n-butanol and 5% Triton X-100. These concentrations of butanol and Triton effectively solubilized the membrane, as determined with ^{14}C-palmitate-labeled membrane fragments. As shown in Table 5, after centrifugation of butanol-Triton treated membrane fragments from Novikoff tumor at 105,000 x g(ave) for 20 hours, DP_{II}, TK, RR and TS were all sedimented. Virtually all of the DP_{II} and

TABLE 4. Sub-fractionation of Novikoff P-4
on a discontinuous sucrose gradient

Fraction	DP II Total units	TK Total units	RR Total units	TS Total units
Soluble (0 - 1.0 M sucrose)	1.4	22	169	250
1.3 M Sucrose	579.0	1797	11,308	1797
2.0 M Sucrose	39.6	100	163	93
Pellet	0.8	7	13	8

The post-microsomal pellet (P-4) from Novikoff tumor was resuspended in 1.3 M sucrose-TKMD. This was underlayered with one-half volume of 2.0 M sucrose-TKMD and overlayered with 1.0 M sucrose-TKMD. Centrifugation was at 25,000 RPM (SW 25.1) for 20 hours. Fractions were carefully removed with a bent-tip pipet. The pellet was resuspended in TKMD buffer. Units of activity are as indicated in Table 1.

TABLE 5.   Distribution of DP II, TK, RR and TS after treatment
of Novikoff P-4 with 2.5% butanol-5% Triton

Fraction	DP II Total units	TK Total units	RR Total units	TS Total units
Novikoff post-microsomal pellet (P-4)	19.7	326.4	1713.3	16.3
P-4 treated with 2.5% butanol-5% Triton and centrifuged at 105,000 x g(ave) for 20 hours				
Pellet	18.3	288.8	914.4	15.4
Supernatant	ND	8.3	ND	1.1

The procedure for treatment of Novikoff P-4 with butanol-Triton
was as described in Methods.   Units of enzyme activity are as indicated in Table 1.

TS activities were recovered in the pellet but some loss of TK and
RR activities occurred.   The latter enzymes may be partially inactivated by the butanol-Triton treatment, since the balance of these
enzyme activities is not recovered in the supernatant.

The electron microscopic observations presented in Fig. 3
indicate that Novikoff P-4 samples contain several classes of particles (viz., small and large circles, and rectangles (Fig. 3B,
arrows)) associated with larger amorphous membrane fragments
(30-100 nm in size) (Fig. 3A, arrow).   These particles measure be-
between 8.5 and 12.0 nm in diameter and are separated from the
membrane fragments after butanol-Triton treatment (Fig. 3B).   Both
the particles and the multi-enzyme activities are present at this
and later stages of purification.   This co-purification of activities
may indicate that separate particles corresponding to separate
enzymes avidly bind to each other or to the same membrane fragments
and once released have similar sedimentation properties.   Further
characterization of the purified enzymes is needed to decide if this
interpretation is correct.

When the particulate material present in the 1.3 M sucrose layer
of the discontinuous gradient is subjected to isopycnic centrifugation in CsCl, DNA polymerase II and thymidine kinase activities
band with the predominant fraction at a density of 1.23 g/cc.
Essentially all of the activities of these enzymes are recovered
after CsCl gradient centrifugation, but ribonucleotide reductase

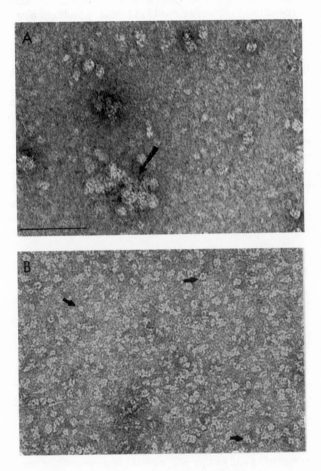

Fig. 3.   Electron microscopy of P-4 from Novikoff tumor.   A) Before
and B) after 2.5% butanol - 5% Triton treatment.   Total protein
concentration of the specimens was 100 μg/ml.   Staining was with
2% UA.   Magnification was 215,000x.   Bar on scale was 0.1 μ.

activity is partially lost during this procedure. After treatment of the membrane fragments with butanol and Triton, DNA polymerase II and thymidine kinase activities band at a density of 1.35 g/cc in CsCl. The activities of these enzymes also coincide during gel-filtration of the butanol-Triton treated fraction on Biogel A-1.5M. This evidence suggests that the enzymes do, in fact, represent a multi-enzyme aggregate. However, additional studies are required to substantiate this.

INCORPORATION OF THYMIDINE INTO DNA BY THE COMPLEX

Using either the membrane fragments or the 8-12 nm particles derived from them, it has been possible to demonstrate sequential reactions that produce (^{3}H)-dTTP from (^{3}H)-thymidine and subsequently incorporate the (^{3}H)-dTTP into DNA (Table 6). The first step of the procedure was to incubate the membrane fragments, or the 8-12 nm particles, for 60 minutes under the conditions used to assay thymidine kinase (Sneider, Potter and Morris, 1969). The production of dTMP, dTDP and dTTP was confirmed by chromatography of aliquots on Whatman #1 paper (Weissman, Smellie and Paul, 1960). The second step of the procedure involved addition of the four

TABLE 6. Incorporation of ^{3}H-Thymidine into DNA

Condition	Acid-insoluble counts/min.
Complete system: steps #1 and #2	5,245
Omit step #1, add [^{3}H]-dT during step #2	90
Omit DNA	75
Omit cold dNTP during step #2	864
*Omit P-4, replace by S-4	0

Procedure

*P-4, post-microsomal pellet; S-4, resulting supernatant (soluble material) after centrifugation for P-4.

Step #1 - Post-microsomal pellet (P-4) or supernatant (S-4) incubated with ^{3}H-thymidine for 60 minutes for thymidine kinase assay (Elford et al., 1970).

Step #2 - Following incubation for Step #1, cold dATP, dGTP, dCTP and dTTP (50 mμmoles) and 60 μg of activated calf thymus DNA were added. Incubation for DNA polymerase assay was then continued for 60 minutes (Baril et al., 1971).

unlabeled deoxyribonucleoside triphosphates and activated DNA to the incubation mixture. Incubation was then continued for 60 minutes for the assay of DNA polymerase.

As shown in Table 6, (^{3}H)-thymidine was incorporated into acid-insoluble material when the post-microsomal pellet (P-4) from Novikoff tumor was carried through steps 1 and 2 of the procedure. The incorporation is dependent on: the preincubation (Step 1); the addition of DNA; the presence of all four deoxyribonucleoside tri-phosphates during Step 2; the post-microsomal pellet (P-4). The post-microsomal pellet (P-4) cannot be replaced by the supernatant (S-4), although this fraction contains thymidine phosphorylating activity (see Table 1).

The acid-insoluble radioactivity from Step 2 is sensitive to DNase but is insensitive to RNase, pronase and exposure to 1 N NaOH at 80°. It, therefore, seems likely that the (^{3}H)-thymidine is incorporated into DNA. These data also suggest that the particles contain all of the components necessary for the sequence of reactions that convert thymidine into dTTP and incorporate the dTTP into DNA.

DISCUSSION AND SUMMARY

Two forms of DNA polymerase, in addition to the mitochondrial enzyme, have been demonstrated in rat liver and hepatomas. Each enzyme has distinct physical and enzymatic properties and we desig-nate the enzymes as DNA polymerases I and II. They are probably similar to the 6-8 S and 3.3 S DNA polymerases that have been observed in other mammalian cells by Chang and Bollum (1971) and others (Weissbach et al., 1971; Sedwick et al., 1972; Smith and Gallo, 1972). The activity of only one form, DNA polymerase II,

TABLE 7. Check that incorporation at Step #2 is into DNA

Condition	Acid-insoluble counts/min.
Product from Step #2 (Table 7)	5,245
Product incubated with:	
DNase I (10 µg) at 37° for 30 min.	728
Pancreatic RNase (20 µg) and T$_1$ RNase	
(7 units) at 37° for 30 min.	4,980
1 N NaOH at 80° for 10 min.	4,945
Pronase (20 µg) at 37° for 30 min.	5,030

changes with proliferation. In normal rat liver we can detect DNA polymerase II only with membrane fragments that are isolated from the post-microsomal supernatant solution (Baril et al., 1971). In regenerating rat liver and hepatomas, however, we find that DNA polymerase II is present in the nucleus, as well as with membrane fragments isolated from the post-microsomal supernatant solution. The activities of DNA polymerase II in the nuclear and post-microsomal membrane fractions change in parallel with the reported rate of DNA synthesis (Bucher and Malt, 1971) throughout liver regeneration (Baril et al., in press).

The nature and origin of the post-microsomal membrane fragments are, at present, unknown. The membrane has no detectable DNA or cytoplasmic membrane markers and has a low RNA to protein ratio of 0.06 (Baril et al., 1970). Identification of the origin of these membrane fragments may provide clues to the physiological localization and function of DNA polymerase II.

We have observed an association of at least four enzymes for DNA synthesis with the post-microsomal membrane fragments from proliferating tissues. These enzymes participate at different levels of DNA biosynthesis in rat liver and tumors. Three of the enzymes, ribonucleotide reductase, thymidylate synthetase and thymidine kinase, function in the production of deoxyribonucleotide precursors. The fourth enzyme is DNA polymerase II whose activity, like those of the other three enzymes, correlates positively with the rate of cell proliferation. The membrane fragments contain essentially all of the ribonucleotide reductase of the cell and all of the DNA polymerase II of the cytoplasm. Thymidine kinase and thymidylate synthetase activities are present with the membrane fragments, as well as the supernatant, S-4 (Baril et al., submitted). However, it is not known whether the same molecular forms of these enzymes are present with both the membrane fragments and the soluble fraction. Multiple forms of thymidine kinase have been reported in Yoshida sarcoma (Hashimoto et al., 1972).

The enzymes remain associated after treatment of the membrane fragments with concentrated salt solution or combined butanol and Triton (Table 3 and 5). Therefore, it seems doubtful that the co-fractionation of these enzymes is fortuitous and the result of electrostatic absorption or aggregation. Electron microscopic observations (Fig. 3) suggest that after the butanol-Triton treatment the enzyme activities are present with 8.5-12 nm particles. The particles, or the membrane fragments from which they are derived, do promote the synthesis of dTMP, dTDP and dTTP from thymidine and the subsequent incorporation of the dTTP into DNA in vitro. It is, therefore, very probable that deoxyribonucleotide kinases are also present with the membrane fragments and the particles derived by butanol-Triton treatment.

It is tempting to speculate from these data that a multi-enzyme complex for DNA biosynthesis exists in eukaryotes. The complex would conceivably contain key enzymes for the production of deoxyribonucleoside triphosphates, as well as the enzymes and factors participating directly in DNA replication. There are precedents for the existence of functional multi-enzyme aggregates that catalyze consecutive steps of a metabolic sequence (Reed and Cox, 1970; Plate et al., 1968; Lue and Kaplan, 1970). The existence of a multi-enzyme system for DNA replication would be consistent with the replicon model for DNA replication and the evidence that DNA replication in prokaryotes is regulated by several genetic loci (Jacob, Brenner and Cuzin, 1963; Karamata and Gross, 1970).

We are, however, aware of alternative explanations for the co-fractionation of the enzymes observed here. The recovery of the membrane fragments from a cytoplasmic fraction (the post-microsomal supernatant solution) may be an artifact of the fractionation procedure. On the other hand, the enzyme aggregate and membrane fragments may represent part of a larger complex that is disrupted during the fractionation. We are presently exploring the latter possibility.

Determination of the physiological functions of the two DNA polymerases and examination of the possibility that DNA polymerase II is part of a multi-enzyme complex for DNA synthesis are important aspects for future study. The supramolecular organization of the DNA synthesizing machinery, in eukaryotic cells especially, is probably of paramount importance with respect to both the mechanism and control of DNA replication.

ACKNOWLEDGEMENT

This research was supported by USPHS Research Grants CA-11265, CA-10441, CA-08800-04 and CA-11976. E. B. and B. B. extend their appreciation to Dr. J. Laszlo for his assistance.

REFERENCES

Baril, E. F., M. D. Jenkins, O. E. Brown and J. Laszlo. 1970. Science. 169:87.

Baril, E. F., O. E. Brown, M. D. Jenkins and J. Laszlo. 1971. Biochemistry. 10:1981.

Baril, E. F. and J. Laszlo. 1971. Advan. Enzyme Regulation. 9:183.

Baril, E., B. Baril and H. Elford. 1972. Proc. Amer. Assoc. Cancer Res. 13:84.

Baril, E. F., M. D. Jenkins, O. E. Brown, J. Laszlo and H. P. Morris. In press. Cancer Res.

Baril, E., B. Baril, R. B. Luftig and H. Elford. Submitted. Proc. Nat. Acad. Sci., U.S.A.

Bucher, N. L. R. and R. A. Malt. 1971. In Regeneration of Liver and Kidney. Little, Brown and Co., Boston. Pp. 55-77.

Chang, L. M. S. and F. J. Bollum. 1971. J. Biol. Chem. 246:5835.

Chang, L. M. S. and F. J. Bollum. This symposium.

DeLucia, R. and J. Cairns. 1969. Nature. 224:1164.

Elford, H. L., M. Freese, E. Passamani and H. P. Morris. 1970. J. Biol. Chem. 245:5228.

Hashimoto, T., T. Arima, H. Okuda and S. Fujii. 1972. Cancer Res. 32:67.

Howk, R. and T. Y. Wang. 1969. Arch. Biochem. Biophys. 133:238.

Jacob, F., S. Brenner and F. Cuzin. 1963. Cold Spr. Harb. Symp. Quant. Biol. 28:329.

Karamata, D. and J. D. Gross. 1970. Mol. Gen. Genetics. 108:277.

Knippers, R. 1970. Nature. 228:1050.

Kornberg, A. 1969. Science. 163:1410.

Kornberg, T. and M. L. Gefter. 1971. Proc. Nat. Acad. Sci., U.S.A. 68:761.

Lomax, M. I. S. and G. R. Greenberg. 1967. J. Biol. Chem. 242:109.

Lue, P. F. and J. G. Kaplan. 1970. Biochim. Biophys. Acta. 220:365.

Moses, R. E. and C. C. Richardson. 1970. Biochem. Biophys. Res. Commun. 41:1557 and 1565.

Plate, C. A., V. C. Joshi, B. Sedgwick and S. J. Wakil. 1968. J. Biol. Chem. 243:5349.

Reed, L. and D. J. Cox. 1970. Biochim. Biophys. Acta. 220:365.

Sedwick, W. D., T. S. Wang and D. Korn. 1972. J. Biol. Chem. 247:5026.

Smith, R. G. and R. C. Gallo. 1972. Proc. Nat. Acad. Sci., U.S.A. 69:2879.

Sneider, T., V. R. Potter and H. P. Morris. 1969. Cancer Res. 29:40.

Weissbach, A., A. Schlabach, B. Fridlender and A. Bolden. 1971. Nature New Biol. 231:167.

Weissman, S. M., R. M. S. Smellie and J. Paul. 1960. Biochim. Biophys. Acta. 45:101.

A EUKARYOTIC DNA-BINDING PROTEIN IN MEIOTIC CELLS

J. Mather and Y. Hotta

Department of Biology, University of California

La Jolla, California

In recent years a number of DNA-binding proteins have been
reported to occur in organisms ranging from viruses (3,9) and
bacteria (1), to flowering plants and mammals (4,6,7,10). In the
case of the gene-32 protein whose function has been closely studied
the importance of its biological role has been clearly established.
Thus, in T_4 infected bacterial cells, $32-T_4$ mutants are deficient
in both replication and recombination (2).

Hotta and Stern have recently provided less direct evidence
for an important biological function of a nuclear DNA-binding protein which occurs in the meiotic cells (microsporocytes) of Lilium.
A similar protein has also been isolated from human, bovine, and
rat spermatocytes. The properties of this partially-purified protein from lilies and rats have been broadly studied. More intensive
studies are being pursued with the aim of clarifying the function
of the protein in the meiotic process.

Generally, the protein is obtained by first isolating nuclei
from meiotic cells. These can be collected at different stages of
meiosis in Lilium by extruding them from anthers present in developing flower buds of predetermined lengths (6). In mammalian systems,
isolated nuclei (11) or whole cells (5) are separated into different
meiotic stages by sedimentation in discontinuous or continuous density
gradients. Nuclei obtained by these procedures are broken by sonication and washed with concentrated salt solution to remove chromatin
and soluble proteins. The residue consists largely of nuclear membranes. The preparation may be further fractionated on a discontinuous sucrose gradient (8) or made 0.1% in DOC to solubilize the
membrane proteins which are then loaded onto a SS-DNA-cellulose
column. The column is eluted stepwise with 0.15-2.0 M NaCl gradient

in the presence of 0.02 M Tris HCl pH 8.1 buffer and 50 mM MgCl$_2$.
The protein elutes at 2.0 M NaCl, 50 mM EDTA. This eluate is concentrated by dialysis and designated as the partially-purified
preparation.

The binding activity of the protein is measured by reacting
it with denatured ^{3}H-thymidine-labeled T7-DNA. The DNA-protein
complex can be separated from unreacted DNA and protein on a
sucrose gradient or retained by the filter. The DNA-protein complex passes the filter, is collected, precipitated with trichloroacetic acid, and collected on a glass-fiber filter.

The protein was found to be localized in the nucleus and to
occur at appreciable levels only in meiotic cells. Moreover, its
occurrence is transient even within meiotic cells. Nuclear activity arises from a barely detectable level to a maximum during the
zygotene-pachytene stages of meiosis and then rapidly disappears.

TABLE 1

Incubation time	µgm DNA through filter/µgm protein
5 sec	.055
15 sec	.055
30 sec	.055
1 min	.065
2 min	.050
10 min	.058
20 min	.058
1 hr	.062
48 hr	.060

T7-DNA binding activity of the partially-purified protein at
various incubation times. The nitrocellulose filter assay was
used. The figures given are averages of duplicates obtained
in two separate experiments 8 µg (5 sec - 2 min) or 18 µg (10
min - 48 hr) protein were mixed with 0.6 µg DNA and the mixture
filtered after the times indicated. The reaction was stopped
by diluting the reaction mix 10-fold with 5 x SSC. No DNA-
protein association occurs at this salt concentration, although
the preformed complex is stable.

The molecular weight, as estimated from sedimentation behavior
and from sephadex chromatography, is approximately 35,000 daltons.
Binding activity is virtually absent in preparations treated with
pronase or chymotrypsin but is unaffected by treatment with RNase
I, DNase I, α-amylase, or phospholipase C. The incorporation of

tryptophan by the protein and its absorption on DEAE at pH 8.0,
indicate that it is not basic, and therefore not a histone (6).
The binding of the protein to DNA shows an absolute requirement
for Mg++, is specific for single-stranded but unreactive toward
native-DNA, and non-specific with respect to the origin of the
DNA. It does not bind to RNA.

The reaction is not enzyme catalyzed and is completed in less
than 5 seconds at 4° C. As shown in Table 1, the amount of complex formed after 48 hours was not appreciably different from that
formed after 5 seconds. It has not proved feasible to measure
shorter reaction times with the present method of assay. The
kinetics of the reaction are obviously complex (Fig. 1 and 2).

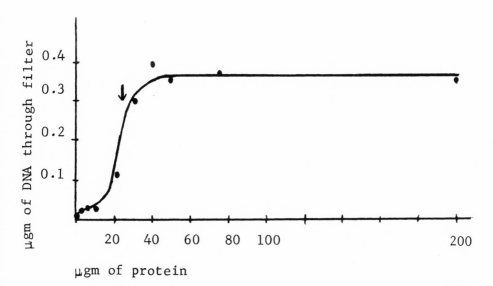

μgm of protein

Fig. 1: Binding activity of the partially purified protein where
         denatured T7-DNA concentration is constant (0.63 μgm DNA/
         reaction) and protein concentration is varied. The re-
         action mix also contains 0.005 M β-mercaptoethanol, 0.2 M
         MgCl$_2$, and distilled water to a constant volume of 0.2 ml.
         Arrows in this and in Fig. 2 point to corresponding DNA/
         protein ratios in the reaction mixture.

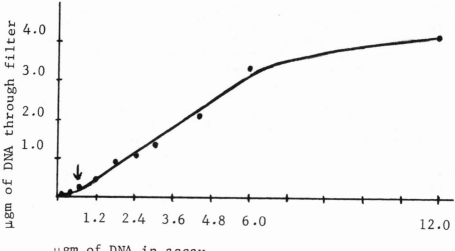

μgm of DNA in assay

Fig. 2: Binding activity of partially purified protein as in
         Fig. 1 except protein concentration is constant and the
         denatured [^{3}H]-T7-DNA concentration is varied. Arrows
         in this and in Fig. 1 point to corresponding DNA/protein
         ratios in the reaction mixture.

The reaction, like that of the gene 32 protein, is not stoichiometric. When the concentration of the ^{3}H-T7-DNA in the reaction
mixture is held constant and the concentration of the protein is
increased, we see a signoidal curve (Fig. 1). There seems to be
a cooperative effect in the binding of the protein to DNA. However, when the protein concentration is held constant and the DNA
concentration is increased (Fig. 2), we see an apparently progressive increase in the amount of DNA bound/μgm of protein corresponding to the increase of DNA in the reaction mixture. That the
observed increase is not due to a passage of free DNA through the
filter may be inferred from the absence of a significant amount
of DNA in the filtrate if protein is excluded from the reaction
mixtures and from the presence of a plateau in the amount of
filterably DNA plotted against the amount of DNA added to the
original reaction mixture (Fig. 2).

        Since all the assays have been made in terms of DNA radio-
activity, the data leave open the question of the minimal amount
of bound protein required for passage of DNA through the filter
or of the range of protein/DNA ratios present among the molecular
complexes in the filtrate. Isopycnic centrifugation of the filtrate reveals a fairly broad range of values (6). Nevertheless,
if the curve in Fig. 1 and 2 are compared closely, it becomes
apparent that conditions of assay may be defined in which the

amount of DNA radioactivity in the filtrate is proportional to the amount of protein present in the reaction mixture. Within these limits the filter technique provides a convenient and relatively simple method for assaying the protein. The complexity of the kinetics almost certainly reflects the cooperative characteristics of the reaction and elucidation of these characteristics is beyond the reach of the filter technique used.

In summary, the DNA-binding protein identified in meiotic cells appears to be a single nuclear membrane-bound species whose activity is restricted to a particular stage of meiosis during which pairing and crossing over are presumed to occur. It is found in a broad range of higher organisms and shows many of the same properties associated with the DNA-binding proteins found in the prokaryotes and known to be associated with replication and recombination. This leads us to believe that the DNA-binding protein discussed above may be involved in the more elaborate process of pairing and crossing over in higher organisms. Although the evidence favoring this conclusion is still circumstantial, the unique properties and intracellular location of the protein as well as its wide phylogenetic distribution would argue for the wisdom of further study.

REFERENCES

1. Alberts, B. M., Amodio, F. S., Jenkins, M., Gutmann, E. B. and Ferris, F. L. CSH Symp. Quant. Biol. 33: 289, 1968.

2. Alberts, B. M. and Frey, L. Nature 227: 1313, 1970.

3. Alberts, B. M., Frey, L. and Delius, H. J. Mol. Biol. 68: 139, 1972.

4. Fox, T. and Pardee, A. B. J. Biol. Chem. 246: 6159, 1971.

5. Go, V. L. W., Veron, R. G. and Fritz, I. B. Can. J. Biochem. 49: 753, 1971.

6. Hotta, Y. and Stern, H. Dev. Biol. 26: 87, 1971.

7. Hotta, Y. and Stern, H. Nature (New Biol.) 234: 83, 1971.

8. Kashnig, D. M. and Kasper, C. B. J. Biol. Chem. 244: 3786, 1969.

9. Oey, J. L. and Knippers, R. J. Mol. Biol. 68: 125, 1972.

10. Tsai, R. L. and Green, H. (submitted for publication).

11. Utakoji, T. In Methods in Cell Physiology, D.M. Prescott (ed) Academic Press, New York, 4:1 (1969).

DNA REPLICATION, THE NUCLEAR MEMBRANE,

AND OKAZAKI FRAGMENTS IN EUKARYOTIC ORGANISMS

Joel A. Huberman

Department of Biology, Massachusetts Institute of

Technology, Cambridge, Massachusetts 02139

THE NUCLEAR MEMBRANE AND DNA REPLICATION

There are two topics which I would like to cover.

The first, dealt with by David Comings (this meeting) previously, is the involvement or lack of involvement of the nuclear membrane in DNA replication. The second topic is the presence and role of Okazaki fragments during DNA replication in eukaryotic cells.

Concerning DNA replication and the cell membrane in eukaryotes, Comings (this meeting) has already talked about evidence (essentially the same as our evidence) that the nuclear membrane is not involved in DNA replication. Such evidence has also been obtained recently by Fakan, Turner, Pagano, and Hancock (1972). On the other hand, others at this meeting have discussed evidence supporting a role for the membrane in DNA replication, and the literature is full of reports supporting such a role. I will not go over the evidence here for association between DNA replication and the membrane in prokaryotes. The evidence is of essentially the same nature as the evidence obtained with eukaryotic organisms. In eukaryotic organisms, there have been a number of experiments which suggest this association. One type of experiment is exemplified by the work of Friedman and Mueller (1969). These investigators pulse-labeled HeLa cells with ^{3}H-thymidine for a short period (several minutes), lysed the cells and then fractioned the DNA in various ways and compared what happened to the pulse labeled DNA with what happened to bulk DNA which was not replicating. They found that the ^{3}H-labeled DNA behaved as if it were attached to some cell component with properties of large size and light density, possibly also having hydrophobic qualities. Friedman and Mueller found that this component could be removed from the pulse-labeled DNA by treatment with heat, alkali, and sonication, but not by protease treatment. They did not try ribonuclease. Friedman and Mueller were rather cautious in their interpretation of these results. People who have examined their data subsequently have put a stronger interpretation on their results, and have assumed that the large component attached to the pulse-labeled DNA is actually nuclear membrane. There is, however, no real evidence for this conclusion.

Another type of experiment has been done by two groups of investigators (Mizuno, Stoops and Sinha - 1971; and O'Brien, Sanyal and Stanton - 1972). In these experiments, cells were pulse-labeled with ^{3}H-thymidine, then nuclei were isolated. The isolated nuclei were separated from each other either on the basis of density in an isopynic sucrose gradient, or on the basis of sedimentation velocity; and in these experiments, too, it was found that the pulse labeled DNA preferentially associated with the fractions containing the nuclear membrane, while the bulk DNA did not. A third kind of experiment that has demonstrated association between replicating DNA and the nuclear membrane used the M band technique as earlier explained by Schachter (this meeting). This has been done in eukaryotic cells by Hanaoka and Yamada (1971) and by Pearson and Hanawalt (1972). Both groups found preferential association of replicating DNA with the M band.

There are also some electron microscope autoradiographic experiments which have suggested association between the nuclear membrane and replicating DNA. David Comings (this meeting) has described his experiments and why he now thinks the earlier experiments were probably wrong. There is a second group; O'Brien, Sanyal and Stanton (1972) which has also done such experiments. The results of their experiments are a bit more difficult to explain. They pulse labeled HeLa cells for one minute with ^{3}H-thymidine, fixed the cells, sectioned them, did autoradiography on the sections, and found after looking in the electron microscope that after this one minute pulse, in all of the cells which they examined, all the grains were located around the nuclear membrane; a very striking configuration of grains. These results were very similar to the pictures Comings (this meeting) showed you of peripheral grain distributions. If this short pulse were followed by one hour chase, then O'Brien et al, (1972) found that the grains were distributed throughout the nucleus. That experiment seems at first rather hard to argue with. I will explain later in this talk why we feel these conclusions are incorrect.

At the time we began our experiments, I was convinced by the previous biochemical experiments I have just described that replication must be taking place on the nuclear membrane. However, I was aware of other experiments on eukaryotic cells which suggest that the nuclear membrane is not associated with the replication point. These are all electron microscopic autoradiographic experiments; and they have been done by the following groups: Blondel

(1968) using two minute pulses and KB (human) cells, Williams and
Ockey (1970), using ten minute pulses in Chinese hamster cells,
and Erlandson and deHarven (1971), using 15 minute pulses in HeLa
cells. All of these groups obtained essentially the same results,
namely that if the cells were pulse labeled in early S phase and
then autoradiographed, the label was found preferentially over the
periphery of the nucleus, near the nuclear membrane. Williams and
Ockey (1970) suggested that this is a reflection of later replication of heterochromatin. Dense chromatin is indeed found preferentially around the nuclear membrane in eukaryotic cells. So
these experiments suggested that the replication point did not
occur on the nuclear membrane while the biochemical experiments I
have described suggested the opposite. It occurred to me that a
possible reconciliation between these conflicting experiments was
as follows. The shortest pulse time used in the autoradiographic
experiments is two minutes. Now in two minutes a considerable
amount of DNA can be synthesized. Using the Cairns technique, of
DNA fiber autoradiography, Huberman and Riggs (1968) measured the
rate of replication per growing point in Chinese hamster cells and
found that the rate of replication can vary between 0.3 and 2.5
microns per minute per growing point. Thus, in two minutes as much
as 5 microns of DNA can be synthesized. But the size of the nucleus
in mammalian cells is only 5 to 10 microns in diameter, so you can
imagine the possibility that DNA might be synthesized at the nuclear
membrane, then stretched out into the interior of the nucleus to
give the impression, after a two minute pulse, that the grains were
spread uniformly throughout the nucleus. In order to rule out that
possibility we simply chose a shorter pulse time. The pulse time
in our experiments was 30 seconds. In 30 seconds, at the most only
1.25 microns of DNA could be synthesized; and that would give us
sufficient resolution to see whether or not the replication was
actually taking place on the nuclear membrane. In our experiment
we used Chinese hamster cells because they are very easy to syn-
chronize. The Chinese hamster cells were pulse labeled with tritiated thymidine for 30 seconds, then trypsinized, fixed, sectioned
and autoradiographed. All of the pictures I will show you are with
unsynchronized cultures of Chinese hamster cells. In the cells of
Figure 1a and 1b, you can see that the grains are distributed
throughout the nucleus. The length of the bar is one micron, and
as you recall we calculated that grains greater than 1.25 microns
away from the nuclear membrane cannot possibly by synthesized on
the membrane. In Figure 1 we can see one disadvantage of using
Chinese hamster cells; the nucleus is unusually irregular in shape.
There are many nuclear invaginations, which causes the major problem that some of the central grains could be due to nuclear membrane
invaginations coming down from above or up from below. This is a
possibility which I think can be ruled out as I will explain later.
As it turns out, HeLa cells have much smoother nuclei and in future
experiments I would be tempted to use them instead.

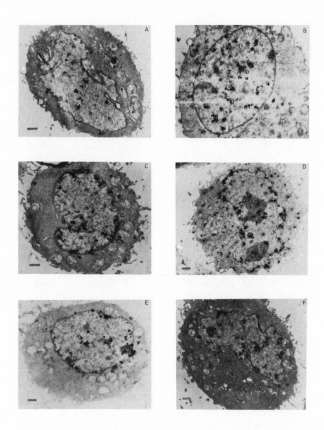

FIGURE 1. Autoradiography of unsynchronized CHO cells exposed to
³H-TdR for 0.5 minutes. a and b, grains distributed over entire nucleus. c and d, grains distributed around nuclear membrane. e, clustered grain distribution. f, mixed grain distribution. CHO cells
were grown on Petri plates in Joklik-modified MEM (Grand Island Biological Co.) supplemented with 7% foetal calf serum and non-essential
amino-acids. Pulse-labeling was performed by first adding 5-fluorour-
idine deoxyriboside (FUDR; Hoffmann-LaRoche) to a final concentration
of 9.1 g ml⁻l to inhibit further biosynthesis of dTTP. After 1 min.,
³H-TdR (51 Ci mmole⁻¹; New England Nuclear) was added to 17µCi ml⁻¹.
After 30 seconds the plates were removed from 37°C, and the medium
rapidly sucked off. The plates were then washed with ice cold iso-
tonic saline containing FUDR at 0.1µg ml⁻¹ (two changes). Cells were
removed from the plates by trypsinization at room temperature in iso-
tonic saline still containing FUDR. The cells were pelleted, then
fixed with glutaraldehyde and OsO₄, dehydrated, embedded in epoxy resin
and cut into gold-purple sections. The sections were mounted on grids
and autoradiographed by the technique of Caro and Van Tubergen (1962)
except that the acetic acid stop bath was replaced with distilled water.
Exposure time was 3.5 months. The bar in each figure represents 1µm.
(From Huberman, Tsai and Deich (1973)

Figure 1c and d show cells with grains distributed preferentially around the nuclear membrane. In Figure 1e a few grains are
located in the middle of the nucleus as well but they happen to be
located over the dense chromatin in the middle of the nucleus and
according to the interpretation of Williams and Ockey (1970) and of
Comings, this is consistent with the late replication of heterochromatin. Figure 1f shows a cell which may be described as a combination of the previous examples. There are grains around the nuclear membrane and quite a few grains in the interior of the nucleus.

The immediate conclusion of these experiments is that the nuclear
membrane is not required for DNA replication. Replication can take
place in the absence of the nuclear membrane. But before we make
that conclusion firm, we have to rule out a few possible artifacts.
One possible artifact is that the grains that we have been seeing
might be background grains. This is ruled out by the strikingly
high ratio of grains over the nucleus to grains over other areas.
In addition, we must prove that the grains are due to real DNA
replication. This we have done in two ways. First, we have looked
at cells from synchronized cultures in G1 or G2 phase. We do not
see grains over their nuclei. Second, the grains can be removed by
DNAse treatments, so the ^{3}H-thymidine giving rise to the grains has
really been incorporated into DNA. A more difficult possibility to
rule out is the possibility that the grains that we see in the center
of the nucleus might be due to invaginations of the nuclear membrane
which bring some of the nuclear membrane within 1.25μ above or below
the plane of the section. A weak argument against that possibility
is that the frequency of invaginations judged from the contour of
the nucleus in these thin sections cannot be great enough to account
for the number of grains seen in the interior. A stronger argument
comes from our studies with synchronized cells. Our data are similar to the data of Comings (this meeting) in that they show that in
early S phase all of the nuclei are generally labeled while in late
S phase nearly all of the nuclei are labeled perpherally. Yet there
is no change in the frequency of nuclear invaginations going from
early S to late S. Therefore, the central grains that we see cannot
be due to invaginations. A further argument against invaginations
is that a few experiments have been done with HeLa cells which have
smoother nuclei (Huberman, Tsai and Deich (1973). These experiments
also indicate that replication can occur in the interior of the nucleus.

We also did experiments with synchronized cells in order to see
during which period of S this peripheral pattern of labeling was
occurring and we used the Colcemidreversal method for synchronization of cells (Stubblefield, 1968). It is an excellent method with
Chinese hamster cells. One obtains a population which is more than
99% pure mitotic cells. This population goes through division in
the first hour after release from mitosis and in our hands more than
50% of the cells divide again within 15 hours. The S phase lasts
from about 3 hours to 14 hours after release from mitosis. We did
30 second pulses at various times after release from mitosis, and
Figures 2 and 3 presents our results in histogram form. We divided

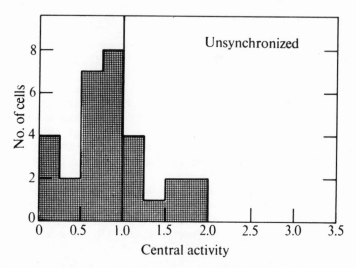

FIGURE 2. Distribution of grains over unsynchronized CHO cells exposed to ^{3}H-TdR for 0.5 minutes. Prints at about x15,000 magnification were made of autoradiograms of individual randomly chosen
cells similar to and including those in Figure 1. A line was drawn
within each nucleus which was at all points 1.25μ away from the nearest nuclear membrane. This line divided the nucleus into a central
area and a peripheral area. Grains were counted over each area,
and the size of each area was measured. In these determinations,
grains lying over the nucleolus and nucleolar areas, were ignored
because the nucleolus has so much less DNA than the rest of the
nucleus. Grains lying outside the nucleus were also ignored.
"Central activity" was calculated as the ratio of the fraction of
grains which were central to the fraction of area which was central.
Thirty nuclei were measured for the histogram. (From Huberman,
Tsai and Deich, 1973)

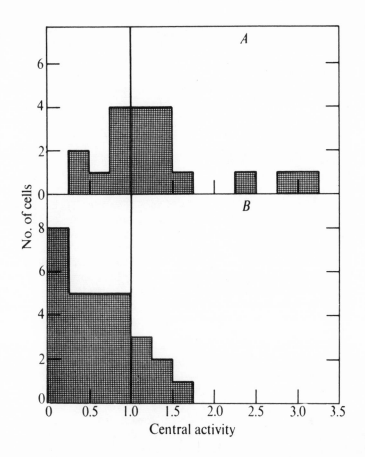

FIGURE 3. Distribution of grains over synchronized CHO cells exposed to ^{3}H-TdR for 0.5 minutes. A, Early S phase (4-6 h after release from mitosis). B, Late S phase (10-12 h after release from mitosis). CHO cells were synchronized by the 'Colcemid' reversal method of Stubblefield (1968). Pulse-labeling (performed at 2 h intervals after release from mitosis), preparation for electron microscopy, and autoradiography were performed as in Figure 1. Exposure time was 2-4 months. Histograms were prepared as in Figure 2. For A, eight nuclei pulse-labeled at 4 h and eleven nuclei labeled at 6 h after mitotic release were used. For B, thirteen nuclei labeled at 10 h and sixteen nuclei labeled at 12 h were used (from Huberman, Tsai and Deich, 1973).

the nucleus into two areas, a peripheral area and a central area. The peripheral area was defined by drawing a line in the interior of the nucleus at all points 1.25 microns away from the nearest membrane. We counted the grains over the central and peripheral areas and determined the area by tracing the nuclei onto paper, cutting out the central and peripheral areas and weighing the papers. Then we determined a quantity called the central activity. Central activity is defined as the ratio of fraction of grains which are central to the fraction of area which is central. When this ratio (the central activity) is 1.0 all the grains are evenly distributed throughout the nucleus. If the ratio is greated than 1.0, the concentration of grains is over the central part of the nucleus. If the ratio is less than 1.0, the distribution of grains is the greatest in the peripheral area. In these Figures we have made histograms of the distribution of central activity in unsynchronized populations (Figure 2), and in early S phase and late S phase (Figure 3). In the unsynchronized populations (Figure 2) there are some cells which show strong peripheral labeling and other cells which show grains throughout the nucleus. In early S phase (Figure 3, top) all of the cells show grains throughout the nucleus. There are no cells showing strong peripheral labeling. In late S phase (Figure 3, bottom) a large fraction of the cells show strong peripheral labeling and the frequency of central labeling is certainly reduced. These are similar to the results of Comings (this meeting), and probably reflect the late replication of heterochromatin around the nuclear membrane.

If it is true that DNA replication can take place anywhere in the nucleus and the DNA does not have to migrate to the nuclear membrane and back again in order to be replicated, then the DNA could be stably located inside the nucleus. Thus, if we were to do a pulse chase experiment, or a longer pulse, we would get exactly the same distribution of grains that we get when we do a short simple pulse. This is the case, as shown in the experiments illustrated by Figure 4.

In the top section one can see the results of pulse labeling in early S phase and chasing with cold thymidine into the late S phase. You can see that even though the cells are now being examined when they are in late S phase, the grains are distributed throughout the nucleus. There is still no peripheral labeling, suggesting that DNA is stably distributed inside the nucleus. In the lower two frames, we compare the results of a half minute pulse, an unsynchronized culture followed by a six hours' chase with cold thymidine, and a ten minutes' pulse of an unsynchronized culture. In both cases one finds generally labeled cells and cells with strong peripheral labeling just as in the case of a half-minute pulse of an unsynchronized culture. Because of the low number of cells scored, the histograms here are rather bumpy. I think the important thing is that one sees both strong peripheral labeling and also general labeling with either a half-minute pulse, a half-minute pulse followed

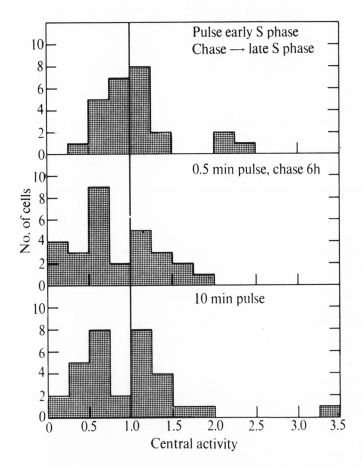

Figure 4. Distribution of grains over CHO cells after pulse-chase or long pulse labeling with ³H-TdR. a, synchronized cells (see Figure 3) were pulse-labeled 6 h after release from mitosis as in Figure 1. Then 30 seconds after addition of label, the medium was sucked off and the plates were washed, then replaced with medium containing 2.5 μg ml of TdR. After 6 h, the cells were collected by trypsinization and prepared for electron microscope autoradiography as in Figure 1. Exposure time was 1.5-4 months. b, Unsynchronized cells were pulse-labeled as in Figure 1 and chased for 6.75 h as in Figure 4a. Exposure time was 3.25 months. c, unsynchronized cells were pulse-labeled as in Figure 1 except that the ³H-TdR was left in contact with the cells for 10 minutes instead of 30 s. Exposure time was 1.5-6 months. In all cases, histograms were prepared as in Figure 2. For a, twenty-four cells were used, whereas for b, twenty-nine cells were used and for c, thirty-two cells were used. (From Huberman, Tsai and Deich, 1973)

by a six hour chase, or a ten minute pulse, again suggesting that
the DNA does not move to the nuclear membrane and back again for
replication.

This finding, that the ten minute pulse gives essentially the
same results as the half minute pulse, also allows us to ask in more
detail whether or not initiation of replication takes place on the
nuclear membrane. So far we have been concerned with whether the
growing point is on the nuclear membrane. We have not considered
whether the initiation at the very beginning of S phase occurs on
the nuclear membrane. Recall that in early S phase (Figure 3, top)
4-6 hours after release from mitosis, cells were labeled throughout
the nucleus after a 30 second pulse. We found in those experiments
with synchronized cells that the early S phase cells had a much
smaller number of grains over each nucleus than the late S phase
cells, suggesting that the late S phase cells have a higher specific
rate of replication. In fact the earlier we went into S, the fewer
grains there were over each nucleus, and that suggested that with a
30 second pulse we might not be able to see the very beginning of S
phase because there would be so few grains that we would not be able
to distinguish real incorporation from background. In order to
examine the very beginning of S phase, we decided to use a longer
pulse (ten minutes) and to do this pulse just two hours after release
from mitosis. At this time, whole cell autoradiographic studies have
shown us that many cells are just entering S phase, while many other
cells have not yet started DNA synthesis (Huberman & Tsai, unpublished
results). We did a ten minute pulse and exposed for 5 months; and
after this long exposure time, we developed the autoradiograms and
found a relatively small number of labeled nuclei because we were
looking at early S phase. The nuclei were divided into three classes
depending upon the number of grains over each nucleus. Cells with
zero to five grains over each nucleus constituted about 1/3 of the
total cells. This is essentially background level, and these cells
were not examined further. The second class were cells with 5-50
grains over each nucleus, again about 1/3 of the total cells. The
third class were cells with 50-200 grains over the nucleus, also
about 1/3 of the total cells. We felt that the cells which were
entering S phase during the ten minute pulse would probably be found
in the second category, the category of just barely significant
labeling over the nucleus. We looked at this category and did statistical analysis as before of the distribution of grains over the
nucleus, and we found no peripheral labeling at all. All of the
cells were generally labeled with the grains distributed throughout
the nucleus. The same also proved to be true of the more highly
labeled cells. Thus, even with a ten minute pulse period we can
detect no sign that initiation occurs at the nuclear membrane.

I would now like to explain the contradictory autoradiographic
results of O'Brien, Sanyal and Stanton (1972). As you recall, this
group pulse labeled HeLa cells for one minute, did electron microscope autoradiography and found label distributed only around the

periphery of the nucleus. They also did a one minute pulse followed by a one hour chase and now found that the label was distributed throughout the nucleus. The question is how to explain these results, which appear to be contradictory to our results. One possible explanation comes from our finding that if we pulse-label cells for 30 seconds we get a certain number of grains over each nucleus after an autoradiographic exposure of a certain time, but if we follow this 1/2 minute pulse with a chase of several minutes to one hour with cold thymidine, then (even though we chase by completely removing the external medium containing ^{3}H-thymidine and washing the cells with medium containing cold thymidine before allowing them to grow again in medium with cold thymidine) the number of grains incorporated over each cell after the chase increases by a factor of three or four. A reasonable explanation is that in a 1/2 minute pulse a considerable amount of H^{3}thymidine gets into the cell and is incorporated into the pool of thymidine nucleotides, but only a fraction of the H^3-thymidine is actually incorporated into DNA. During the subsequent chase a lot more of the H^{3}thymidine which has gotten into the pool can be incorporated into DNA. This may have happened in O'Brien et al's experiments. In addition, as I mentioned earlier, we find many more grains over late S phase cells than over early S phase cells when both have been exposed to a pulse of H^3-thymidine lasting 30 seconds. Therefore, I predict that if O'Brien et al had exposed the autoradiographs of pulse-labeled cells for a longer time they would have seen generally labeled cells from early S phase as well as the more obvious peripherally labeled cells from late S phase. If our explanation of the increase in grain quantity after chasing is correct, then during the one hour chase of O'Brien et al, more radioactivity would have been incorporated into both early S phase and late S phase cells, with the result that after their short autoradiographic exposure they could now see generally labeled cells from the early S phase.

The next question is how to explain the biochemical experiments which show association between nuclear membrane and DNA replication. I think Schachter (this meeting) offered one possible explanation by saying that single stranded DNA can bind to the M-band. It is clear that replicating DNA has a different structure from non-replicating DNA and one of the differences is that there is a certain amount of single-stranded character in replicating DNA. In any case, there are certainly different proteins associated with the replicating point (DNA polymerases, other factors) which are not associated with bulk DNA. It seems quite possible that the differences in structure could be such that after cells are broken open, preferential adsorption of replicating regions to membranes and other cellular constituents takes place, which would cause the replicating regions to fractionate differently from the bulk DNA. Because we now think that the biochemical experiments that have been done with mammalian cells, which we previously thought showed attachment of the replicating region to the membrane, have been

incorrectly interpreted, and since the biochemical experiments
that have been done with prokaryotes were essentially the same in
character, perhaps they should be reexamined. With that challenge,
I conclude the section of this presentation concerning the role of
the nuclear membrane in DNA replication in eukaryotic cells.

Okazaki Fragments in Eukaryotes

The first question which we addressed ourselves to is whether
or not there are Okazaki fragments (Okazaki et al, 1968) in eukaryotic cells, because in the literature of this subject there are about
a dozen papers and only about half of these papers (Painter and
Schaefer, 1969; Schandl and Taylor, 1969; Kidwell and Mueller, 1970;
Nuzzo, Brega and Falaschi, 1970; Sato, Tankaka and Sugimura, 1970;
Hyodo, Koyama and Ono, 1971) come to the conclusion that there are
fragments as short as Okazaki's. One paper in particular (Nuzzo,
Brega and Falaschi, 1970) obtained sucrose gradient patterns very
much like Okazaki's, but in most of the others there is some discrepancy between their results and Okazaki's findings, ranging from
finding no strands as short as Okazaki's after pulse labeling mammalian cells (Paoletti et al, 1967; Tsukada et al, 1968; Habener,
Bynum and Shack, 1969; Berger and Irvin, 1970; Hyodo, Koyama and
Ono, 1970) to finding very short strands of nucleotides about ten
nucleotides long, as well as various other kinds (Schandl and Taylor,
1971; Schandl, 1972).

We have been trying to investigate Okazaki fragments in our laboratory for the last seven or eight months. And during the course
of our preliminary experiments, we were able to reproduce all of the
findings in the literature. Some of the findings were clearly due,
however, to certain experimental problems. These problems can
result in loss of Okazaki fragments or can result in finding oligonucleotides much shorter than Okazaki fragments. Loss of fragments
may come about because Okazaki fragments frequently can be dissoci-
ated from the replication fork during lysis of the cells, coming
off as single strand pieces (Okazaki et al, 1968). Such single
strand pieces can be selectively lost from DNA during purification
by phenol extraction. Denatured DNA, which the Okazaki fragments
actually are, as they come off the replication fork, can be adsorbed
to the interface of the phenol extraction (Fakan, Turner, Pagano and
Hancock, 1972). Thus, if you phenol-extract replicating DNA and just
take the aqueous phase and run that on a sucrose gradient, you may
obtain the result in which you see no Okazaki fragments. In addition, Okazaki fragments are short-lived in mammalian cells, as in
bacteria. Pulses longer than one minute may fail to resolve them.

The second problem, finding oligonucleotides much shorter than
Okazaki fragments has been reported by Schandl & Taylor (1971). At
first, we also felt we had isolated very short oligonucleotides which
sedimented at about 2S! We think now that this finding is in error,

that the apparent oligonucleotides are not oligonucleotides but
rather represent the triphosphate pool. By a number of commonly
used analytical techniques, it is possible to adsorb a small fraction
of TTP onto filters during tests for acid precipitability. Thus,
you may get some counts in your experiment by adsorption of small
fractions of the TTP pool. How can you tell, if you pulse label
and lyse the cells and pass the lysate through a sucrose gradient
and you get a peak of material at about 2S, whether it is due to
selective adsorption from the TTP pool? The fact is that when we
take this material and try to estimate its size by chromotography
on columns or on paper, we find it has the size of TTP. We can
detect no significant amount of oligonucleotides 2-15 nucleotides
long.

We have done experiments on Okazaki fragments with mammalian
cells (Chinese hamster cells and HeLa cells) and with the slime
mold, Physarum polycephalum. In both systems we find fragments very
similar to those of Okazaki. Today I will confine my attention to
Physarum.

I would like to explain why we first started using Physarum.
Physarum is a slime mold, one of two slime molds which are now becoming popular in microbiological research. The second is Dictyostelium. These two are completely different organisms, and it is
quite by coincidence that they are both called slime molds. They
both form slimey plasmodia which have a certain amount of motility.
In the case of Physarum, the plasmodium is a single syncytium containing millions of nuclei, but in the case of Dictyostelium the
plasmodium is an aggregate of many cells, each containing a single
nucleus. Dictyostelium goes through interesting developmental
changes in the various stages of its life cycle and for this reason
has been used widely in studies of developmental biology. Physarum
is unusual in having a syncytial plasmodium in which all of the
nuclei replicate their DNA in natural synchrony; there is no need
to use drugs to achieve synchrony. If one measures the rate of DNA
synthesis by Physarum as a function of time during the cell cycle,
one finds that during the G2 phase of the cell cycle the rate is
very low. At the time of mitosis, which lasts just a few minutes,
this rate jumps up to a maximum value, remains at the maximum value
for one hour to one hour and a half, and then gradually decreases
for the next four hours, returning to the G2 level, (Braun, Mittermayer and Rusch, 1965). The bulk of the nuclear DNA which is not
nuclear satellite is synthesized during the S period. It is easy to
determine when a plasmodium is in S phase by cutting off little
pieces of plasmodium at various times and examining them micro-
scopically. It is possible to observe characteristic changes when
the plasmodium is entering and leaving mitosis, and one knows that
S phase lasts for 3-4 hours after mitosis.

We started using Physarum because we were interested in the

question of whether or not there is RNA serving as a primer for DNA synthesis in eukaryotic organisms. Arthur Kornberg will present (this meeting) some of the evidence his group has accumulated suggesting that RNA is a primer for DNA synthesis in the phages M13 and φX174. Those findings as well as findings from other groups (especially Sugino, Hirose and Okazaki, 1972) influenced us to begin work with eukaryotic cells, and one of the approaches which we thought might work was an approach first used in the Kornberg lab, demonstrating a covalent linkage between ribonucleotides and deoxynucleotides by a sort of nearest neighbor analysis, a transfer of labeled phosphate from a deoxy- to a ribo- nucleotide   (Wickner et al, 1972; Schekman et al, 1972).

In order to use this approach one needs an in vitro system or other system capable of incorporating deoxyribonucleoside phosphates into DNA. There are some in vitro systems described in the literature for mammalian cells which in many ways do seem to replicate DNA in a fashion similar to in vivo replication. We tried labeled phosphate transfer experiments in such isolated nuclei, but the results were rather unreproducible. We needed a more reproducible system, and we thought of the possibility of using Physarum, which offers such a large single cell that we could easily micro-inject deoxyribo-nucleoside triphosphates directly into the cell. Once the tri-phosphate got past the barrier of the cell membrane it should be incorporated directly into the DNA. So we tried micro-injection of deoxyribonucleoside triphosphates and we found that triphosphates are taken up by the DNA in Physarum. At that point we did not know whether or not there were Okazaki fragments. So we went back to using ^{3}H-thymidine labeling of Physarum plasmodia in order to char-acterize the Okazaki fragments and other species of Physarum DNA replication.

Let us see what happens when we pulse-label the Physarum plas-modium in S phase with tritiated thymidine for 30 seconds or 20 minutes. For historic reasons not particularly relevant to following the putative RNA primer, we used the following procedure. We first lysed the pulse labeled plasmodia. Lysing Physarum nuclei is more difficult than lysing mammalian nuclei: the nuclear membranes are tough and require high concentrations of Sarkosyl, followed by rolling in the cold for 45 minutes. The lysate of the plasmodium is applied to the top of a neutral sucrose gradient and sedimented for two and a half hours at 25,000 rpm as shown in Figure 5. We used two radioisotopes here. The ^{14}C thymidine was used to label the plas-modia for several generations so that bulk, old DNA would be thor-oughly labeled. The ^{14}C thymidine was removed before adding the ^{3}H-thymidine pulse label. The ^{3}H-thymidine pulse label was present for 30 seconds (Figure 5a) or 20 minutes (Figure 5b). Then the plas-modium was removed and lysed with Sarkosyl. One part forms a pellet and a large fraction of the counts, both ^{14}C and ^{3}H are in that pellet. The remainder of both isotopes is at the top of the neutral

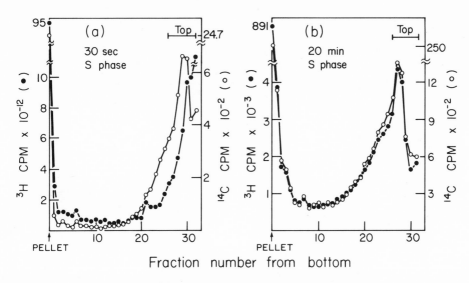

Fraction number from bottom

FIGURE 5. Sedimentation of pulse-labeled <u>Physarum</u> DNA through neutral sucrose gradients. Physarum microplasmodia were grown in shaker
culture medium and macroplasmodia were formed by fusion of microplasmodia and then grown on filter papers supported above the culture
medium by glass beads, as described by Holt and Gurney (1969). The
macroplasmodia were uniformly labeled with ^{14}C-thymidine (0.3µCi/15
ml medium) for 32 to 38 hours after fusion. Pulse labeling was carried out during the third postfusion S phase. 5-fluorodeoxyuridine
was added to a final concentration of 3µg/ml for 5 minutes before
labeling with ^{3}H-thymidine. For pulse labeling with ^{3}H-thymidine,
the plasmodium on the filter paper was blotted on another filter paper
and then cut into sections which were floated on drops (1 ml) of culture medium containing ^{3}H-thymidine (150µCi/ml) and 5-fluorodeoxy-
uridine (3µg/ml) at 20°C. The pulses were terminated and the plasmodia were lysed by modification of the method of Sonenshein & Holt
(1968). The plasmodia were immediately scraped with a blunt scalpel
into 1.5 ml ice-cold 0.1M EDTA, 0.1M tris, PH 8.1. An equal volume
of 15% (v/v) Sarkosyl NL30 (Geigy Chemicals) was added. The lysates,
in test tubes, were rotated for 45 minutes at 60 rpm, around a circle of 20 cm diameter in a vertical plane, in the cold room at 2-4°C.
Lysates were centrifuged through neutral sucrose density gradients
(33ml of 5-20% sucrose in 2mM EDTA, 5mM potassium phosphate, PH 7.5)
in polyallomer tubes for 2.5 hours at 5 C and 20,000 rpm in the Spinco
SW 27 rotor. 1.2 ml fractions were collected from the top of the
tube. 0.1 ml aliquots were dried onto filter paper discs (#895-E,
Schleicher & Schuell, Inc., Keene, N.H.), followed by a 1 hour wash
in cold 10% TCA and 5 one hour washes in cold 5% TCA and finally by
2 five minute washes with 95% ethanol. The filter discs were then

Legend to Fig. 5 (cont'd).

dried and counted in a scintillation counter. Pellets were resuspended in 1.2 ml of 99% formamide and counted as above. The ^{3}H-thymidine pulse was in early S phase for (a) 30 seconds and (b) 20 minutes. The indicated top fractions were separately pooled for further analysis. The recovery of ^{14}C and ^{3}H label was 99%-100%.

sucrose gradient. Note that the ^{3}H label at the top of the neutral sucrose gradients sediments more slowly than the ^{14}C label after a short pulse (30 seconds - Figure 5a), implying that this ^{3}H label is in separate structures from the bulk DNA. It is this slowly-sedimenting ^{3}H-labeled DNA which, as I shall describe below, contains attached RNA.

Fractions containing the ^{3}H label at the top of the neutral sucrose gradients (indicated in Figure 5) were pooled, denatured by incubation in 90% formamide at $37°$ for 30 minutes, and then sedimented through 0-15% sucrose gradients in 99% formamide. Under these conditions, DNA is totally denatured (Marmur and Ts'o, 1961), so that sedimentation should be a true measure of single strand length for linear strands. In addition, since alkali is not used, RNA is not hydrolyzed.

The data of Figure 6 show that the ^{14}C-labeled material pooled from the top of the neutral sucrose gradients consists of single strands which sediment heterogeneously, with two peaks near the top of the formamide-sucrose gradient. On the other hand, the ^{3}H-labeled DNA from a 30 second pulse (Figure 6a) sediments entirely as very short strands, at about the same position as the slower of the ^{14}C peaks. When the pulse time is extended to 20 minutes (Figure 6b) the ^{3}H-labeled DNA sediments more rapidly, at approximately the position of the faster ^{14}C peak.

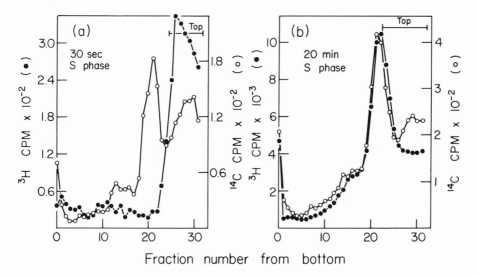

FIGURE 6. Sedimentation of pooled top fractions from Figure 5 through formamide-sucrose gradients. Selected fractions from the neutral sucrose gradients in Fig. 5 were pooled, brought to 90% formamide, heated at 37°C for 30 minutes (Sugino, Hirose and Okazaki, 1972), concentrated by evaporation under vacuum, and then centrifuged through 0-15% sucrose gradients in 99% formamide at 25°C and 25,000 rpm for 20 hours in the SW 27 rotor. 1.2ml fractions were collected and 0.1ml aliquots were taken onto filter discs which were washed and counted as in Fig. 5. The ^{3}H-thymidine pulse was for (a) 30 seconds and (b) 20 minutes. The recovery was 90-100%.

In order to test for possible attached RNA, the slowly sedimenting material from the formamide-sucrose gradients (fractions indicated in Figure 6) was pooled, dialyzed, and then centrifuged in isopycnic Cs_2SO_4 gradients along with added ^{14}C-labeled HeLa cell ribosomal RNA and ^{14}C-labeled single-stranded Physarum DNA. As shown in Figure 7, the density of a portion of the ^{3}H-thymidine labeled DNA was greater than that of the marker DNA, the difference in density being much more pronounced for the material labeled for 30 seconds (Fig. 7a) than for the material labeled for 20 minutes (Fig. 7b). This density difference could be partially abolished by pretreatment of the ^{3}H-thymidine labeled DNA strands with agents

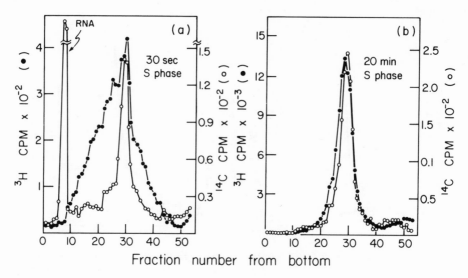

FIGURE 7. Isopycnic centrifugation of slowly-sedimenting pulse-labeled DNA chains. The indicated fractions from the formamide su-crose gradients shown in Fig. 2 were pooled, dialyzed at 0-4°C against 10mM tris, 1mM EDTA, pH7.4, for 18 hours, and then centrifuged in Cs_2SO_4 equilibrium density gradients (Sugino, Hirose, and Okazaki, 1972). Cs_2SO_4 gradients were centrifuged in a Beckman SW 50L rotor for 48 hours at 36,000 rpm and 15°C. Ten drop fractions were col-lected directly onto filter discs through the needle by puncturing at the bottom of the tube. Drying, washing of the discs with TCA and counting were as described above. In (a) marker ^{14}C-labeled HeLa cell ribosomal RNA was added. The ^{14}C marker for the density of single-stranded DNA is ^{14}C overnight-labeled Physarum DNA (de-natured by formamide). (a) 30 second pulse, from Fig. 6a, (b) 20 minutes pulse, from Fig. 6b. The recovery of ^{14}C label was about 100% and the recovery of ^{3}H label was about 88%.

capable of hydrolyzing RNA. As shown in Fig. 8a and b, alkaline hydrolysis resulted in nearly complete loss of detectable density difference. On the other hand, treatment with pancreatic RNase (Fig. 8c) caused only partial loss of density difference under

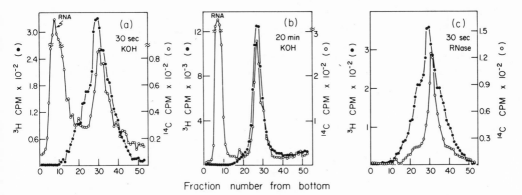

FIGURE 8. Effect of hydrolysis by KOH or RNase A on the density of pulse-labeled DNA chains. Pancreatic RNase A digestion was carried out at 37°C for three hours in 1.0M NaCl, 0.1M EDTA, pH 7.0, at a RNase A concentration of 1 mg/ml. The KOH digestion was in 0.3M KOH at 37°C for 18 hours. KOH was removed by dialysis against the appropriate buffer before running the Cs_2SO_4 equilibrium density gradients. Conditions for Cs_2SO_4 contrifugation were like those of Fig. 7. (a) 30 second pulse from Fig. 6a, hydrolysis by KOH; (b) 20 minute pulse from Fig. 6b, hydrolysis by KOH, (c) 30 second pulse from Fig. 6a, hydrolysis by RNase A. ^{14}C-labeled HeLa ribosomal RNA was added to (a) and (b) after hydrolysis, but was added to (c) before hydrolysis.

conditions where the marker ribosomal RNA was completely degraded.

These facts suggest that the material causing the density shift includes RNA. If it is only RNA, one can calculate from the position of the ^{3}H label between the RNA and DNA markers in Fig. 7a. that some of these short chains contain more RNA than DNA!

Confirmation of a covalent linkage between RNA and DNA in these short chains will come when we complete the ^{32}P-transfer, injection experiments I described above. These experiments are now in progress in our laboratory.

Note added in proof: Because I thought it best to preserve as much as possible of the informal style of my original talk in this revision, I have removed my coworkers from the responsibilities of authorship. I wish to emphasize, though, that most of the work using electron microscopic autoradiography was done by Alice Tsai and Robert Deich, while Henry Horwitz, Robert Minkoff and Fay Feldman carried out the work on Okazaki fragments in mammalian cells, and Anwar Waqar did the experiments on Okazaki fragments in Physarum polycephalum.

REFERENCES

1. Berger, H., Jr., & Irvin, J.L. (1970) Proc. Nat. Acad. Sci. USA 65: 152.
2. Blondel, B. (1968) Exptl. Cell Res. 53: 348.
3. Braun, R., Mittermayer, C. & Rusch, H.P. (1965) Proc. Nat. Acad. Sci. USA 53: 924.
4. Caro, L.G. & Van Tubergen, R.P. (1962) J. Cell Biol. 15: 173.
5. Erlandson, R.A. & deHarven, E. (1971) J. Cell Sci. 8: 353.
6. Fakan, S., Turner, G.N., Pagano, J.S. & Hancock, R. (1972) Proc. Nat. Sci. USA 69: 2300.
7. Friedman, D.F. & Mueller, G.C. (1969) Biochim. Biophys. Acta 174: 253.
8. Habener, J.F., Bynum, B.S. & Shack, J. (1969) Biochim. Biophys. Acta 195: 484.
9. Hanaoka, F. & Yamada, M. (1971) Biochem. Biophys. Res. Comm. 42: 647.
0. Holt, C.E. & Gurney, E.G. (1969) J. Cell Biol. 40: 484.
11. Huberman, J.A. & Riggs, A.D. (1968) J. Mol. Biol. 32: 327.
12. Huberman, J.A., Tsai, A. & Deich, R.A. (1973) Nature 241: 32.
13. Hyodo, M., Koyama, H. & Ono, T. (1970) Biochem. Biophys. Res. Comm. 38: 513.
14. Hyodo, M., Koyama, H. & Ono, T. (1971) Exp. Cell. Res. 67: 461.
15. Kidwell, W.R. & Mueller, G.C. (1970) Biochem. Biophys. Res. Comm. 36: 756.
16. Mamur, J. & Ts'o, P.O.P. (1961) Biochim. Biophys. Acta 51: 32.
17. Mizuno, N.S., Stoops, C.D. & Sinha, A.A. (1971) nature New Biology 229: 22.
18. Nuzzo, F., Brega, A. & Falaschi, A. (1970) Proc. Nat. Acad. Sci. USA 65: 1017.

19. O'Brien, R.L., Sanyal, A.B. & Stanton, R.H. (1972) Expt. Cells Res. 70: 106.

20. Okazaki, R., Okazaki, T., Sakabe, K., Sugimoto, K., Kainuma, R., Sugino, A. & Iwarsuki, N. (1968) Cold Spring Harbor Symp. Quant. Biol. 33: 129.

21. Painter, R.B. & Schaefer, A. (1969) Nature 221: 1215.

22. Paoletti, C., Dutheillet-Lamonthezie, N., Jeanteur, P. & Obrenovitch, A. (1967) Biochim. Biophys. Acta 149: 435.

23. Pearson, G.D. & Hanawalt, P.C. (1971) J. Mol. Biol. 62: 65.

24. Sato, S., Tankaka, M. & Sugimura, T. (1970) Biophys. Acta 209: 43.

25. Schandl, E.K. & Taylor, J.H. (1969) Biochem. Biophys. Res. Commun. 34: 291.

26. Schandl, E.K. & Taylor, J.H. (1971) Biochim. Biophys. Acta 228: 595.

27. Schandl, E.K. (1972) Cancer Res. 32: 726.

28. Schekman, R., Wickner, W., Westergaard, O., Brutlag, D., Geider, K., Bertsch, L.L. & Kornberg, A. (1972) Proc. Nat. Acad. Sci. USA 69: 2691.

29. Sonenshein, G.E. & Holt, C.E. (1968) Biochem. Biophys. Res. Commun. 33: 361.

30. Stubblefield, E. (1968) In Methods in Cell Physiology, Vol III, ed. by D.M. Prescott. Academic Press, New York, 1968, pp. 25-43.

31. Sugino, A., Hirose, S. & Okazaki, R. (1972) Proc. Nat. Acad. Sci. USA 69: 1863.

32. Tsukada, K., Moriyama, E., Lynch, W.E. & Lieberman, I. (1968) Nature 220: 162.

33. Wickner, W., Brutlag, D., Schekman, R. & Kornberg, A. (1972) Proc. Nat. Acad. Sci. USA 69: 965.

34. Williams, C.A. & Ockey, C.H. (1970) Exptl. Cell Res. 63: 365.

IS THE NUCLEAR MEMBRANE INVOLVED IN DNA REPLICATION?

David E. Comings and Tadashi A. Okada

Department of Medical Genetics, City of Hope National

Medical Center, Duarte, California

In 1963 Jacob, Brenner and Cuzin proposed the replicon concept for bacterial DNA and on the basis of some later modifications
suggested that the replicating point was attached to the bacterial
cell membrane. Since basic mechanisms in prokaryotes have a certain probability of being similar to basic mechanisms in eukaryotes,
I was interested in examining the question of whether membrane
attachment played a role in DNA replication in mammalian cells.
We were primarily interested in whether DNA replication might be
initiated at the nuclear membrane since a few autoradiographic
studies had already suggested that replication could take place
throughout the nucleus, implying that in mammalian cells the replication point did not always remain at the membrane. These previous
studies, however, were mostly with 30 minute pulse labels. To examine this question human amnion cells were synchronized first with
excess thymidine for 24 hours. The excess thymidine was then washed
out and the cells allowed to pass through the S period for 10 hours
and then further synchronized by exposure to amethopterin for 14
hours. The cultures were then exposed to tritiated thymidine for
10 minutes and after appropriate fixation electron microscope autoradiography performed. In these studies the label was primarily
around the nuclear membrane and periphery of the nucleolus (Comings
and Kakefuda, 1968). This pattern was only occasionally seen in
unsynchronized cells labeled for 10 minutes. We felt that the most
likely interpretation of these results was that DNA replication was
being initiated at the nuclear membrane. On the basis of these
and other studies which showed that chromatin was attached to the
nuclear membrane (DuPraw, 1965) and some studies by Bovieri at the
first part of the century which indicated that chromosomes maintained fairly constant positions inside the nucleus, it seemed
that there were several possible levels of order in regard to the

arrangement of interphase chromatin (Comings, 1968). The lowest
order would be simply the attachment of chromatin at numerous
sites to the inner surface of the nuclear membrane. A second
level of order would be that the chromosomes might bear some non-
random relationship to each other, the most obvious relationship
being that of some degree of homologous pairing. A third degree
of order would be that the sites on the chromatin which were
attaching to the membrane might be non-random, one possibility
being that the attachment sites are the initiation points (repli-
cators) of all or of some replicons. Although the final level of
order would be that the chromatin attachment would be at specific
sites on the nuclear membrane, there was no evidence that this
was the case.

Attachment of Chromatin to the Nuclear Membrane

        By whole mount electron microscopy we have found that meta-
phase cells frequently have pieces of nuclear membrane remaining
attached to them at random sites up and down the chromosome
(Comings and Okada, 1970a). It was frequently possible to see
areas where two sister chromatids were sharing a piece of nuclear
membrane suggesting that the daughter DNA molecules were attached
close together on the same piece of nuclear membrane. In prometa-
phase it was possible to see the chromosomes peeling away from
many attachment sites onto the inner side of the nuclear membrane
(Comings and Okada, 1970a). In quail meiotic cells interesting
configurations were seen which suggested that the chromatin was
converging in many different directions to the annuli of the nuclear
membrane (Comings and Okada, 1970b). If interphase chromosomes are
so intimately associated with the nuclear membrane one would expect
that during mitosis the chromosomes would condense onto the under
side of the nuclear membrane, leaving a relatively chromatin-free
nuclear center. This has indeed been observed in Muntjac and other
types of mammalian cells (Comings and okada, 1971).

        It has been known for many years that meiotic chromosomes are
attached to the nuclear membrane at the telomeres. That is, the
ends of the bivalents or the ends of the synaptonemal complexes
are firmly attached by a plate-like structure to the inner surface
of the nuclear membrane (Comings and Okada, 1970c; Moses, 1969).
This probably plays some role in aiding the initiation of pairing
during meiosis (Comings and Okada, 1970d).

        There is some evidence both for and against the proposition
that in mammalian cells there is a certain degree of somatic pairing. This somatic pairing may be greater in the spermatogonal
cells prior to meiosis than it is in other somatic cells. Recent
studies in the Muntjac have provided evidence that the chromosomes
are not randomly arranged in the nucleus (Heneen and Nichols, 1972).
The unique thing about this study was that in the Muntjac there

are a very small number of chromosomes and it is possible to in-
dividually identify them without exposing the cells to either
colchicine or hypotonic media. This study suggested that the non-
random arrangement of chromosomes is very easily destroyed by
these two treatments, probably accounting for why it is not picked
up in most studies.

DNA Replication and the Nuclear Membrane

In recent years there have been a number of biochemical studies
which have suggested that the replication point in eukaryotes remains at the nuclear membrane (see Comings, 1972, for review).
EM autoradiographic studies by Ockey (1972) using Chinese hamster
cells which had been synchronized by mitotic selection and FUdR,
suggested that DNA synthesis was associated with the membrane but
this was attributed to an artifact due to prolonged unbalanced
growth. Because of these varying results we decided to re-examine
the question and this time switched our cell system to Chinese
hamster cells since they are ideally suited for synchronization.
They grow rapidly, with a cell cycle of approximately 12 hours, and
it is easy to obtain large amounts of metaphase cells by shaking.
To obtain further synchrony to the beginning of S period we used
hydroxyurea which Amaldi (1972) has shown stops DNA replication
at the beginning of S period. He also demonstrated that FUdR
and amethopterin does not stop DNA replication but only slows it
down.

To investigate the question of whether the replication point
is associated with the nuclear membrane we exposed the cells to
tritiated thymidine for 30 seconds, 2 min, 4 min, 10 min and 20
min and then did EM autoradiography (Comings and Okada, 1973). So
that we would not be biased in scoring we simply scored on the
basis of whether there was a grain touching the nuclear membrane
or not and this was converted to percent of grains touching the
nuclear membrane. The mean of 30 seconds was 53, at 2 min 54,
at 5 min 57, at 10 min 59 and at 62 percent. Thus, if anything,
exactly the opposite occurred of what one would expect if the
replication point was at the nuclear membrane. If that were the
case one would expect the greatest degree of membrane-associated
grains in the very brief pulses and progressively less with the
longer pulses. This seemed to clearly indicate that the replication point was not remaining at the nuclear membrane. The fact
that slightly more than 50 percent of the grains were nuclear-
associated is simply a reflection of the fact that there is more
DNA condensed under the membrane than there is in the center of
the cell. The next question we investigated was whether the grains
were predominately membrane-associated late in S. For this experiment synchronized cells were labeled at 3 hrs, 5 hrs, 7 hrs,
9 hrs and 11.5 hrs after the onset of mitosis. Here the percentage
of grains at the nuclear membrane was 22% at 3 hrs, 30% at 5 hrs,
48% at 7 hrs, and 56% at 9 hrs and it then dropped down a little

bit to 47% at 11.5 hrs probably because some of these cells had already undergone division and were actually back at the early part of the cell cycle again. This was clear evidence that the membrane-associated DNA was predominately found among the late replicating DNA as would be expected with peripherally located heterochromatic DNA (Williams and Ockey, 1970).

The final question to investigate was whether there was any membrane-associated grains at the very beginning part of the S period. To do this cells were first synchronized by mitotic selection and then exposed to hydroxyurea for several hours. To begin labeling, hydroxyurea was removed and the cells labeled with tritiated thymidine for 4 min. In five different experiments the percent of grains touching the nuclear membrane ranged between 9 and 12%. This seemed very conclusive evidence that DNA synthesis was not being initiated at the nuclear membrane. There are several possibilities to account for the previous studies, one is that the prolonged exposure to amethopterin had allowed the cells to progress into late S when replication is membrane-associated, and the other is that prolonged exposure to amethopterin itself may have caused some type of abnormal replication pattern (Ockey, 1972). These studies then seemed to indicate that: (1) the replication point does not remain associated with the nuclear membrane; (2) peripheral label is seen predominately when late replicating membrane-associated heterochromatin is undergoing DNA synthesis.

A final question that can be legitimately asked is whether DNA replication might be controlled at the nuclear membrane in heterochromatin but not in euchromatin. For example, one could make the following case. It could be suggested that units of DNA replication and units of function are similar and that attaching the replicator to the nuclear membrane would prevent RNA synthesis by this segment of DNA and thus prevent priming for DNA replication and at the same time prevent transcription by RNA polymerase. This could be a potential explanation for the often observed correlation between genetic inactivation and late DNA replication. The one observation which makes this seem unlikely is the fact that heterochromatin is also seen to be clustered around the nucleolus and the nucleolus has no membranous structure. This would seem to indicate that heterochromatin can be genetically inactive and can undergo late DNA replication without being membrane-associated.

This leads us to the conclusion that the nuclear membrane appears to play no role in DNA replication in mammalian cells.

REFERENCES

1. Amaldi, F., Carnevali, F., Leoni, L. and Mariotti, D. The

replicon origins in Chinese hamster cell DNA: I. Labeling procedure and preliminary observations. Exp. Cell Res. <u>74</u>: 367-374, 1972.

2. Comings, D. E. The rationale for an ordered arrangement of chromatin in the interphase nucleus. Am. J. Human Genet. <u>20</u>: 440-460, 1968.

3. Comings, D. E. The structure and function of chromatin, in <u>Advances in Human Genetics</u>, vol. 3, pp 237-431, Harris and Hirschhorn (eds), Plenum Press, New York, 1972.

4. Comings, D. E. and Kakefuda, T. Initiation of DNA replication at the nuclear membrane in human cells. J. Mol. Biol. <u>33</u>: 225-229, 1968.

5. Comings, D. E. and Kakefuda, T. Association of nuclear membrane fragments with metaphase and anaphase chromosomes as observed by whole mount electron microscopy. Exp. Cell Res. <u>63</u>: 62-68, 1970a.

6. Comings, D. E. and Okada, T. A. Association of chromatin fibers with the annuli of the nuclear membrane. Exp. Cell Res. <u>62</u>: 293-302, 1970b.

7. Comings, D. E. and Okada, T. A. Whole mount electron microscopy of meiotic chromosomes and the synaptonemal complex. Chromosoma <u>30</u>: 269-286, 1970c.

8. Comings, D. E. and Okada, T. A. Mechanism of chromosome pairing during meiosis. Nature <u>227</u>: 451-456, 1970d.

9. Comings, D. E. and Okada, T. A. Condensation of chromosomes onto the nuclear membrane during prophase. Exp. Cell Res. <u>63</u>: 471-473, 1971.

10. Comings, D. E. and Okada, T. A. DNA replication and the nuclear membrane. J. Mol. Biol. <u>75</u>: (in press) 1973.

11. DuPraw, E. J. The organization of nuclei and chromosomes in honeybee embryonic cells. Proc. Nat. Acad. Sci. USA <u>53</u>: 161-168, 1965.

12. Heneen, W. K. and Nichols, W. W. Non-random arrangement of metaphase chromosomes in cultured cells of the Indian deer, <u>Muntiacus muntjak</u>. Cytogenetics <u>11</u>: 153-164, 1972.

13. Jacob, F., Brenner, S. and Cuzin, F. On the regulation of DNA replication. Cold Spring Harbor Symposium on Quant. Biol. <u>28</u>: 329-348, 1963.

14.   Moses, M. J.   Structure and function of the synaptonemal complex.   Genetics <u>61</u>: 41-51, 1969.

15.   Ockey, C. H.   Distribution of DNA replicator sites in mammalian nuclei. II. Effects of prolonged inhibition of DNA synthesis. Exp. Cell Res. <u>70</u>: 203-213, 1972.

16.   Williams, C. A. and Ockey, C. H.   Distribution of DNA replicator sites in mammalian nuclei after different methods of cell synchronization.   Exp. Cell Res. <u>63</u>: 365-372, 1970.

DNA REPLICATION AND THE CONSTRUCTION

OF THE CHROMOSOME

Harold Weintraub

Medical Research Council Laboratory of Molecular

Biology; Cambridge, CB2 2QH England

It was previously proposed that the normal synthesis of DNA in higher cells was dependent on <u>concurrent</u> synthesis of histone[1],[2]. Evidence for this was based on the fact that four relatively specific experimental manipulations of the rate of histone synthesis led to characteristic alterations in the rate of DNA synthesis. A possible insight into the way histone might function during DNA replication came from the finding that added histone has the capacity to remove nascent DNA from a presumptive replication complex when tested in a cell free system. In continuing these studies on the coupling between DNA synthesis and chromosome assembly, I will present further evidence that (a) nascent DNA is associated with a complex, possibly some "nuclear organelle", and (b) newly made histone enters the chromosome almost exclusively at the growing fork of DNA replication

THE ASSOCIATION OF NASCENT DNA WITH A NUCLEAR COMPLEX

The basic assay to be used in this communication involves the treatment of isolated chick erythroblast nuclei with pancreatic DNase in the presence of Mg^{+2}. Others have used analogous assays as a probe to the structure of nuclear chromatin[3-8]. For pancreatic DNase, the digestibility of DNA in nuclei is a function of the dose of DNase treatment, and the extent to which histone covers the DNA[5-7].

Figure 1 shows the DNase sensitivity of newly-replicated DNA in isolated nuclei from pulse-labeled cells. The sensitivity is monitored as a function of replication time during a continuous exposure to 3H-TdR. Similar curves apply to chromatin. The data indicates that cell samples taken at 30 seconds after addition of label are

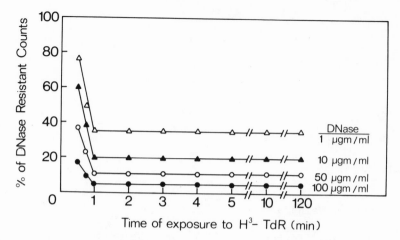

FIGURE 1. Sensitivity of pulse-labeled DNA to DNase. Primitive
chick erythroblasts were obtained from the blood islands of White
Leghorn Hen eggs after 4 days of incubation at 37° C. The cells
were incubated in vitro with H3-TdR (50 µc/ml; 15 Ci/mM) for increasing periods of time. Nuclei were obtained by suspending the
washed cells in cold RSB (0.0IM NaCl; 0.0IM Tris-HCl, pH 7.2; 0.003M
$MgCl_2$) containing 0.5% Nonidet P40. The nuclei were resuspended in
0.25M sucrose containing 5mM Na-phosphate buffer, pH 6.7 and 3mM
$MgCl_2$. Nuclei (10^6/ml) were digested with pancreatic DNase (Sigma)
as described by Mirsky (1972). Digestion was at 22° C for 30 minutes at a DNase concentration of 20 µgm/ml. At the end of the
digestion, the remaining DNA was precipitated with 5% TCA and carrier,
washed and counted. All TCA precipitable counts could be recovered
from the pelleted nuclei after digestion.

relatively resistant to DNase when compared to samples taken after
longer exposures to label. As will be shown later, these in turn
are much more resistant than free DNA. High DNase concentrations
eventually digest over 90% of the resistant DNA labeled for 30 seconds. This indicates that the label is probably in DNA. It is un-

likely that some single-stranded character of pulse-labeled DNA is responsible for its resistance since pancreatic DNase prefers single-stranded DNA. The resistance of pulse-labeled DNA to DNase may offer an opportunity to identify those nuclear proteins associated with the replication complex.

When pulse-labeled nuclei are digested with other enzymes in

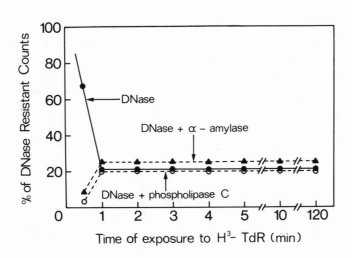

FIGURE 2. Increased sensitivity to DNase for pulse-labeled DNA incubated in the presence of alpha-amylase and phospholipase C. Conditions of the experiments were identical to those described in Figure 1. Alpha-amylase was obtained from Sigma and used at 200 μgm/ml. Phospholipase C was obtained from Calbiochem and used at 200 μgm/ml.

addition to DNase, it is found (Figure 2) that both alpha-amylase
and phospholipase C can make the resistant, 30 second pulse much
more sensitive to DNase.   Both enzymes have no affect on the digestion of DNA labeled for longer periods.   Additon of histone
(Figure 3) and in particular, the lysine rich histones, inhibits
the digestion of DNA in these nuclei.   This has also been demonstrated by Mirsky et al. (1972) who have shown that the protection
of DNA in nuclei stems largely from histones.   In contrast to the
selective effects of alpha-amylase and phospholipase C, Figure 3
also shows that low levels of pre-digested trypsin or pronase affect

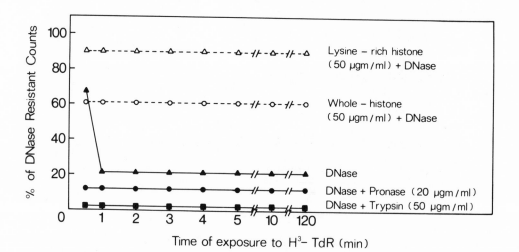

FIGURE 3.   Decreased sensitivity in the presence of histone.   Conditions were the same as in Figure 1.   All chemicals were obtained
from Sigma.   The figure also shows that pronase and trypsin sensi-
tize DNA in nuclei to digestion by DNase.

the digestion of both the short and long pulses to about the same extent.

The following conclusions can be drawn from these experiments: 1) Pulse-labeled DNA exists in a relatively DNase resistant form. 2) Within two minutes, pulse labeled DNA is incorporated into a structure that has the same sensitivity to DNase as normal chromatin. This indicates that histone is bound to new DNA within two minutes of synthesis.   3) Commercial preparations of both alpha-amylase and phospholipase C make the resistant DNA more accessible to DNase.

It was previously[1] shown that alpha-amylase and phospholipase C also destroy the characteristics of pulse-labeled DNA that cause it to sediment to low densities on CsCl and cause it to be extractable in phenol. The present studies with DNase, performed under much milder conditions, are consistent with these other observations. Rigorous documentation of the association of pulse-labeled DNA with a lipo-polysaccharide containing matrix must, however, await further purification of these enzymes.

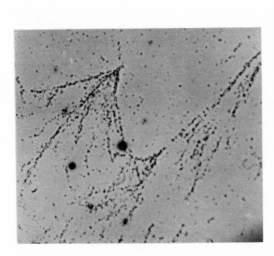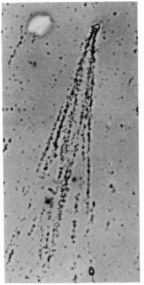

PLATE 1.   Association of newly-replicated DNA with a replication complex.   DNA autoradiography was performed as described by Lark et al. (1971).   Cells were pulsed with H3-TdR (200 μCi/ml; 15 Ci/mM) for 15 minutes.   Isolated nuclei were washed and 300 dropped onto an albumin coated glass slide.   One drop of 1% SDS was added and the lysate pulled across the slide with the side of a glass pipette.   The slide was dried, fixed in 5% TCA, washed with ethanol, dried, and processed for autoradiography.   Exposure was for 3 months.

Plate 1 is a DNA autoradiograph that may eventually yield some insight into the nature of the replication complex. Cells were labeled for 15 minutes with high specific activity H3-TdR; nuclei were isolated and the cells dropped onto an albumin coated slide. They were then lysed by the addition of SDS and autoradiography done as described by Lark et al. (1971)[9]. The plate shows many newly replicating DNA strands radiating from a single point. A very tentative conclusion from these pictures is that several replication forks share the same replication complex. K. Lark has found similar structures in HeLa cells (personal communication). Clearly, much work remains to be done before an unequivocal interpretation of these structures can be made. Recent electron microscopic studies[10] have demonstrated the presence of stained polysaccharide associated with DNA. These "cyclomeres" may be related to the replication complex described here.

THE ASSEMBLY OF CHROMATIN IN THE PRESENCE OF CYCLOHEXIMIDE

When protein synthesis is inhibited by cycloheximide, DNA synthesis is inhibited almost immediately[2,11-18]. In most cells, however, some DNA synthesis continues. For chick erythroblasts, this residual synthesis is about half the control rate for 30-45 minutes. We have previously shown that DNA replication under these conditions is, in all respects tested, no different from control synthesis except that the rate of chain elongation is less. The inhibition in elongation was attributed to a requirement for continued histone synthesis, while the residual synthesis was thought to be maintained by the recycling of pre-replicative histone to post-replicative DNA.

The obvious problem posed by the finding that some DNA synthesis continues in the absence of histone synthesis is whether this DNA is ever made into chromatin. Figure 4 shows the DNase sensitivity as a function of time of synthesis in medium containing cycloheximide. Also shown are comparable curves for synthesis in control conditions and in the presence of FUdR, where external thymidine is rate limiting and overall synthesis is about 30% of the control rate. In the presence of either cycloheximide or FUdR, pulse-labeled DNA remains relatively resistant for a longer period of time. This is consistent with the known decrease in elongation that occurs during these treatments. For longer pulses, cycloheximide DNA is about twice as sensitive to this concentration of DNase as either control or FUdR-DNA. The sensitivity remains consistant for over 30 minutes. This indicates that histone from other areas of the DNA does not migrate to the sensitive regions generated in the presence of cycloheximide. Addition of cycloheximide after removal of the label showed no additional sensitivity of the labeled DNA to DNase.

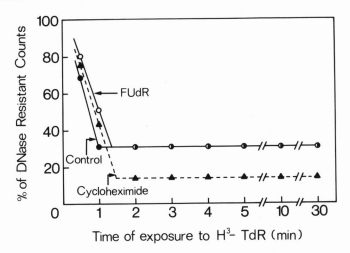

FIGURE 4. Increased sensitivity of DNA synthesized in the absence of protein synthesis. Conditions were the same as in Figure 1. Cycloheximide (Sigma) was at 50 uM. FUdR was at 10^{-5}M. Protein synthesis was inhibited by over 98%; DNA synthesis was inhibited by 70%.

Figure 5 is a dose response curve of DNase for free DNA, and nuclear DNA generated in the presence and absence of cycloheximide. Cycloheximide DNA is seen to be intermediate in sensitivity between free DNA and control, nuclear DNA. If it is assumed that half of the DNA generated in the presence of cycloheximide is free and the other half constructed as normal chromosomal DNA, then the resulting curve is shown by the open triangles in Figure 5. Such a curve follows the curve for chromatin generated in the presence of cycloheximide. This data is consistent with the idea that DNA generated in the presence of cycloheximide. This data is consistent with the idea that DNA generated in the presence of cycloheximide contains regions of free DNA (as measured by its ability to be protected from DNase) and regions of DNA covered by histone. The dose response curve of normal

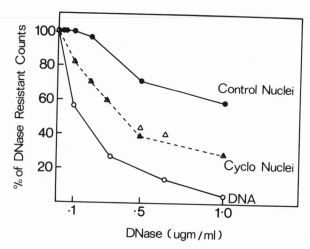

FIGURE 5. "Free" DNA and normal DNA synthesized in the presence of cycloheximide. Conditions were the same as in Figure 4, except digestion was for 15 minutes. Incubation was for 15 minutes in the presence of cycloheximide. Pure erythroblast DNA was isolated from CsCl gradients and used at 50 ugm/ml. The open triangles give the reconstructed curve assuming half of the DNA is "free" and the other half constructed to give the same DNase sensitivity as control chromatin.

chromatin shows two phases, one rapid and one slow. This has been noted in kinetic studies of others. It is likely from the correspondence between the reconstructed curve and the cycloheximide curve that whatever the sub-structure that confers these phases, it is preserved in the presence of cycloheximide for half of the DNA. The data in Figure 5 gives no insight into how the free and covered regions are distributed on the two daughter strands. Preliminary evidence indicates, however, that in the absence of protein synthesis, one free strand is generated while the other daughter strand is covered normally by histone. This histone is presumably recycled from the original parental template. Lastly, the

repair of the free DNA can be monitored after removal of cycloheximide. Such experiments show that free DNA becomes associated
with histone within 10 minutes after protein synthesis resumes.

The results in Figures 4 and 5 are compatible with a model in
which (a) newly replicated histone is present very near the growing
point (b) old histone is recycled from parental to at least one
daughter strand (c) new histone combines with new DNA almost exclusively at the growing point, and (d) histone does not slide along
the DNA in the intact cell.

Recent studies monitoring the replication of polyoma and SV40
in the presence of puromycin have demonstrated a very rapid inhibition by half in the rate of replication[19,20]. Under these conditions, "free" viral DNA is produced as opposed to the normal situation in which most DNA becomes rapidly associated with protein.
Given the parallel between the observations with polyoma and SV40
and those with chick erythroblasts, it may follow that cellular
histone synthesis is also required for the synthesis of viral DNA.
Lastly, many of the conclusions from this study rest on data obtained
from cells treated in the presence of cycloheximide. Although the
action of the drug is readily reversible, very different processes
may be occurring in cells incubated under normal conditions.

DNA REPLICATION AND THE CONSTRUCTION OF THE CHROMOSOME

Previous work[1] implicated histone in the synthesis of eukaryotic DNA. The present study demonstrates that newly made histone is
present very near the growing point. This is consistent with the
rapid inhibition of DNA synthesis observed after the addition of
cycloheximide. The present study also shows that some histone
becomes recycled in the presence of cycloheximide. This is consistent with the fact that there is a residual amount of DNA synthesis (50%) in the absence of protein synthesis. Proof for the
involvement of histone in DNA synthesis awaits the demonstration
that the proper concentration of these basic proteins (or one of them,
or a modified form of one of them) stimulates the rate of DNA chain
elongation in a nuclear system. The recent report that cadaverine
stimulates nuclear DNA synthesis[21] may be related to this point as
is the finding that histone stimulates the Qβ replicase[22].

Two additional problems remain totally unsolved. What is the
mechanism of the histone requirement for DNA synthesis? And why
should such a requirement exist? The most tempting answer to the
second question is that the coupling exists in order to preserve
some relationship between DNA synthesis and the structure of the new
chromosome; however, it is very difficult to reconcile this proposal
with the indication from this work that some "free" DNA is generated
in the absence of histone synthesis. Clearly, the reason for this

coupling will prove to be more complex than at first imagined. In particular the role of other proteins[23] in directing the assembly of the chromosome at the growing point will have to be examined.

With respect to the first question -- the mechanism for the coupling -- it is possible that free DNA inhibits the DNA replicase. The free DNA generated in the absence of histone synthesis might then feedback to inhibit the synthesis of other DNA. Any mechanism that removes the free DNA would then allow the replication fork to proceed at the normal rate. Alternatively, the finding that viral DNA, generated in the absence of protein synthesis, is relatively "uncoiled"[20] might imply a mechanism for cellular DNA synthesis whereby post-replicative supercoiling (induced by histone association) pulls pre-replicative DNA through the replication complex. In either case, it is possible that modification of newly-made histone[24] is required for the proper interactions with newly-made DNA, while phosphorylation[25] of old histone may be involved in its proper recycling at the growing point.

REFERENCES

1. Weintraub, H.,(1972) Nature 240: 449.
2. Weintraub, H. and Holtzer, H., (1972) J. Mol. Biol. 66: 13.
3. Clark, R.J. and Felsenfeld, G., (1971) Nature New Biol. 229: 101.
4. Murray, K., (1969) J. Mol. Biol. 39: 125.
5. Mirsky, A.E.,(1971) Proc. Nat. Acad. Sci. USA 68: 2945.
6. Mirsky, A.E. and Silverman, B. (1972) Proc. Nat. Acad. Sci. USA 69: 2115.
7. Mirsky, A.E., Silverman, B. and Panda, N. (1972) Proc. Nat. Acad. Sci. USA 69: 3243.
8. Billing, R.J. and Bonner, J. (1972) Biochim. Biophys. Acta. 281, 453.
9. Lark, K.G., Consigli, R. and Toliver, A. (1971) J. Mol. Biol. 58: 873.
10. Engelhardt, P. and Pusa, K. (1972) Nature New Biol. 240: 163.
11. Brega, A., Falaschi, A., deCarli, L. and Pavan, M. (1968) J. Cell Biol. 36: 484.
12. Bennett, L.L., Smither, D. and Ward, C.T. (1964) Biochim. Biophys. Acta. 87: 60.
13. Brown, R.F., Umeda, T., Takai, S. and Lieberman, I. (1970) Biochim. Biophys. Acta. 209: 49.
14. Ensminger, W.D. and Tamm, I. (1970) Virology 40: 152.
15. Grollman, A.P. (1968) J. Biol. Chem. 243: 4089.
16. Young, C.W. (1966) Mol. Pharm. 2: 50.
17. Taylor, E.W. (1965 Exp. Cell Res. 40: 316.
18. Storrie, B. and Attardi, G. (1972) J. Mol. Biol. 71: 177.
19. Bourgax, P. and Bourgaux-Ramoisy, D. (1972) Nature 235: 105.
20. White, M. and Eason, R. (1973) Nature New Biol. 241: 47.
21. Liebermann, I. (1973) Biochim. Biophys. Acta., in press.

22. Kuo, C.H. and August, J.T. (1972) Nature 239: 134.
23. Paul, J. and More, I.R. (1972) Nature 239: 134.
24. Louie, A.J. and Dixon, G.H. (1972) Proc. Nat. Acad. Sci. USA 69: 1975.
25. Balhorn, R., Oliver, D., Hohmann, P., Chalkley, R. and Granner, D. (1972) Biochemistry 11: 3915.

THE REPLICATION OF SIMIAN VIRUS 40 DNA

Arnold J. Levine, Rudolf Jaenisch,[1] Allen Mayer,[2]

Arup Sen and Ronald Hancock[3]

Princeton University Department of Biochemical Sciences

Princeton, New Jersey, U.S.A.

I would like to divide the topic of SV40 DNA replication into three parts: 1) the isolation and characterization of SV40 replicative intermediates, 2) the mechanism of formation of SV40 dimeric DNA and 3) the characterization of an SV40 specific nucleoprotein-complex found in infected cells.

We chose to study SV40 DNA replication for four reasons: 1) SV40 DNA has been very well characterized by Vinograd and his collaborators (Vinograd and Lebowitz, 1966; Bauer and Vinograd, 1968; Radloff, Bauer and Vinograd, 1967) and by Crawford (Crawford and Black, 1964; Crawford, 1969). They have shown that SV40 DNA is a closed circular double stranded DNA with superhelical turns and has a molecular weight of about 3×10^6 daltons. Several different techniques are available for the isolation of this DNA in pure form (Bauer and Vinograd, 1968; Radloff et al., 1967). 2) The size and shape of SV40 DNA is about the same as that found in ϕX174 RF. These two similar DNA molecules replicate, however, in very different environments. SV40 DNA replicates in the nucleus of African green monkey kidney cells while ϕX174 RF duplicates itself in the cytoplasm of E. coli. This then, represents an interesting problem in compar-

---

[1]Present address: The Salk Institute, LaJolla, California, U.S.A.

[2]Present address: Columbia University, New York, New York, U.S.A.

[3]Present address: ISREC, Lausanne, Switzerland

ative biochemistry of DNA replication. What is the effect of these
two different environments on the replication of these similar (in
size and shape) molecules? 3) The molecular weight of SV40 DNA is
about 3×10^6 daltons and one would expect that this DNA should
contain only about 6-10 genes. Of this number several gene
products (probably 2-3) are employed as coat proteins (Estes, Huang
and Pagano, 1971; Girard, Marty and Suarez, 1970). One might
expect then, that SV40 would borrow from the eukaryotic cell a
number of replication functions required for its duplication
process. We can then anticipate in such a study, to learn something
about the eukaryotic DNA replication mechanisms. 4) SV40 DNA can
integrate into the cellular genome (Sambrook et al., 1968; Hirai,
Lehman and Defendi, 1971) and there are good reasons to believe that
this viral DNA undergoes recombination events (matings) with the
mammalian chromosome (Aloni et al., 1969; Lavi and Winocour, 1972;
Tai et al., 1972). Again, it might be expected that such a
capability is derived from viral and/or cellular functions employed
in the replication cycle of this DNA. If this is correct, a study
of SV40 DNA replication could lead to a better understanding of the
process of recombination in eukaryotic cells.

The system we employ to study these events is as follows:
Confluent monolayer cultures of African Green Monkey kidney cells
(AGMK) are infected with SV40 (Levine, Kang and Billheimer, 1970)
at time zero. A one step growth curve for virus production and
viral DNA synthesis is presented in Fig. 1. Infectious virus (PFU)
begins to appear about twenty to twenty-four hours after infection
and continues to increase for some sixty to seventy hours. By 15-
18 hours post infection viral DNA synthesis is first detected and
continues to increase so that by 30-40 hours this rate is maximal.
Most of the experiments presented here have been performed between
30-40 hours after infection.

In order to isolate SV40 replicative intermediates, infected
AGMK cells were labeled with ^{3}H-thymidine for varying lengths of
time. The pulse-labeling period was stopped by washing the cells
with ice cold phosphate buffered saline (Levine et al., 1970)
followed by addition of 0.6% sodium dodecyl sulfate to lyse the
cells. The lysate is then gently scraped off of the petri dish
surface and NaCl (final 1M) is added. When this extract is kept
at 4°C for 6-24 hours the large molecular weight DNA precipitates
out of solution with most of the SDS. Low molecular weight DNA
(viral, mitochondrial and some cellular fragments) remains
soluble and can be isolated after centrifugation. This procedure
was first described by Hirt (1967). For short labeling times
(1 minute to 1 hour) mitochondrial DNA and cellular fragments
represent only a very small percentage (at 30-40 hours after infection) of the labeled DNA found in the 1 M NaCl-SDS soluble fraction
(1-4%). For longer labeling times further procedures must be

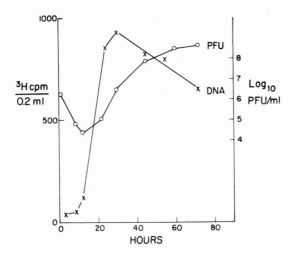

Fig. 1. One step growth curve for SV40 replicating in African Green
Monkey Kidney cells.

Monolayer cultures of AGMK cells were infected with 100 PFU/cell
of SV40 at time zero. At various times after infection a petri dish
was harvested and titered for infectious virus production (PFU). A
second set of petri dishes were labeled with ^{3}H-thymidine for one
hour and viral DNA was selectively extracted (Hirt, 1967) and sedimented through neutral sucrose gradients. The rate of viral DNA
synthesis was determined from the level of cpm sedimenting at 21S.

o-o PFU/ml
x-x ^{3}Hcpm/0.2ml

employed to purify SV40 DNA.

The first question we asked was "What does the replicating intermediate of SV40 DNA look like?" In order to get at this question we labeled infected cells with ^{3}H-thymidine for 2.5, 5.0, 10.0 and 20.0 minutes, selectively extracted the viral DNA and centrifuged it through neutral sucrose gradients (Levine et al., 1970). Fig. 2 shows the results of such an experiment. In each sucrose gradient a ^{32}P-labeled 21S mature viral DNA marker was included in the gradient. For short pulse-labeling times one observes a component labeled with ^{3}H-thymidine sedimenting slightly ahead of the 21S marker at about 25S. This component continues to build up until about 10 minutes of labeling when one sees a shoulder appear that co-sediments with mature viral DNA. As the pulse length increases (20 minutes, Fig. 2 or 1 hour) (Levine et al., 1970) the 25S DNA remains relatively constant in amount while the 21S DNA increases linearly (Levine et al., 1970). DNA–DNA hybridization experiments have shown that the 25S DNA is viral specific (Levine et al., 1970). The kinetics of labeling of 25S DNA suggest it may be a precursor (replicating intermediate) of mature 21S DNA. This was confirmed by pulse-labeling infected cells with ^{3}H-thymidine for one minute followed by a chase period with unlabeled thymidine for one hour (Levine et al., 1970). In this experiment (Levine et al., 1970) (not shown here) more than 90% of the ^{3}H-label found in the 25S peak after a one minute pulse shifted into the 21S region of the sucrose gradient after the one hour chase. This result strongly suggested that the 25S viral DNA was a precursor of mature 21S DNA.

The results of these experiments (Fig. 2) were surprising for a number of reasons. When the same experiment was performed with ΦX174 RF (Sinsheimer, Knippers and Komano, 1968) and H. Kaerner (This symposium), one observed replicating intermediates of ΦX174 RF that sediment much more heterogeneously in a neutral sucrose gradient, from the 16S to 25S regions of the gradient. The SV40 replicating intermediates seem to be sedimenting faster (at 25S) and more homogeneously than expected. The fast sedimenting position of 25S DNA could be due to RNA or protein complexed with the replicative intermediate. To see if this was correct we pronase treated, phenol extracted and RNase treated this DNA before sedimentation in neutral sucrose (Levine et al., 1970). None of these treatments alone or in combination altered the sedimentation pattern of replicating SV40 DNA isolated after SDS treatment (Levine et al., 1970). We concluded that the structure of the replicating DNA itself was responsible for the unusual fast sedimentation rate in neutral sucrose (Levine et al., 1970).

There are two possibilities then why SV40 replicative intermediate would sediment faster than expected: 1) either the mass of the DNA was larger than anticipated (Levine et al., 1970) or 2) the

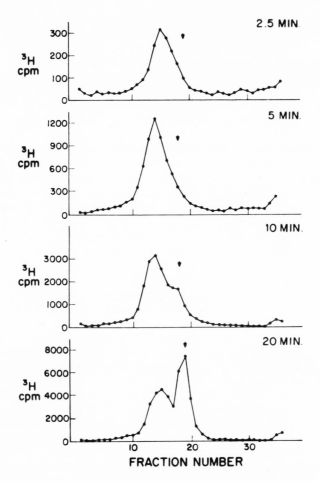

Fig. 2.   Sedimentation of SV40 DNA synthesized in short pulses with
[3]H-thymidine.

    Viral DNA synthesized during 2.5, 5.0, 10.0 and 20.0 minutes of
labeling with ^{3}H-thymidine was centrifuged through 5-20% neutral
sucrose gradients at 40,000 rpm for 3 hours in an SW 50.1 rotor.
The arrow is the position of a ^{32}P-labeled 21S sedimentation marker.

    o-o [3]Hcpm

shape of this DNA was more compact than we expected it to be (Jaenisch, Mayer and Levine, 1971; Sebring et al., 1971; Mayer and Levine, 1972). Although we didn't know it at the time we did this experiment, essentially both those explanations are correct (Jaenisch et al., 1971; Sebring et al., 1971; Mayer and Levine, 1972). That is to say, most of the replicative intermediates we find by this procedure appear to be almost completed replicating molecules (Levine et al., 1970; Mayer and Levine, 1972) (almost two times the size of mature SV40 DNA) and 2) those replicating molecules that have not yet almost completed their replication have a more compact shape than expected due to their enhanced tertiary structure (Jaenisch et al., 1971; Sebring et al., 1971).

Let us first address ourselves to the unusual shape of SV40 replicative intermediates. To get at this problem, we pulse-labeled SV40 infected cells for 3.5, 7.0, 15.0 and 120 minutes. The selectively extracted viral DNA was then analyzed on CsCl-ethidium bromide (EtBr) equilibrium gradients (Jaenisch et al., 1971). This result is presented in Fig. 3. The arrow (I) shows the position of mature closed circular ^{32}P-labeled viral DNA and the arrow (II) the position of relaxed circular viral DNA. For short pulse-labeling times (3.5 and 7.0 minutes) the majority of the ^{3}H-cpm were distributed at both an intermediate (with regard to the positions of form I and II DNA) and relaxed density position. With continued labeling the mature form of viral DNA (I) continued to build up (see 15 and 120 minute labeling times). Viral DNA (labeled for 3.5 or 7.0 minutes) banding at the intermediate and light density positions in this gradient both sediment at about 25S in neutral sucrose gradients (Mayer and Levine, 1972) while viral DNA banding with component I (dense position) sedimented at 21S (Mayer and Levine, 1972). This experiment (Jaenisch et al., 1971) suggested that EtBr-CsCl equilibrium centrifugation fractionated the SV40 replicative intermediate into roughly two classes (intermediate and light density). Both intermediate and light density SV40 DNA were purified by repeated centrifugations in EtBr CsCl. The resultant fractions were then prepared for examination in the electron microscope (Jaenisch et al., 1971). Fig. 4 (lower line) shows the two classes of SV40 replicative intermediates found in these preparations. In the light density region of the EtBr-CsCl gradient we found replicative intermediates (Fig. 4, lower right) (Jaenisch et al., 1971) of the type previously observed by Cairns in E. coli (1963), as well as with lambda (Tomizawa and Ogawa, 1968; Schnös and Inman, 1970; Inman and Schnös, 1971), Polyoma (Hirt, 1969; Bourgaux, Bourgaux-Ramoisy and Seiler, 1971), SV40 (Jaenisch et al., 1971) and other circular DNA's. These molecules contain two circles of equal length (presumed replicated regions) attached by a common region of DNA. One circle plus this common region is equivalent to the size of mature SV40 DNA (1.66μ) (Levine et al., 1970; Jaenisch et al., 1971). The intermediate density DNA (Fig. 4)(lower left three

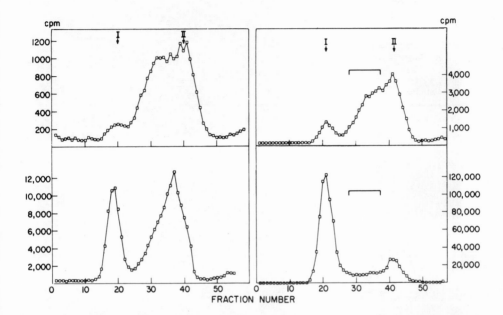

Fig. 3.   Ethidium bromide - CsCl equilibrium centrifugation of SV40
DNA synthesized in short pulses with ^{3}H-thymidine.

Viral DNA synthesized during a 3.5, 7.0, 15 and 120 minute
labeling period with ^{3}H-thymidine was selectively extracted (Hirt,
1967) and centrifuged to equilibrium in an EtBr (100 ug/ml) CsCl
(1.56 gms/cc) gradient.   The arrows are a closed circular marker
(I) and a relaxed circular marker (II) labeled with ^{32}P in SV40
DNA.

Top right - 3.5 minute exposure to ^{3}H-thymidine
Top left - 7.0 minute exposure to ^{3}H-thymidine
Lower right - 15.0 minute exposure to ^{3}H-thymidine
Lower left - 120.0 minute exposure to ^{3}H-thymidine

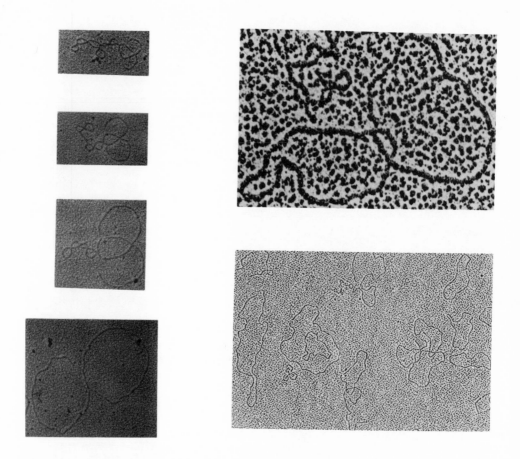

Fig. 4. Electron Micrographs of Replicating SV40 DNA (lower pictures) and Dimeric SV40 DNA.

SV40 replicating and dimeric DNA were isolated as described in the text and prepared for microscopy as described (Jaensch et al., 1971; Jaenisch and Levine, 1971a).

Top left – a catenated dimer
Top right – a mixture of circular and catenated oligomers and
    monomeric DNA.
Lower left three pictures – Early replicative intermediates
    from the intermediate density region of an EtBr-CsCl gradient.
Lower right – a late replicative intermediate from the light
    density region of an EtBr-CsCl gradient.

pictures) contained two circles (loops) of equal length (presumed replicative region) attached to a common region of DNA that contained superhelical turns. One circle plus the superhelical region of DNA was approximately the size of mature SV40 DNA (Jaenisch et al., 1971). Similar results have been obtained by Sebring et al. (1971) and Bourgaux and Bourgaux-Ramoisy (1972). In order for the replicating molecules to have superhelical turns it must behave as a closed circular molecule. One representation of such a replicative inter- mediate (shown in Fig. 7a) would be a molecule where the template strands remain as continuous closed polynucleotide circles (solid line, Fig. 7a) and the newly replicated linear fragments of progeny DNA (dashed lines, Fig. 7a) would be hydrogen bonded to the separated strands of the closed circular template molecules. Such a molecule would behave as partially closed circular in the unreplicated region (bind less EtBr) and partially relaxed (bind more EtBr) in the replicated region. This could give rise to its intermediate density in EtBr CsCl gradients.

If indeed the template strands of such molecules are closed circular then one might expect that treating these replicative intermediates with pancreatic DNase would result in a break in the phosphodiester backbone of DNA of the template strands (in the un- replicated region of the molecule) resulting in a shift in the density of such molecules (to the lighter side) in EtBr-CsCl. To test this SV40 DNA was isolated from the intermediate density region of EtBr-CsCl gradients (as in Fig. 3). One half of this DNA was treated with pancreatic DNase (Jaenisch et al., 1971) for nine minutes and both fractions were then centrifuged to equilibrium in EtBr-CsCl gradients. The results of this experiment are presented in Fig. 5. The top portion of Fig. 5 contains a ^{32}P-labeled SV40 DNA marker which is distributed in the closed circular (dense) and relaxed (light) regions of the equilibrium gradient. The ^{3}H-labeled intermediate density replicative forms continue to band at the intermediate density in EtBr-CsCl upon repeated centrifugations. Treating this sample with pancreatic DNase converts the ^{32}P-labeled component I marker to a relaxed position and shifts the density of the replicative intermediate to the lighter density region. This experiment confirms the closed circular nature of the template strands in the SV40 replicative intermediate. Identical conclusions have been reached by sedimenting these repli- cative intermediates in alkali (Sebring et al., 1971; Mayer and Levine, 1972), varying concentrations of EtBr (Mayer and Levine, 1972) and at high or low salt concentrations (Mayer and Levine, 1972). All the evidence is consistent with the structure shown in Fig. 7a.

In all our experiments where pancreatic DNase was employed to break a phosphodiester bond and shift the density of replicative intermediate in EtBr-CsCl, the replicating SV40 DNA relaxed with DNase banded one tube heavier in density than the ^{32}P-labeled relaxed

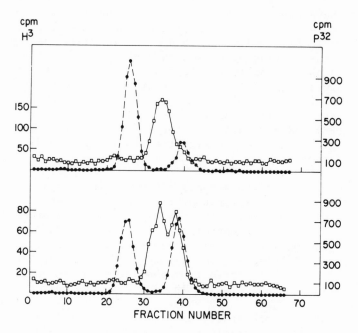

Fig. 5. An EtBr-CsCl gradient profile of intermediate density replicative forms before (top) and after (bottom) treatment with pancreatic DNase.

The intermediate density DNA from Fig. 3, upper left was isolated and rebanded 2 times in EtBr-CsCl. One half of this DNA was treated with pancreatic DNase (10^{-3} ug/ml) for 9 minutes.

Upper gradient: No DNase
  o-o ^{32}P-SV40 DNA marker form I + II
  □-□ ^{3}H-labeled replicative intermediate

Lower gradient: With DNase
  symbols are the same

viral DNA (Fig. 5). This may be due to single stranded regions in
the replicative intermediates. Such single stranded regions were
first detected by the tight binding of SV40 replicating DNA to
benzylated-napholyated DEAE cellulose columns (Levine et al., 1970).
These single stranded regions can be visualized in the electron
micrograph shown in Fig. 4. In these micrographs we observed a
discontinuity (single stranded region) at both replicating forks.
Bourgaux demonstrated single stranded regions at both replicating
forks of polyoma replicative intermediate employing the single
strand specific enzyme from Neurospora (Bourgaux and Bourgaux-Ramoisy,
1971). Single stranded regions have been interpreted as regions of
active DNA synthesis (Inman and Schnös, 1971) and so these data
suggested that SV40 and polyoma replicate bidirectionally. Nathans
and Danna (1972) and Fareed and Salzman (1972a) have shown by
completely different techniques that SV40 replicates bidirectionally
from a unique origin (Nathans and Danna, 1972; Fareed and Salzman,
1972a; Thoren, Sebring and Salzman, 1972).

This then is the available evidence that early replicative
intermediates (10-60% replicated) have an unusual tertiary structure
(superhelical turns) and sediment faster in neutral sucrose gradients (Fig. 2) than expected. There is yet another reason why
replicative intermediates of SV40 sediment more rapidly (25S) than
expected. When we look at the distribution, with regard to the
extent of replication, of replicative intermediates isolated by the
1M NaCl-SDS (Hirt, 1967) procedure we find many more almost completed
molecules (larger in size) than those just starting replication
(see Fig. 7b).

To demonstrate this, the following experiment was performed.
SV40 infected AGMK cells were labeled with ^{3}H-thymidine for 15
minutes at 36 hours after infection. The selectively extracted
viral DNA was spun to equilibrium in a CsCl-EtBr gradient. The
results of this gradient are presented in Fig. 6. The regions in
the gradient marked A, B, C and D were isolated and centrifuged
through an alkaline sucrose gradient. The newly labeled linear
progeny strands were thus separated from the closed circular template
strands and sedimented as a function of their number average molecular weight in alkaline sucrose gradients. We calculated the number
average molecular weight of the linear DNA fragment for each class
(A, B, C, D) of molecules from the EtBr-CsCl gradient. Fig. 6
shows that the most dense fraction (A) was an average of 18% of one
genome length, the next fraction (B) was on the average 32% replicated, fraction C was 58% replicated while fraction D (lightest
density) about 80% completed. This demonstrates that the position
of the replicative intermediate in an EtBr-CsCl gradient is determined
by the degree of replication of the molecule. The lightest density
molecules are almost completed replication while the intermediate
density molecules are just beginning replication. This experiment

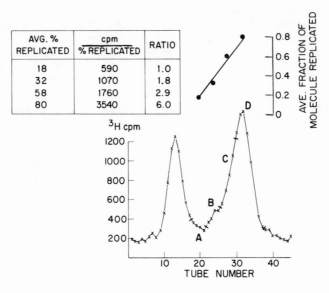

Fig. 6. An EtBr-CsCl gradient analysis of the distribution of
replicative intermediates by size and amount isolated by the selective
extraction procedure.

Infected cells were labeled with ^{3}H-thymidine for 15 minutes
and viral DNA was centrifuged to equilibrium in an EtBr-CsCl gradient. Fractions A, B, C and D were sedimented through alkaline
sucrose gradients to determine the size of the newly made daughter
strands (sizes shown in upper right insert). The relative amounts
of each of these four size classes of replicating molecules is
calculated in the upper left insert.

x-x   ^{3}H-cpm

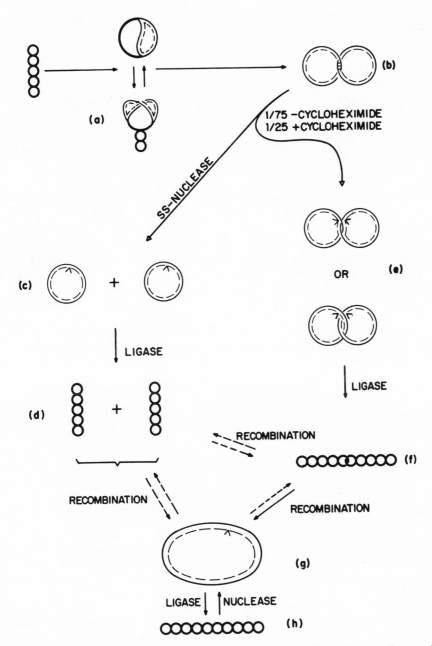

Fig. 7. A model for the replication of SV40 DNA and the formation
of catenated and circular dimers.

Details are explained in the text.

also gave us the opportunity to ask, on a molar basis which replicative intermediates are found in largest amounts? Do we find more D type molecules than A type (see Fig. 6)? To calculate this one divides the degree of replication found in each fraction into the total number of counts per minute found in each fraction. This normalizes for the fact that larger DNA molecules contain more ^{3}H-thymidine and gives us the ratio of progeny strands found in each class of replicated molecules. When this is done (Fig. 6) one finds that the ratio of molecules just beginning replication (class A, 18% replicated) is 1/6th the level found in the 80% replicated region of the gradient (class D). Over several experiments this number (80% replicated molecules/20% replicated molecules) varied between 5-10. This experiment then demonstrates that there is an excess of almost completed molecules (almost 2 times the size of SV40) in the replicative pool as isolated by the 1M NaCl-SDS procedure.

Fareed and Salzman (1972b) have shown that the replicative intermediate of SV40 contains 4S fragments that are probably intermediates in polynucleotide chain growth. This does not alter the interpretation of this experiment. For a 15 minute pulse-labeling period employed in this experiment these 4S fragments represent less than 2% of the total labeled polynucleotide strands. In addition one could only expect a discontinuous replication mechanism to shorten the chain length and yield the opposite result found in these experiments.

The available evidence indicates that there is an unusual distribution of molecules in the replicative pool and that most replicative intermediates have almost completed replication (Fig. 7b). There could be two explanations for this. One, it is artifact of the extraction procedure. We may be losing early replicative molecules perhaps in the 1M NaCl-SDS pellet with host DNA. We have looked in this pellet and do not find these molecules (Levine et al., 1970). Or, two, it could be that this represents the real distribution of replicating molecules and that this unusual result reflects kinetic events during replication.

Fig. 7 presents a model for SV40 DNA replication that is consistent with the experiments presented here. The solid lines represent the template strands while the dashed lines represent the newly synthesized progeny linear polynucleotide chains. Initiation of SV40 DNA replication requires at least one viral protein (Tegtmeyer, 1972) coded for by the mutants TSA-7 or TSA-30 (White and Eason, 1971; Mayer, 1972). The replicative intermediate containing closed circular template strands and linear newly synthesized progeny strands (Fig. 7a) can be isolated at the intermediate density region of EtBr-CsCl gradients (Figs. 3 and 6). In order for replication to proceed semiconservatively further strand separation requires a break in the unreplicated region (superhelical region) of this

molecule (Sebring et al., 1971; Mayer and Levine, 1972). The
relaxed replicative intermediates are rarely observed (Jaenisch et
al., 1971; Sebring et al., 1971) indicating that the equilibrium
favors the closed circular superhelical replicative intermediate.
Replication proceeds until a form observed in Fig. 7b is produced.
This intermediate has two interlocked circles with linear fragments
of 80-96% completed chains (Mayer and Levine, 1972). This intermediate (late replicative form) represents the majority form
observed in the replicating pool. Two lines of evidence indicate
that early replicative intermediates (7a) replicate at a faster rate
than do late replicative intermediates (7b). First, short pulses
with ^{3}H-thymidine (see Fig. 3) preferentially label early replicative intermediates that band at an intermediate density in EtBr-
CsCl. Up to seven minutes of labeling is required to label uniformly
all the replicating molecules in the pool (Levine et al., 1970;
Mayer and Levine, 1972). Only after seven minutes of exposure to
isotope is mature 21S closed circular and supercoiled DNA (7d) synthesized at a linear rate (Levine et al., 1970; Mayer and Levine,
1972).

There could be two reasons why early replicative forms propagate
their polynucleotide chains at a faster rate than late replicative
forms: 1) A break in the phophodiester backbone in the unreplicated
region of these intermediates is required to allow replication to
proceed. The target size of the unreplicated region of SV40 DNA
becomes smaller as replication proceeds. If such a break was rate
limiting during SV40 DNA replication, then early replicative forms
would replicate faster (be attacked by this nuclease more often)
than late forms. This notion assumes a nonspecific nuclease action
(Mayer and Levine, 1972). 2) Toward the end of replication there
might be a rate limiting step for the separation of two closed
circular molecules (7b) into two relaxed circles (7c). This would
also yield the unusual distribution of SV40 intermediates observed
here (Fig. 6).

The model presented in Fig. 7 makes certain predictions.
Failure to separate the two replicating SV40 circles (7b) should
result in the formation of a catenated dimer (7e, f). Thus an
error in the replicating mechanism (failure to separate the replicating circles) should yield the catenated dimeric form of DNA.
Circular dimers (7g, h) could then be formed by recombination of
two monomers or from the catenated dimer. Dr. Jaenisch in our
laboratory set out to test this notion.

The basic question is, how are catenated and circular dimers
formed in SV40 infected cells? The experiments required the following format: first we must isolate and demonstrate the existence
of such dimeric DNA in SV40 infected cells, second we must devise a
method for separating and quantitating catenated and circular dimeric

DNA and third we must introduce replication errors (by using cycloheximide) and follow the resultant formation of dimers.

Dimeric SV40 DNA was isolated by repeated sedimentation of
closed circular DNA through alkaline sucrose gradients at pH 12.2
(Jaenisch and Levine, 1971a and 1972). Two classes of dimers were
observed in the electron microscope, circular dimers (Fig. 4, top
right) and catenated dimers (Fig. 4, top left). The length distribution of these molecules indicated they were two times the size of
SV40 (Jaenisch and Levine, 1971a and 1972) and such molecules were
infectious (Jaenisch and Levine, 1971a and 1972). On a mass basis,
each dimer had a relative infectivity, when compared to monomers of
0.5 (Jaenisch and Levine, 1971a and 1972).

Fig. 8 shows the sedimentation of dimeric SV40 DNA through an
alkaline sucrose gradient at pH 12.2. The faster sedimenting peak
contains SV40 dimers while the slower sedimenting DNA represents
SV40 monomers. When infected cells are labeled with ^{3}H-thymidine
(for varying periods of time) the number of cpm found in purified
dimer DNA divided by the total number of cpm found in monomeric
closed circular DNA gives us the percentage of dimers synthesized
in that labeling period (Jaenisch and Levine, 1971a and 1972).

In order to separate and quantitate the proportion of catenated
and circular dimers found in these cells, the equations shown in
Fig. 9 were derived with the aid of Dr. N. Sueoka. The basic principle of this method is as follows: component I catenated or circular dimers band in the dense position (heavy) of an EtBr-CsCl
gradient (Fig. 9, I). Treatment of this DNA with pancreatic DNase
introduces one break in the DNA backbone. The circular dimer (one
continuous circle) then relaxes and shifts its position in the
EtBr-CsCl gradient to the least dense (light, Fig. 9, III) position
in the gradient. The catenated dimer is made up of two interlocked
independent circles, so that at low DNase concentrations only one
circle is relaxed and this molecule bands at an intermediate density
(Fig. 9, II) position in EtBr-CsCl. Since pancreatic DNase treatment of these DNA's approximates a random breaking process we have
used a derivative of the Poisson distribution to determine the
levels of catenated dimers (Bo) found in a preparation of SV40
dimeric DNA. A more complete discussion of this technique can be
found in Jaenisch and Levine (1973).

These experiments demonstrate that catenated and circular dimers
can be isolated from infected cells and that circular and catenated
dimers can be separated and quantitated. The prediction we would
like to test is that catenated dimers should arise from replication
errors (failure to separate two interlocked and replicating circles).

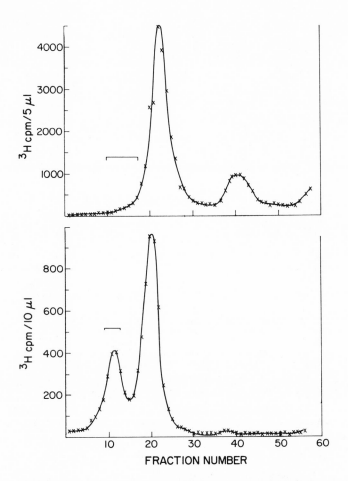

Fig. 8.   Sedimentation of SV40 Monomeric and dimeric DNA through alkaline sucrose gradients.

Selectively extracted viral DNA labeled with ^{3}H-thymidine for 45 minutes and sedimented through an alkaline sucrose gradient (pH 12.2) for 9 hours at 41,000 rpm using an SW 41 rotor (top gradient). The fractions indicated by the bars in the top gradient were collected, neutralized, ethanol precipitated and rerun on a second alkaline sucrose gradient (bottom). The bar above the leading peak in the second gradient is the dimer fraction.

x-x   ^{3}H-cpm

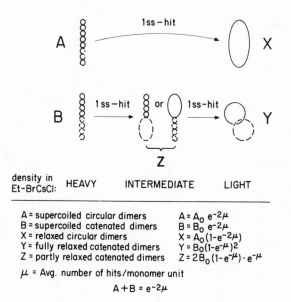

Fig. 9.   Schematic diagram of the method for determining the level of circular and catenated dimers found in a preparation of dimeric SV40 DNA.

Purified SV40 dimers are incubated with pancreatic DNase for various time intervals and the average number of polynucleotide chain breaks per monomer unit are determined (μ) from the distribution of radioactive DNA in an EtBr-CsCl gradient. The fraction of dimers that are catenated (B_0) can then be determined (Jaenisch and Levine, 1973). Note that $A_0 + B_0 = 1$.

To determine the level of dimers found in infected cells, ^{3}H–thymidine was used to label these cells for long periods of time (20–90 hours and 20–105 hours after infection). For these long labeling times we found about 1.5% of the total viral DNA synthesis was in the dimeric form and 10% of these dimers were catenated dimers (Table 1, lines 1 and 2). When shorter pulse-labeling times with ^{3}H-thymidine were employed (4 hours, 1 hour or 45 minutes) the percentage of dimers synthesized remained constant (1.25-1.6%) but the proportion of these dimers in the catenated form increased in an indirect proportion to the pulse length. Thus the percentage of catenated dimers found in the purified preparations of dimeric DNA labeled for 4 hours, 1 hour or 45 minutes was 25%, 43% and 48% respectively (Table 1). The shorter the pulse length the greater the percentage of catenated dimers. Since longer labeling times (70 hours) yielding 10% catenated dimers while short labeling times (45 minutes) yielded 48% catenated dimers, these experiments indicated that catenated dimers were unstable and decayed to an alternate form. To test this, infected cells were labeled for 45 minutes with ^{3}H-thymidine and a portion of the culture was chased in the absence of labeled thymidine for 4 hours. During the 45 minute pulse 1.6% of the total viral DNA synthesized was dimeric DNA and 48% of these dimers were in the catenated form. After a 4 hour chase period, 1.4% of the viral DNA was dimeric DNA while 22% of these dimers were in the catenated form (Table 1). Thus catenated dimers are formed with a high frequency and decay, with a half-life of about 3.7 hours, to an alternative form.

TABLE 1. The percentage of viral DNA present as dimers and the fraction of these dimers found as catenated molecules

Expt. no.	Time of labeling post infection	cycloheximide treatment	% of total DNA present as dimers	% of dimers present as catenates
1	20 – 90 hrs.	–	1.4	10
2	20 – 105 hrs.	–	1.5	10
3	37 – 41 hrs.	–	1.25	25
4	36 – 37 hrs.	–	1.25	43
5	36 – 36.83 hrs.	–	1.6	48
6	36 – 36.83 hrs. plus 4 hr. chase	–	1.4	22
7	36 – 37 hrs.	–	1.25	43
8	36 – 37 hrs.	+	4.8	67

These data are consistent with the hypothesis that ^{3}H-thymidine enters SV40 DNA via the replicative pool (incorporated into replicating molecules). In one in every seventy-five rounds of replication the interlocked replicating circles of SV40 DNA (Fig. 7b) fail to separate (yielding 1.5% dimers) and catenated dimers are formed (highly labeled). This form of SV40 dimer is unstable and decays probably to circular dimers or to monomeric DNA or both (Jaenisch and Levine, 1973). Thus the shorter the pulse length the greater the frequency of detecting catenated dimers.

If this notion is correct then introducing replication errors should increase the frequency of dimers and increase the proportion of dimers found in the catenated form. When cycloheximide is added to SV40 infected cells, SV40 DNA replication is inhibited. There is a requirement for protein synthesis both at the initiation step (7a) and during the last stages in separating the replicating circles (7b). Evidently inhibiting protein synthesis causes an imbalance in the levels of proteins involved in different stages of replication (initiation, chain propagation and separation of the two interlocked circles) and these imbalances can lead to replication errors.

What is the effect of cycloheximide treatment on the frequency of dimer formation and the proportion of catenated dimers? When SV40 DNA is replicated in the presence of cycloheximide the frequency of dimers produced increases about 3 fold (from 1.3 to 4.8% dimeric DNA) and a larger percentage of dimers are in the catenated form (67% instead of 43%) (see Table 1). These experiments are consistent with the expectation that catenated dimers are formed by failure to separate the two replicating circles at the end of SV40 replication and that protein synthesis is required for an efficient separation of these two replicated molecules.

Two other independent lines of evidence support the notion that dimers are formed from replication errors. One, viral DNA synthesized prior to cycloheximide treatment (^{32}P-labeled) does not participate in the enhanced synthesis of dimers (^{3}H-labeled) observed after cycloheximide treatment (Jaenisch and Levine, 1971a and 1972). One might expect that if an SV40 dimer arose by recombination events that SV40 monomers synthesized before cycloheximide treatment (^{32}P-labeled) would participate in dimer formation. Instead a small percentage of the monomers synthesized before cycloheximide treatment (^{32}P-labeled) enter the replicating pool as templates and are labeled with ^{3}H-thymidine during the cycloheximide treatment. What is observed is that ^{3}H-labeled DNA (replicated after cycloheximide is added) is enriched in these dimers (over ^{32}P-labeled viral DNA made before cycloheximide is added) (Tegtmeyer, 1972). This shows replication must occur to yield high levels of dimers. Second, SV40 tetramers are consistently observed in higher levels than SV40

trimers. It is likely that a monomer (98.5% of the total DNA) and
a dimer (1.5%) would recombine more frequently to give a trimer
than would two dimers recombine to yield a tetramer. On the other
hand once a monomer replicated to yield a dimer (by a replication
error) the dimer could replicate to give a tetramer. The fact that
SV40 dimers can replicate has been shown previously (Jaenisch and
Levine, 1971b). Thus one would predict that more tetramers than
trimers should be present if SV40 dimeric DNA was formed by an
aberrant mode of DNA replication and not by recombination (Jaenisch
and Levine, 1973).

Fig. 7 outlines the events occurring in SV40 infected cells
with regard to the proposed mechanism of dimer formation. Replicating SV40 intermediates (7a and b) normally result in a separation
of the two interlocked replicated circles (7b) into monomeric DNA
(7c and d). Once in every 75 rounds of replication these replicating circles fail to separate (7e) and form catenated dimers (7e, f).
The frequency of replicative errors (7b) can be increased 3 fold by
cycloheximide treatment resulting in a higher proportion of dimers
that are catenated (7e). Catenated dimers are unstable and decay
(with a half life of 3.7 hours) to either monomers (7d) or circular
dimers (7g and h). The limitations of quantitation in these
experiments do not permit us to state that all catenated dimers are
formed by replication errors. It is possible that a minority of
catenated dimers may be formed by recombination events. It is also
possible that some circular dimers may be formed from replication
errors. We believe Fig. 7 represents the major pathway of dimer
formation in infected cells.

Table 2 compares the results I have just reviewed for the
formation of SV40 dimers with those of Benbow, Eisenberg and Sinsheimer (1972) who, employing different techniques and ϕX174 grown

TABLE 2.   The mechanisms postulated for the generation of
           oligomeric forms of closed circular DNA

	Type of dimer	
Virus	Circular	Catenated
SV40	The majority from recombination of the catenated dimer.	The majority from replication errors.
ϕX174[a]	The majority from a replication mechanism.	Some from recombination events.

(a) Benbow, Eisenberg and Sinsheimer, 1972.

in E. coli, have studied the mechanism of formation of φX174 dimeric DNA. We have concluded that most SV40 catenated dimers arise from replication errors and circular dimers arise from recombination events. For φX174, they have concluded that most circular dimers arise from replication errors while some catenated dimers arise from recombination events. These opposite conclusions however are consistent with the known suggested modes of replication of these two viruses. SV40 employs a closed circular intermediate where newly synthesized DNA is a shorter than unit length fragment. Failure to separate the two replicating circles results in a catenated dimer. φX174 replicates by a rolling circle mechanism (Dressler, 1970) where the growing newly made strands are longer than unit length. Failure to separate the parental or template strand from the newly synthesized progeny strand (replication error) yields a continuous DNA strand that circularizes into a circular dimer. Thus the mode of replication, rolling circle or interlocked closed circular replicative intermediates, determines the type of dimer that is formed by replication errors.

Now let us shift into the third topic of this presentation. In all the work I have described so far SV40 DNA has been isolated by the 1M NaCl-SDS selective extraction procedure (Hirt, 1967). Clearly this procedure removes all or most of the proteins from this DNA. The mature viral DNA and SV40 replicative intermediates might be expected to be associated with cellular and viral proteins. To get at the more natural template for SV40 DNA replication we have employed a more gentle cell lysis procedure first described by Green, Miller and Hendler (1971) for polyoma and White and Eason (1971) for SV40. Instead of lysing infected cells with SDS and treatment with 1M NaCl, one lyses the cells with triton x-100 and employs 0.2M NaCl concentration. Large molecular weight cellular DNA is selectively removed by a fast centrifugation step (Green et al., 1971; White and Eason, 1971).

In the first experiment we labeled SV40 infected cells with ^{3}H-thymidine for 1 hour. One half the cultures were lysed by the 1M NaCl-SDS procedure (Hirt, 1967) while the other half were lysed by the triton x-100-0.2M NaCl procedure (Green et al., 1971). Each sample was then sedimented through a linear 5-20% neural sucrose gradient (0.2M NaCl) at 40,000 rpm for 2 hours. Fig. 10 shows both these gradients. The 1M NaCl-SDS lyses procedure yields 21S viral DNA that cosediments with a mature viral DNA marker. The 0.2M NaCl-triton x-100 procedure yielded a heterogeneous peak of viral DNA sedimenting about 44S and no viral DNA in the 21S position of the gradient. The recovery of ^{3}H-viral DNA in the 0.2M NaCl-triton x-100 procedure is about 50-80% of that observed with 1.0M NaCl-SDS. These results are identical to those reported by Green et al. (1971) and White and Eason (1971).

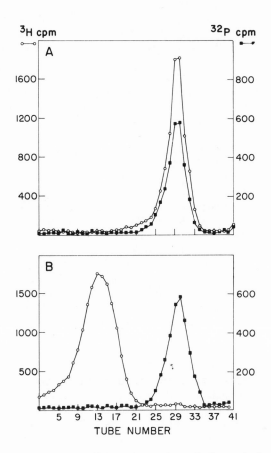

Fig. 10. Sedimentation of SV40 DNA isolated from infected AGMK cells by the 1M NaCl-SDS lysis procedure (Hirt, 1967) (top graph) or the 0.2M NaCl-Triton x-100 lysis procedure (Green et al., 1971) (bottom graph).

SV40 DNA (1 hour labeling with ³H-thymidine) was extracted from infected AGMK cells by the procedure of Hirt (1967) (top) or that of Green et al. (1971) (bottom). The DNA was sedimented through a neutral sucrose gradient at 40,000 rpm for 2 hours in an SW 50.1 rotor.

o-o ³H-cpm
□-□ ³²P-cpm from a labeled 21S sedimentation marker.

Employing those techniques discussed earlier (EtBr-CsCl gradients and neutral sucrose gradients) we could show that SV40 replicative intermediates were enriched at the leading edge of the 44S peak (Mayer, 1972). A DNA negative mutant of SV40 (tsA-7) did not yield a 44S DNA-protein complex at the nonpermissive temperature. This shows that 100% of the DNA in this 44S complex is viral DNA.

We next asked the question, how much protein is associated with SV40 DNA in the 44S complex and what is the origin (host or viral) of this protein? To determine this we isolated the 44S complex and crosslinked the DNA and protein by formaldehyde fixation (Hancock, 1970). The resultant DNA-protein complex was run in CsCl gradients to determine the DNA to protein ratio. Fig. 11 shows the results of such an experiment. Two peaks of ^{3}H-thymidine were observed. The major peak (95-99.5%) had a density of 1.47 gms/cc while the minor component (0.5-5%) had a density of 1.67 gms/cc. The DNA to protein ratios were calculated with the aid of the following values: 1) SV40 DNA has a buoyant density in CsCl of 1.700 gms/cc, 2) The density in CsCl of proteins associated with chromatin is 1.275 gms/cc (Hancock, unpublished) and SV40 empty shells band at 1.300 gms/cc in CsCl (Koch et al., 1967; Ozer, 1972). Both these numbers were employed as protein standards to give the limits of DNA to protein ratios found in the 44S complex. The calculations (Brutlag, Schlehuber and Bonner, 1969) indicate that the protein to DNA ratio of the 1.47 gms/cc component is between 0.82 and 0.95 while that of the 1.67 component is 0.04-0.06.

When uninfected cellular ^{14}C-labeled proteins were mixed with ^{3}H-labeled 44S complex from infected cells less than 1% of the uninfected cell protein bound nonspecifically to the 44S, 1.47 gms/cc density component. White and Eason (1971) as well as Green et al., (1971) have also concluded that the proteins found in the 44S complex were not derived by nonspecific attachment after cell lysis.

We can now calculate that the molecular weight of the 44S (1.47 gms/cc) complex is about 5.5-5.9 x 10^6 daltons, the protein contributing about 2.5-2.9 x 10^6 daltons. If the 44S complex had a superhelical configuration (the DNA sediments at 21S after dissociation with SDS) a 5.5-5.9 x 10^6 dalton complex should sediment at 27-29S (Hudson, Clayton and Vinograd, 1968). Since it sediments at 44S the complex must be folded more compactly than free closed circular and supercoiled DNA.

In summary, we have been able to isolate the replicative intermediate of SV40 DNA. It contains closed circular template strands and linear fragments of newly made DNA hydrogen bonded to the template strands. Initiation requires at least one viral protein (Tegtmeyer, 1972; Mayer, 1972) and proceeds from a unique

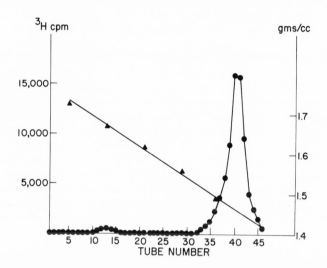

Fig. 11. A CsCl equilibrium gradient of the SV40 nucleoprotein
complex fixed prior to centrifugation with formaldehyde.

The 44S nucleoprotein complex was isolated as in Fig. 10.
After dialysis of the nucleoprotein it was reacted with 6% formaldehyde at 41°C for 24 hours. The reaction mixture was dialysed
and banded to equilibrium in CsCl (1.57 gms/cc) by centrifuging the
sample at 38,000 rpm for 48 hours in a 5 pinco SW 50.1 rotor.

o-o   ³H-cpm
Δ-Δ   density, gms/cc

origin (Nathans and Danna, 1972; Fareed and Salzman, 1972; Thoren et al., 1972). Replication is bidirectional and probably involves the discontinuous synthesis of 4S fragments which are then added to the growing linear fragment (Nathans and Danna, 1972; Fareed and Salzman, 1972). A break in the phosphosiester bond of the unreplicated region of the intermediate is required several times throughout replication to allow chain propagation and template strand separation to proceed (Sebring et al., 1971; Mayer and Levine, 1972). Almost completed replicative forms are isolated in greater quantity than newly started replicative intermediates. At the end of replication the two interlocked replicated circles must separate at a unique point (terminus) to yield two monomers. In 1 out of 75 rounds of replication this termination process fails, probably yielding a catenated dimer. Separation of the two progeny replicative intermediates requires protein synthesis to occur efficiently. Catenated dimers once formed are unstable and decay probably to circular dimers or monomers.

SV40 DNA isolated from infected cells by the triton x-100-0.2M NaCl lysis procedure is associated with about 2.5-2.9 x 10^6 daltons worth of protein. The replicative intermediates of SV40 DNA sediment slightly faster than this nucleoprotein-complex (>44S) and probably contain similar proteins in addition to those proteins required for its replication. It should be most useful to further characterize the replicative intermediates with their associated proteins.

ACKNOWLEDGMENTS

We thank S. Arwood and A. K. Teresky for their excellent technical assistance. The advice of Dr. N. Sueoka is gratefully acknowledged in helping to formulate the equations for analysis of circular and catenated dimers. This research was supported by grants CA 11049-04 and CA 12068-02 from the National Cancer Institute and American Cancer Society grant E 591. We thank the Whitehall Foundation for the generous use of their equipment.

REFERENCES

Aloni, Y., E. Winocour, L. Sachs and J. Torten. 1969. J. Mol. Biol. 44:333.

Bauer, W. and J. Vinograd. 1968. J. Mol. Biol. 33:141.

Benbow, R. M., M. Eisenberg and R. L. Sinsheimer. 1972. Nat. New Biol. 237:141.

Bourgaux, P. and D. Bourgaux-Ramoisy. 1971. J. Mol. Biol. 62:513.

Bourgaux, P., D. Bourgaux-Ramoisy and P. Seiler. 1971. J. Mol. Biol. 59:195.

Bourgaux, P. and D. Bourgaux-Ramoisy. 1972. J. Mol. Biol. 70:399.

Brutlag, D., C. Schlehuber and J. Bonner. 1969. Biochemistry. 8:3214.

Cairns, J. 1963. J. Mol. Biol. 6:208.

Crawford, L. V. and P. H. Black. 1964. Virology. 24:388.

Crawford, L. V. 1969. Adv. in Virus Research. 14:89.

Dressler, D. 1970. Proc. Nat. Acad. Sci., U.S.A. 67:1934.

Estes, M. K., E. S. Huang and J. S. Pagano. 1971. J. Virol. 7:635.

Fareed, G. C. and N. P. Salzman. 1972a. Fed. Proc. 31:442.

Fareed, G. C. and N. P. Salzman. 1972b. Nat. New Biol. 238:274.

Girard, M., L. Marty and F. Suarez. 1970. Biochem. Biophys. Res. Commun. 40:97.

Green, M. H., H. J. Miller and S. Hendler. 1971. Proc. Nat. Acad. Sci., U.S.A. 68:1032.

Hancock, R. 1970. J. Mol. Biol. 48:357.

Hancock, R. Unpublished data.

Hirai, K., J. Lehman and V. Defendi. 1971. J. Virol. 8:708.

Hirt, B. 1967. J. Mol. Biol. 26:365.

Hirt, B. 1969. J. Mol. Biol. 40:141.

Hudson, B., D. A. Clayton and J. Vinograd. 1968. Cold Spr. Harb. Symp. Quant. Biol. 33:435.

Inman, R. B. and M. Schnös. 1971. J. Mol. Biol. 56:319.

Jaenisch, R. and A. J. Levine. 1971a. Virology. 44:481.

Jaenisch, R. and A. J. Levine. 1971b. J. Mol. Biol. 61:735.

Jaenisch, R. and A. J. Levine. 1972. Virology. 48:373.

Jaenisch, R. and A. J. Levine. 1973. J. Mol. Biol. 73:199.

Jaenisch, R., A. J. Mayer and A. Levine. 1971. Nat. New Biol. 233:72.

Koch, M. A., H. Becht, F. A. Anderer, H. D. Schlumberger and H. Frank. 1967. Virology. 32:503.

Lavi, S. and E. Winocour. 1972. J. Virol. 9:309.

Levine, A. J., H. S. Kang and F. E. Billheimer. 1970. J. Mol. Biol. 50:549.

Mayer, A. J. 1972. Ph. D. Thesis, Princeton University.

Mayer, A. J. and A. J. Levine. 1972. Virology. 50:328.

Nathans, D. and K. Danna. 1972. Nat. New Biol. 236:200.

Ozer, H. L. 1972. J. Virol. 9:41.

Radloff, R., W. Bauer and J. Vinograd. 1967. Proc. Nat. Acad. Sci., U.S.A. 57:1514.

Sambrook, J., H. Westpal, P. R. Srinivasan and R. Dulbecco. 1968. Proc. Nat. Acad. Sci., U.S.A. 60:1288.

Schnös, M. and R. B. Inman. 1970. J. Mol. Biol. 51:161.

Sebring, E. D., T. J. Kelly, Jr., M. M. Thoren and N. P. Salzman. 1971. J. Virol. 8:478.

Sinsheimer, R. L., R. Knippers and T. Komano. 1968. Cold Spr. Harb. Symp. Quant. Biol. 33:443.

Tai, H. T., C. A. Smith, P. A. Sharp and J. Vinograd. 1972. J. Virol. 9:317.

Tegtmeyer, P. 1972. J. Virol. 10:591.

Thoren, M. M., E. D. Sebring and N. P. Salzman. 1972. J. Virol. 10:462.

Tomizawa, J. and T. Ogawa. 1968. Cold Spr. Harb. Symp. Quant. Biol. 33:533.

Vinograd, J. and J. Lebowitz. 1966. J. Gen. Physiol. 49:103.

White, M. and R. Eason. 1971. J. Virol. 8:363.

A NOVEL FORM OF DNA APPEARING AFTER INFECTION OF PERMISSIVE CELLS BY SV40 CONTAINING BOTH VIRAL AND REITERATED HOST SEQUENCES

W. Waldeck, K. Kammer, G. Sauer

Institut für Virusforschung

Deutsches Krebsforschungszentrum

Heidelberg, West Germany

INTRODUCTION

Covalently linked SV40 and cellular RNA sequences have been described in productively infected cells (Jaenisch, 1972). In agreement with this observation integration of SV40 DNA into the host DNA of productively infected cells has been reported (Harai and Defendi, 1972). It was furthermore shown that integration of host DNA sequences into the mature SV40 superhelix occurs under conditions of high multiplicities of infection when virus pools are used which were serially passaged undiluted (Lavi and Winocour, 1972). This phenomenon could be explained by assuming integration of viral DNA into the DNA of permissive cells followed by excision of viral and of some host DNA sequences and circularization of these molecules (Lavi and Winocour, 1972; Tai et al., 1972).

We have employed similar conditions of infection, i.e., high multiplicities of serially passaged SV40, and we have investigated the host DNA for the presence of integrated SV40 DNA sequences. We present evidence in this report for the existence in productively infected cells of a new form of DNA consisting of viral nucleotide sequences covalently linked to particular reiterated cellular DNA sequences. The isolation and purification of this DNA has made it possible to investigate tumor virus genes in an integrated state.

MATERIALS AND METHODS

Infection and Radioactive Labeling of Cells

CV-1 cells were grown as previously described (Sauer, 1971) and were infected with SV40 strain Rh 911 (Girardi, 1965). Cells were infected with either diluted or undiluted virus pools. SV40 was used at a multiplicity of 0.5 plaque forming units (PFU)/cell. The virus had been plaque purified three times and was propagated by infecting CV-1 cultures at a multiplicity of 0.1 PFU/cell and harvesting the virus 5 days after infection. Second, infection of CV-1 cultures with undiluted SV40 was carried out by using SV40 preparations which had been passaged 3 times undiluted on CV-1 cells. Owing to the loss of infectivity after undiluted passaging of SV40 (Uchida, Watanabe and Kato, 1966), this mode of infection with undiluted virus is equivalent to a multiplicity of 0.06 PFU/cell, although the actual concentration of physical particles is increased by 10^3-fold over the concentration of physical particles in wild type SV40 pools as revealed by electron microscopic particle counts. The cell cultures were exposed after the period of virus adsorption (1h) to ^{14}C-thymidine (0.20 μCi/ml medium) throughout the period of infection, which was terminated at 48 h after adsorption. Uninfected confluent CV-1 cells were labeled with ^{3}H-thymidine (5 μCi/ml) for 48 h.

Isolation of DNA

The DNA from infected cultures was isolated by selective extraction (Hirt, 1967). The NaCl-dodecyl-SO_4-supernatant was subjected to two cycles of phenol extraction (66% phenol, freshly distilled twice, in 0.05 M Tris, 0.2 M Na-trichlor-acetate, pH 8.15). After removal of the phenol the preparation was treated for 1 h at 37°C with 10 μg/ml of pancreatic RNase previously heated for 15 min. at 90°C, re-extracted with phenol, precipitated with 2 x the volume of undenatured ethanol, and then resuspended in 0.1 x SSC. The DNA from uninfected CV-1 cells was extracted according to the method described by Marmur (1961).

Equilibrium Centrifugation

The conditions for CsCl-EtBr equilibrium centrifugation (Radloff, Bauer and Vinograd, 1967) were as follows: Spinco rotor 50 Ti, 40,000 r.p.m., 64 h, 20°C, 1.56 g $CsCl/cm^3$, 150 μg EtBr/ml in 0.02 M Tris, 0.02 M EDTA, pH 7.0. The EtBr was removed from the DNA preparation with propanol-2. After centrifugation the tubes were punctured and 4 drop fractions were collected. 0.02 ml aliquots were pipetted on Whatman glass filter pads, dried and counted in a Packard liquid scintillation counter. The fractions comprising the denser and the lighter band in CsCl-EtBr gradients were pooled separately.

Velocity Centrifugation
Alkaline velocity sedimentation of DNA took place in preformed
CsCl gradients (SW 50.1, 18 h, 36,000 r.p.m., initial density 1.56
g/cm^3, pH 12.9). The DNA samples were layered on top of the preformed gradients and centrifuged for 150 min. at 40,000 r.p.m. The
density was determined by refractive index measurement according to
the values indicated in the Handbook of Biochemistry with Selected
Data for Molecular Biology (Sober, 1970).

DNA:DNA Hybridization
DNA·DNA hybridization was carried out by a modification of the
method previously described (Lavi and Winocour, 1972). The labeled
DNA samples were sonicated with a Branson sonifier (step 8, 10 sec)
in 0.1 x SSC, heated to 100°C for 10 min, rapidly cooled in an ice
bath and then adjusted to 2 x SSC and 50% formamide to give a
reaction volume of 2 ml. Hybridization took place in Tricarb vials
at 37°C for 3 days. The hybridization reaction was terminated by
rinsing the filters in 5-8 beakers, each containing 250 ml 2 x SSC.
The filters were washed by filtration (3 x 15 ml buffer through
Millipore filters).

RESULTS

Accumulation of 30 S DNA After Infection
Permissive CV-1 cells were productively infected with diluted
plaque purified SV40 or with undiluted SV40 which had been serially
passaged undiluted. The infected cells were labeled with ^{14}C-
thymidine throughout the period of infection for 48 h and then the
viral DNA was selectively isolated and separated from the macromolecular host cell DNA (Hirt, 1967). Infection with diluted SV40
reveals in neutral CsCl-EtBr gradients a prominent peak in the
position of the superhelical SV40 DNA component I and a smaller peak
at the lighter density (Fig. 1a). However, upon infection with
undiluted SV40 which had been serially passaged undiluted, the
pattern is completely reversed. Now the peak in the lighter density
is much larger than the peak which comprises the superhelical
molecules (Fig. 1b). Since infection with undiluted SV40 passaged
serially undiluted appears to increase the homology between cellular
and circular SV40 DNA (Lavi and Winocour, 1972; Tai et al., 1972),
the possibility arises that this is accomplished by the ability of
these particular virus stocks to "excise" large cellular DNA to
smaller fragments, containing both cellular and viral nucelotide
sequences. Such "excision" would render host DNA non-precipitable
during selective extraction thus leading to the formation of a
pronounced peak in CsCl-EtBr equilibrium gradients at the lighter
position where one expects relaxed circular and linear duplex DNA
molecules to band (Fig. 1b).

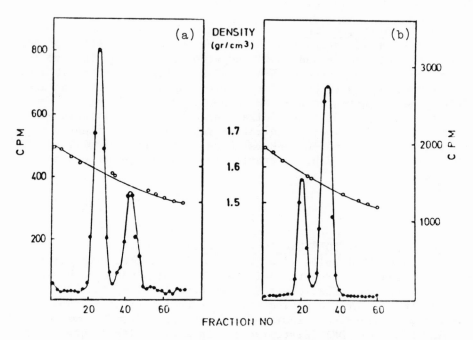

Fig. 1.   CsCl-EtBr equilibrium centrifugation of DNA selectively
isolated from cells infected with diluted and undiluted SV40.   Confluent CV-1 cells were infected either (a) with wild type SV40 at
0.5 p.f.u./cell or (b) with undiluted SV40 which had been passaged
three times undiluted on CV-1 cells (0.06 p.f.u./cell).   Non-adsorbed
virus was removed by washing the cultures with Hank's balanced
salt solution (BSS) after 1 h and the cultures were labeled with
[14]C-thymidine (0.20 μCi/ml medium) for 47 h.   Then the DNA was
selectively isolated (Hirt, 1967) and centrifuged to equilibrium.

To further characterize these molecules, the DNA selectively isolated from CV-1 cells infected with diluted SV40 (Fig. 2a) and undiluted SV40 (Fig. 2b) was subjected to a velocity sedimentation through preformed alkaline CsCl gradients. In Fig. 2b the sedimentation patterns of SV40 DNA component I and component II which originated from component I are superimposed for comparison. It can be seen that the DNA obtained from the NaCl-dodecyl-SO_4-supernatant in both cases contains (Figs. 2a, b) DNA cosedimenting with marker component I DNA and DNA which sediments more slowly than component I, although faster than component II. It may be concluded from a comparison of these data that infection of cells with undiluted SV40 facilitates the yield of those DNA species which sediment more slowly than component I but faster than relaxed component II. Using the sedimentation constant of linear single stranded SV40 DNA component II in alkaline gradients (16S) as a reference, we calculate that the bulk of this denatured DNA sediments at 30 S and has a molecular weight of about 10^7 daltons. This corresponds to 6 to 7 times the length of single stranded 16 S SV40 molecules. Hereafter this material will be termed "30 S DNA."

Purification of 30 S DNA by Velocity Centrifugation

The 30 S DNA was further analyzed by fractionating a neutral CsCl-EtBr equilibrium gradient (Fig. 1b) into four equal portions and sedimenting these fractions through preformed alkaline CsCl gradients (Fig. 3). The denser band of the equilibrium gradient contains almost exclusively superhelical form I DNA and some faster sedimenting oligomeric DNA (Jaenisch and Levine, 1971; Rush, Eason and Vinograd, 1971). The lighter band, however, does not contain, as one might expect, mainly relaxed form SV40 DNA component II; rather it consists of the material termed "30 S DNA." The only known configuration of SV40 DNA that has been demonstrated to both band in the light band of CsCl-EtBr gradients and also sediment in alkaline pH faster than component II are the covalently closed parental template DNA strands released from replicative intermediates (Sebring et al., 1971; Jaenisch, Mayer and Levine, 1971). These molecules have a sedimentation coefficient of 26 S at neutral pH (Sebring et al., 1971; Levine, Kang and Billheimer, 1970). However, such replicative intermediates should not contribute to a marked extent to the 30 S DNA, because replicative forms of SV40 DNA comprise only a rather small portion of the selectively extracted DNA (Levine et al., 1970; Hirt, 1969) and examination of this material in the electron microscope revealed predominantly linear structures. Only rarely was it possible to find circular molecules (unpublished). Replicative intermediates are normally revealed (because of their relatively short half-life) by short-pulse-labeling, while in the experiments described here long term labeling has been employed. It appears, therefore, that the bulk of the 30 S DNA consists either 1) of hitherto undescribed SV40 molecules or 2) of DNA molecules that are of cellular origin, or

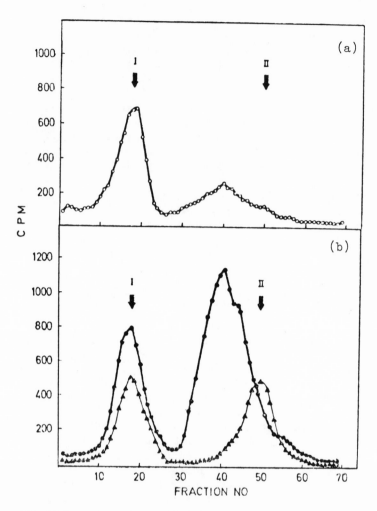

Fig. 2. Alkaline velocity sedimentation of DNA selectively isolated
from cells infected with diluted SV40 and with undiluted SV40 (which
had been serially passaged undiluted). The samples (0.35 ml) had
been exposed to 0.05 ml of 1 N NaOH for 10 min prior to centrifugation (a) DNA from the preparation used in Fig. 1a. (b) DNA from
the preparation used in Fig. 1b. Sedimentation profile of ^{3}H-SV40
DNA component I and II (▲) is superimposed. The marker ^{3}H-DNA was
purified by CsCl-EtBr equilibrium centrifugation and alkaline
velocity sedimentation of the denser band. The peak fraction of the
fast sedimenting peak in alkaline represents SV40 DNA component I.
Component II originated from component I by storage at 4°C of the
DNA for 2 months.

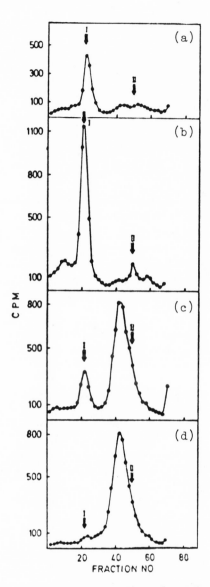

Fig. 3. Alkaline velocity sedimentation of various DNA fractions obtained after CsCl-EtBr equilibrium centrifugation. The two bands depicted in Fig. 1b were divided into four equal portions and after removal of the dye and dialysis against 0.1 x SSC (4000 x the volume) the DNA was sedimented through alkaline CsCl gradients. (a) dense portion of denser band, (b) light portion of denser band, (c) denser portion of the lighter band, (d) light portion of the lighter band. The arrows indicate the peak positions of SV40 DNA marker components I and II.

3) of molecules comprising both viral and cellular DNA sequences.

Homology of 30 S DNA with both Cellular and Viral DNA

To test whether 30 S DNA is homologous to DNA of cellular origin, DNA·DNA-hybridization tests were performed (Table 1). The 30 S DNA preparation was divided into two portions which were hybridized, after denaturation, to both CV-1 and SV40 DNA immobilized

TABLE 1. Homology of 30 S DNA with both cellular and SV40 DNA

DNA in solution	Input per reaction (c.p.m.)*	DNA on filter[+] (5 µg)	% of input bound to filter
30 S DNA	7700	CV-1	17.2
	7700	SV40	12.9
SV40	8755	SV40	80.50
	5940	no DNA	0.70
	8032	T4	1.05
	8910	CV-1	1.32
CV-1	8011	CV-1	9.0
	8825	T4	0.25
Light band (Fig. 1b)	13750	Light band[‡]	21.02
	13750	T4	0.31

*Specific activity of the different DNA samples was as follows: 30 S DNA (fractions 35 to 38, Fig. 2b): 3650 c.p.m./µg; SV40 DNA purified similarly as described in the legend to Fig. 2a had a specific activity of 3700 c.p.m./µg. ^{3}H-CV-1 DNA: 2300 c.p.m./µg. Preparation of the DNA obtained from the light band of the CsCl-EtBr equilibrium gradient (spec. act. 3650 c.p.m./µg) was as described (Fig. 1b).

[+]Unlabeled DNA samples were sonicated and denatured in 0.1 x SSC as described in Materials and Methods. After loading the filters (Millipore HAWP02400) with DNA they were incubated at 80°C for 4 h to immobilize the DNA.

[‡]Unlabeled DNA from the light band of an CsCl-EtBr equilibrium gradient was prepared as described (Fig. 1b) except that isotopic labeling was omitted.

on filters.   It can be seen that the material displays considerable
homology to both host and viral DNA.   The homology to host cell DNA
(17.2%) is somewhat greater than the homology to SV40 DNA (12.9%).
Since contamination of this material with free SV40 DNA component II
is unlikely and since these DNA molecules were sedimenting as a sharp
peak through preformed alkaline gradients, these hybridization tests
suggest covalent linkage of SV40 DNA to host cell DNA.   The conclusion that this 30 S DNA is a new kind of linear duplex DNA consisting
of both host cell and viral DNA sequences covalently linked to each
other is substantiated by a number of other experiments.

It may be seen from a comparison of various controls included
in Table 1 that SV40 DNA self-anneals most efficiently.   Up to 80%
of the labeled denatured viral DNA used in the reaction can be bound
to the filter, while blank filters (containing T4-phage DNA or host
cell DNA) bind very little viral DNA, the background being close to
one percent.   However, self-annealing of host cell DNA is rather
inefficient (9%).   Self-annealing of the DNA which is derived from
the light band shown in Fig. 1b (30 S DNA) to unlabeled DNA of the
same kind is much more efficient (21%) than self-annealing of host
cell DNA.   Thus, this particular type of DNA has properties other
than pure viral or pure host cell DNA.   While host DNA preparations
display an inefficient self-annealing capacity, the 30 S DNA anneals
very efficiently with host DNA (17.2%) as shown in the first line of
Table 1.   This phenomenon can be explained by the presence of particular host cell DNA sequences consisting of reiterated nucleotide
sequences (Britten and Kohne, 1968) which occur frequently and which
are covalently linked to the viral DNA.   These properties allow this
host DNA to anneal exceptionally well to the bulk of host DNA
immobilized.

This conclusion, together with the conclusion that host and
viral DNA sequences are covalently linked to each other, gains
much support from an entirely different experimental approach, the
results of which are summarized in Table 2.   If 30 S DNA preparations contained large amounts of viral-host DNA molecules, it should
be possible, by selective hybridization, to isolate such molecules,
provided they were not artificially degraded to smaller fragments
prior to hybridization assay so as to permit maintenance of linkage
between both viral and cellular DNA sequences.   Therefore, we have
hybridized (without previous ultrasonic treatment) the DNA preparation which was used in Fig. 1b and which contains various forms of
DNA, including superhelical viral DNA with cellular or viral DNA
immobilized on filters.   After completion of the hybridization
reaction the filters were gently washed, since vigorous washing
procedures (sucking buffer through the filters on a millipore
filtration apparatus as described in Table 1) removes large amounts
of the hybridized undegraded long DNA molecules (this is shown in
a control experiment described in the legend to Table 2).   Gentle

TABLE 2. Rehybridization of DNA to either cellular or SV40 DNA which was hybridized to either SV40 or cellular DNA

DNA in solution*	Input per reaction (c.p.m.)	DNA on filter (5 μg)		c.p.m. remaining on filter after elution†	c.p.m. eluted from filters‡	% of input bound to filter
DNA used in Fig. 1b	43000+	(a)	CV-1	115	4120	-
	29000+	(b)	SV40	92	4533	-
SV40 + 5 μg unlabeled CV-1 DNA	16200+		CV-1	55	98	-
DNA eluted from CV-1 filter (a)	2010		CV-1	-	-	21
	2021		SV40	-	-	12.2
DNA eluted from SV40 filter (b)	2210		CV-1	-	-	18
	2250		SV40	-	-	22

*DNA-DNA hybridization was carried out as described in Table 1 except that DNA samples in solution used in selective hybridization experiments (+) were denatured by heating to 90°C for 5 min to minimize degradation. Degradation by sonication was omitted.

†After elution the filters were washed in 100 ml of 2 x SSC, dried and the radioactivity was determined.

‡After completion of the hybridization reaction a modified washing procedure was employed in the case of selective hybridization (+). Sucking buffer through the filters during the washing procedure as described in Table 1 largely removes the hybridized DNA molecules which were not previously degraded by sonic treatment. DNA used in Fig. 1b (input 15640 c.p.m.) was annealed to filters with 5 μg of immobilized CV-1 DNA similar to the reaction described in the first line of Table 2. After vigorous washing (as described in Materials and Methods) a comparatively small amount of DNA could be eluted from the filters (490 c.p.m.) while 104 c.p.m. remained bound to the filter after elution. Therefore filters were rinsed in 3 x 100 ml of 2 x SSC and transferred into new Tricarb vials containing 1 ml of the elution buffer described by Lavi and Winocour (1972). The eluted DNA was diluted to give a concentration of 50% formamide and rehybridized after sonication as described in the legend to Table 1.

washing of the filters proves to be sufficiently efficient in removing non-specifically bound heterologous DNA sequences as evidenced
by the control experiments in Table 2. Less than one percent of
heterologous DNA sequences (SV40 DNA that was subjected to sonic
treatment and then mixed with unlabeled CV-1 DNA) remained attached
to the filters (with CV-1 DNA) after gentle washing and could be
eluted. On the other hand, large amounts of homologous DNA remain
bound after gentle washing to the filters as evidenced by the
data in the first two lines in Table 2.

The DNA specifically hybridized was subjected to sonic treatment after elution and reannealed to both CV-1 and SV40 DNA
immobilized on the filters. Again the results clearly show that
DNA which was hybridized due to its specificity to host cell DNA
(indicated by (a) in Table 2) also contains large amounts of nucleotide sequences which are homologous to viral DNA. The same is true
for the reciprocal experiment indicated by (b) in Table 2.

DISCUSSION

The existence of linear duplex DNA molecules in productively
infected cells which contain both SV40 and host DNA sequences covalently linked to each other and which have an average of 6 to
7 times the length of linear SV40 DNA molecules has been described
in this paper. The appearance of such molecules can be enhanced by
infection of cells with virus serially passaged undiluted. The
mechanism accounting for this phenomenon is unknown. However, the
suggestions made by others (Lavi and Winocour, 1972; Tai et al.,
1972) in the case of enhancement of homology between circular viral
DNA and host DNA may also apply to the situation described herein
by us.

Our observations suggest integration of the tumor virus DNA
into the host cell DNA during the course of a productive cycle of
infection. The covalent linkage of viral to host DNA may result
from a recombination event (e.g., breakage and reunion). A second
rather remote interpretation of these data may be that in an as yet
unknown way host DNA sequences are de novo synthesized onto linear
SV40 strands without involving breakage and reunion.

Experiments are now underway which should determine if linkage
of pre-existing cellular DNA to viral DNA can occur and if viral DNA
can also be covalently bound to the high molecular weight DNA of
permissive cells (i.e., DNA too large to be isolated by the selective
extraction used here). The significance of integration of the viral
DNA into the DNA of transformed cells supposedly is that it allows
perpetuation of the transformed state from cell to cell by continuous
synthesis of particular viral gene products from the integrated

viral genes (Dulbecco and Eckhart, 1970). Whether integration of viral DNA into the host DNA of productively infected cells proves to be a necessary event is not known. Integration may have some bearing on the regulation of transcription of viral genes (Sauer, 1971) or the integration step may be important in as yet unknown way for the replication of viral DNA.

We have found that reiterated host cell DNA sequences (Britten and Kohne, 1968) are covalently linked with the viral DNA. Our data suggest that linkage of viral DNA to host DNA does not occur at random. Experiments are now under way to determine the nature of the reiterated bases. This may help to answer the important question of the site of integration of tumor virus genes. The investigation of integrated tumor virus DNA has been severely hampered thus far by the small number of viral genome equivalents present in the DNA of transformed cells. The study described here permits, owing to the apparently large number of viral DNA sequences involved, a detailed biophysical analysis of tumor virus genes in an integrated state.

ACKNOWLEDGMENTS

We thank Drs. Allan Fried and Alan Kolber for discussions and Miss Dorothea Beutelman for technical assistance. This work was supported by the Deutsche Forschungsgemeinschaft.

REFERENCES

Britten, R. J. and D. E. Kohne. 1968. Science. 161:529.

Dulbecco, R. and W. Eckhart. 1970. Proc. Nat. Acad. Sci., 67:1775.

Girardi, A. J. 1965. Proc. Nat. Acad. Sci., U.S.A. 54:445.

Harai, K. and V. Defendi. 1972. J. Virol. 9:705.

Hirt, B. 1967. J. Mol. Biol. 26:365.

Hirt, B. 1969. J. Mol. Biol. 40:141.

Jaenisch, R. and A. Levine. 1971. Virology. 44:480.

Jaenisch, R., A. Mayer and A. Levine. 1971. Nature. 233:72.

Jaenisch, R. 1972. Nature. 235:46.

Lavi, S. and E. Winocour. 1972. J. Virol. 9:309.

Levine, A. J., H. S. Kang and F. E. Billheimer. 1970. J. Mol.
    Biol. 50:549.

Marmur, J. 1961. J. Mol. Biol. 3:208.

Radloff, R., W. R. Bauer and J. Vinograd. 1967. Proc. Nat. Acad.
    Sci., U.S.A. 57:1514.

Rush, M. G., R. Eason and J. Vinograd. Biochim. Biophys. Acta.
    228:585.

Sauer, G. 1971. Nature. 231:135.

Sebring, E. D., T. J. Kelly, M. M. Thoren and N. P. Salzman. 1971.
    J. Virol. 8:478.

Sober, H. A., Ed. 1970. Handbook of Biochemistry with Selected
    Data for Molecular Biology. Verlag Chemie, Weinheim.

Tai, H. T., C. A. Smith, P. A. Sharp and J. Vinograd. 1972.
    J. Virol. 9:317.

Uchida, S., S. Watanabe and M. Kato. 1966. Virology. 28:135.

MEMBRANE GROWTH AND CELL DIVISION IN E. COLI

Adam Kepes and Françoise Autissier

Laboratoire des Biomembranes, Institut de Biologie
Moléculaire
75005 Paris, France

Our work is concerned with the topology of membrane growth in E. coli.

The question arose from the replicon theory of Jacob, Brenner and Cuzin for prokaryotes, having no mitotic apparatus, what mechanical device could pull apart the two DNA complements after DNA replication. The only anatomically differentiated mechanical device which could perform the task of pulling apart the DNA in bacteria is the membrane. To do this membranes should grow locally somewhere between the insertion points of the 2 DNA complements in order to facilitate segregation, unless some tangential movement directed by an elaborate (contractile ?) mechanism is at work. Many previous reports concluded that membrane was not growing locally but in some diffused manner. Our experiments led to the opposite conclusion. The method which we have used was to look for inducible permeases which are deposited in the cell membrane, after further growth of the bacteria without inducer, generating new membrane devoid of this marker. After cell division the marker should be divided between the daughter cells, and assuming several reasonable simple patterns of localized growth we should find a limited number of cells which have inherited parental membrane. This number of cells depends on the model devised but at some point the cells should produce only progeny with membrane which was not present in the ancestor. The technical problem was to select from a population heterogeneous with respect to permease those individuals which did or did not contain the permease. After some trial and error we found the simplest to use a penicillin selection technique the principle and the practice of which is absolutely classical in the hands of geneticists. When a carbon source taken up via an inducible permease in a population of bacteria, heterogeneous with respect to permease, those bacteria with no permease will not grow and survive while the permease positives, which grow will be lysed by penicillin. Usually, the substrate is also an inducer, therefore those bacteria which do not grow at first will be induced shortly and they will ultimately lyse. In order to produce quick lysis of growing cells, we exposed the cells to an EDTA treatment for one minute, according to Leive, after which penicillin produced lysis very quickly. The early experiments have been done with lactose permease which we have extensively studied in our laboratory. Cells were induced overnight completely with IPTG. These are wildtype cells which have permease, beta-galactosidase, and transacetylase. After start of the experiment, the cells were transferred to the original medium containing no inducer, grown for 1, 2 or 3 generation times and after every one of these periods an aliquot was submitted to the short EDTA treatment. The cells were transferred to the original medium containing lactose as sole carbon source and penicillin.

Figure 1 shows the results of such an experiment. A non-induced population treated with lactose and penicillin lyses after 30 or 40 minutes because lactose has induced the permease at that time. Bacteria which were fully induced (labeled Go because they have made zero generations since the induction) are lysed immediately. The start of the lysis is immediate and terminated after

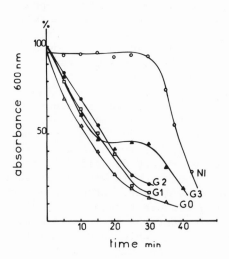

Fig. 1. Time course of lysis of populations of E. coli 3000 after various times of deinduction. Fully induced bacteria were centrifuged and washed with Medium 63, then transferred into Medium 63 glucose Bl at 37°. After 0, 1, 2, 3, generation times (G0, G1, G2, G3), EDTA treatment and lysis in Medium 63 penicillin lactose were performed. Non-induced bacteria were used as control (NI).

15 minutes. The same is true for G1 and G2. At the third generation there is fast but partial lysis, a plateau and a second lysis.
About 50% of the cells survive more than 30 min after 3 generations
and this means that after three generation times half of the cells
no longer contain permease. This observation was repeated for other
markers which allowed us to generalize the phenomenon. Some of
these markers offer special advantages. One of the inducible markers which has a special advantage is the melibiose permease. In
order for melibiose permease to be a faithful marker we must begin
with bacteria which are genetically lac permease minus. These are
grown on melibiose as inducer and carbon source. They are deinduced in a medium where glycerol is the carbon source and for the
test we expose them to melibiose and penicillin. We have the
choice of lysing at 30 degrees where melibiose will de novo induce
melibiose permease in those individuals which don't have the permease, or to lyse at 30°. Melibiose permease is not synthesised
at 39 degrees. Melibiose can still serve as a carbon source when
preformed permease is present but not as inducer for permease.

Indeed when fully induced bacteria exposed to melibiose and
penicillin at 30° they are lysed quickly, and non-induced bacteria
lyse after a lag. In contrast this incubation takes place at 39°
the non-induced bacteria fail to lyse because there is no de novo
induction of melibiose permease.

Figure 2 illustrates the results of de-induction. The fully

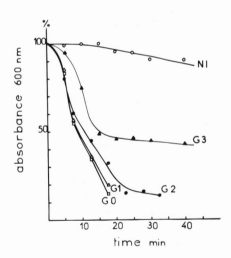

Fig. 2. Time course of lysis at 39° of populations of E. coli
300 P induced and deinduced for Mel-permease, 0, 1, 2, 3 generation
times (G0, G1, G2, G3).

induced bacteria lyse quickly. One and two generations after deinduction the cells still lyse quickly. After three generations
half of the population lyses quickly, the other half survives
indefinitely. This is a special advantage because we have time to
compare the resulting population to the initial population. It
allows for instance to measure the permease content, which is the
basis for the selection, by comparing TMG uptake in the populations
during the lysis which occurs after de-induction. On figure 3
which represents the lysis of the 3-generation-old deinduced bacteria (about 50% lysis). TMG uptake was plotted per unit volume of
suspension taking for 100% the uptake before the start of the
lysis. The uptake is reduced to 5% during the fast step of lysis.
Therefore the 50% survivors contain very little permease. Together
with the test of thermal sensitivity, it is clear that permease
served as the selective agent to discriminate against half the
population having the parental membrane. Figure 3 also shows the
specific activity of β-galactosidase in the same experiment. (Melibiose also induced β-galactosidase). We find that β-galactosidase
is distributed evenly in the 2 parts of the population.

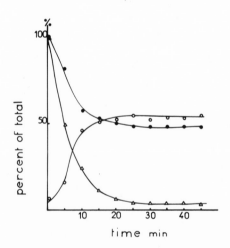

Fig. 3. Fate of Mel-permease and β-galactosidase during segregation. During the lysis at 39° of a 3 generation times deinduced
population, besides absorbance readings at 600 nm (•-•), samples
were removed and treated with penicillinase (10 μg/ml). After centrifugation, Mel-permease content was measured in the survivors, resuspended in Medium 63 with chloramphenicol, by the initial velocity
of uptake of ^{14}C thiomethyl-β-D-galactosidase at 25°, (Δ-Δ). β-galactosidase was estimated in the supernatant by the rate of O-nitro-
phenyl-β-D-galactoside hydrolysis (0-0). 100% control was measured
after toluene treatment of the original population.

Making a calibration curve with artificial mixtures of induced
and non-induced bacteria submitted to the penicillin test, one can
determine the proportion of induced and non-induced cells in a
mixture. Using this test we were able to explore the time course
of appearance of permease negative bacteria during the deinduction
period (figure 4). These bacteria have been deinduced and they are
growing exponentially as indicated in the upper straight line.
The part of the population which is permease-positive is calculated
from the above calibration curve, i.e., the part which lyses within
the first 30 minutes, increases parallel to the total population
until the second generation. From the second generation the number
of permease-positive bacteria stopped increasing and stayed constant
until the 4th generation time. Beyond this time, permease is diluted so that the test is no longer efficient. Permease-negatives
(the difference between the two curves) appear suddenly at 2 generation times and their rate of increase is so fast, that any explanation besides their production by the permease positives would be
absurd.

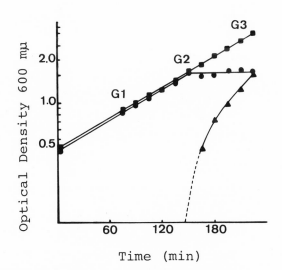

Time (min)

Fig. 4. Segregation kinetics of lactose permease. The deinduction
medium contained 4 g/l glycerol as carbon source. Each aliquot
during deinduction was submitted to the penicillin treatment and
permease-positive and permease-negative population was calculated
from absorbance readings after 30 min according to a calibration
curve made with artificial mixtures. ■ - , absorbance of total population; ▲-▲, absorbance after 30 min penicillin treatment (permease-
negative population); ●-●, absorbance of permease-positive population
calculated by difference.

Figure 5 shows that when deinduction is done in a medium where glucose is the carbon source (in the previous one glycerol was the carbon source) the segregation time is somewhat delayed, it occurs at 2.3 or 2.4 generations after the deinduction. This suggests a correlation between segregation time and number of nuclei. The number of nuclei is higher in bacteria grown on rich medium, glycerol being a rather poor medium and glucose being intermediate. Segregation occurs the earliest with glycerol-grown bacteria, in a middle range with glucose, and one generation later with nutrient broth.

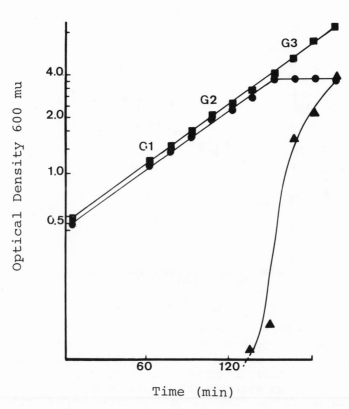

Fig. 5. Segregation kinetics of lactose permease. The experiment was identical to that described in Fig. 4, except that glucose was the carbon source during deinduction.

Figure 6 illustrates a simple model to account for this result. The model is inspired by the separation of nuclei as it has been suggested by Jacob et al., where a median growing zone of membrane growth appears. In the upper row are the cells which are fully induced at the time when deinduction starts. Time runs downwards. New membrane is inserted in the middle and permease is pushed to the poles. The first cell division produces two daughter cells, both of which still have permease. The fate of only one is followed because the other is symmetrical. The cell at the left produces one permease plus and one permease minus cell after two generation times.

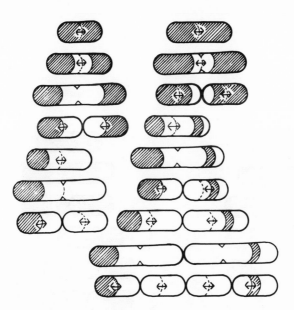

Fig. 6. Distribution of parental membrane among the descendants in bacteria with a single median growing zone. Parental marker (a permease) is the hatched area. Growing zone in dotted line, when distinct from the border between parental and new membrane. Arrows indicate the elongation of new membrane. Left, a cell deinduced just after the last cell division. Right, deinduction of a cell midway between two cell divisions. If a is the time elapsed since the last cell division, and T the generation time (0 < a < T) the cell will yield four positive and four negative segregants after a time of 2T + a. All cell clones will include 50% negative segregants after 3T.

Since bacteria are not synchronized and the cells have various
sizes at the time of deinduction, at the right is the example of a
bacterium which is midway between two cell divisions. After the
first division, giving two bacteria having permease, only one is
followed. At the second division all descendents still have permease. At the third division two bacteria out of the four represented have no permease. This is the reason that we have only 50%
segregation after 3 generation times. Of course, this is not the
only model which can explain these facts and figure 7 shows a
model derived from the work of Donachie for his 2 unit cell.

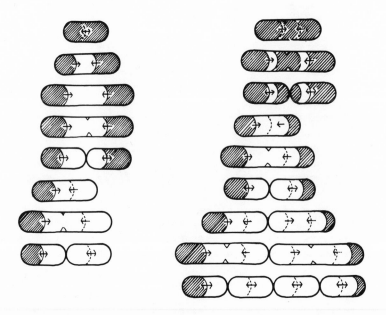

Fig. 7. Distribution of parental membrane among descendants in
bacteria with two unidirectional growing zones moving away from the
equator. (Two unit cell model of Donachie). Symbols as in fig. 6.

The model proposed by this author for the one unit cell does
not fit our data. In the 2 unit cell model a bacterium has two
growing zones which are at constant distance from the 2 poles and
which deposit new membrane only toward the equatorial zone. Therefore the growing zones are themselves moving out from the middle of
the cell and every consequence is exactly the same as in the previous model. The minimal size bacterium gives segregation at 2
generation times and the intermediate size bacterium only at the
third cell division. But there is also a third model which nobody
proposed but which can be imagined. (figure 8). This is a bacterium which has growing zones at two ends depositing new membrane
at the poles. Like before, the bacterium which starts as a newly
divided cell at the time of de-induction produces a segregation
after 2 generation times. The bacterium which is of middle size
segregates at the third generation. The negative bacteria will be
situated at the two ends of the chain, unlike in the other models.

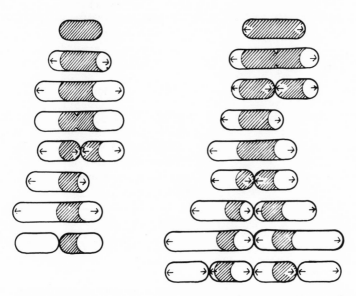

Fig. 8. Distribution of parental membrane among descendants in bacteria with two polar growing zones. Symbols as in Fig. 6 and 7.

Since experiments with chain formation have not been performed, no decisive choice among the different models of membrane growth is possible presently but all results suggest that there are only a small number of growing zones in the membrane. If the membrane grew diffusely the permease would be diluted among all the progeny, and this dilution would continue to the limit of detection. Melvin Cohn and Horibata measured lactose permease with a super-sensitive technique (the maintenance technique). Bacteria which have a very small amount of permease respond to a subliminal dose of inducer and remain induced indefinitely. Using this technique these authors found that permease negative cells did not appear after deinduction until the 7th generation. To reconcile this finding with our results we conclude that the level of detection by the penicillin technique is not as sensitive as by the maintenance technique. Bacteria which have 2 to 5% maximal complement of permease would respond as permease negatives in this test whereas they would respond as permease positives to the hypersensitive test. Therefore with the hypersensitive test every diffusion of a few molecules of permease or a few particles of membrane between the two halves of the original cell will be detected as giving permease positive progeny. It is certain that there are some randomizing processes in the bacterial membrane. We suggest that these randomizing processes are not as fast as the rate of membrane growth. Nevertheless in order to minimize their influence we took care not to introduce any dead time into our experiments.

We have experimented with several inducible markers, by growing the cells with a carbon source and then growing them in the deinducing media containing a different carbon source usually glycerol or glucose, and using the original carbon source during the penicillin step. We found that the segregation after three generation times was common to quite a number of carbon sources. Table I gives the list of those which produced a heterogeneous population and every time this occurred at the third generation. One membrane marker merits special mention. Instead of a transport system, the inducible membrane-bound anaerobic nitrate reductase was examined. This enzyme appears in the membrane when bacteria are grown anaerobically on glucose and nitrate as nitrogen source. After this induction we deinduced in glucose with ammonia as the nitrogen source for 1, 2, and 3 generations and then we submitted these populations to a double test. The first test was similar to all the previous ones. We used again nitrate as the only nitrogen source during the penicillin test and only those cells which contained nitrate reductase could lyse. The results were similar to all the previous results. However, nitrate reductase is also known to reduce chlorate to chlorite. Chlorate is highly toxic agent which kills bacteria so that they cannot be lysed by penicillin. Figure 9 shows the results which we obtained with the chlorate test. These bacteria were placed in a medium containing glucose, ammonia and chlorate for the penicillin test. Those which have been fully

TABLE I

THE DETERMINATION OF GROWTH PATTERN OF CELL ENVELOPES BY VARIOUS METHODS

Species	Year	Method utilized	Envelope component tested	Observation or conclusion	Reference
Streptococcus	1964	Fluorescent antibody. Chain	Surface layer	Equatorial growth	8
E. coli	1964	Fluorescent antibody. Chain	Surface layer	Polar, subpolar and subcentral growing zones	16
E. coli	1966	Fluorescent antibody. Chain	Surface layer	Random dilution	17
E. coli	1966	[^{3}H]Thymidine label. Chain	Nucleus	Non-random distribution	9
B. subtilis	1966	[^{3}H]Thymidine label. Chain	Nucleus	Permanent attachment after one replication cycle	9
B. subtilis	1966	TeO_3^{2-} reduced to tellurium needles. Chain	Respiratory chain	Polar segregation	14
E. coli	1968	[^{3}H]Thymidine label. Chain	Nucleus	Random distribution	10
B. Subtilis	1968	[^{3}H]Thymidine label. Chain	Nucleus	Random distribution	10
E. coli	1970	Direction of elongation.	Surface	Growing zone moving from new pole to equator (1-unit cell)	19
		Penicillin spheroplast formation	Murein sacculus	Growing zones moving away from cell equator (2-unit cell)	19
E. coli	1971	[^{3}H]Thymidine. Chain autoradiography	Nucleus*	Random distribution	11
		[^{3}H]Oleic acid. Chain autoradiography	Phospholipid (membrane + wall)	Random distribution	11
		[^{3}H]DAP**. Chain autoradiography	Murein sacculus	Random distribution	11
		[^{3}H]Glycerol. Single-cell autoradiography	Phospholipid†	Random distribution	11
B. subtilis	1971	Basal body of flagellum (synthesis heat sensitive)	Membrane	Median growing zone	15
E. coli	1971	T6 page receptor after conjugation	Cell wall	Segregation after 2 h. 50% after four generations	18
E. coli	1971	Bromostearic acid label. Isopycnic centrifugation	Membrane	Intermediate density	12
E. coli (minicell forming)	1971	Distribution between cells and minicells of envelope protein label and permeases	Envelopes Membranes	Random distribution	32

* Four generations after [^{3}H]thymidine labeling, the distribution deviates from randomness (Fig. 9 of ref. 11).

** DAP, meso-diaminopimelic acid

† After 2 h (approx. three generation times), the [^{3}H]glycerol label is missing in 33% of the cells, while a Poisson distribution would predict only 24% (see Fig. 12 of ref. 11). Figures are compatible with 40% bacteria with an average of 0.4 silver grain, and 60% bacteria with an average of 2.2 silver grains. This example illustrates well the possible shortcomings of the statistical methods.

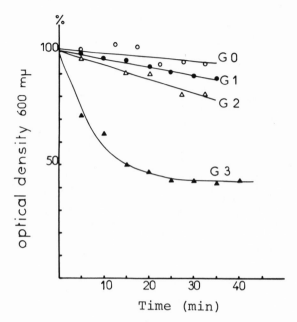

Fig. 9. Segregation of nitrate reductase. E. coli K12, strain
3300, was grown anaerobically overnight on mineral medium with
nitrate as nitrogen source and glucose as carbon source. The cells
were deinduced by anaerobic growth on ammonium-glucose medium.
Cells deinduced for 0, 1, 2, and 3 generation times were submitted
to penicillin lysis in the medium used for deinduction supplemented
with 10 mM $KClO_3$.

induced were killed and not lysed. One and two generations later
all were still killed but not lysed. After three generations half
of the population was lysed because it has not been killed by
chlorate. Here the unlysed population possesses the· markers and
the lysed population has lost the markers. The two methods of
screening furnish the two complementary halves of the population
in unlysed form.

With the use of any of these selective markers we could label
various membrane constituents and determine the fate of the radioactivity in the 2 halves of the heterogeneous population.

Figure 10 shows the distribution of ^{14}C diaminopimelic acid
using melibiose as the selective marker resulting at G3 in lysis
of half the cells. The label from diaminopimelate drops in two

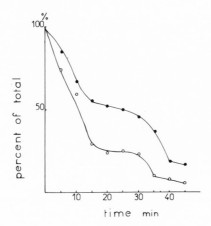

Fig. 10. Distribution of DAP among progeny. A culture of E. coli
autotrophic for DAP and lysine was grown for 6 generation times in
medium containing IPTG. ^{14}C DAP was added one hour before the end
of the induction period. After washing, the bacteria were grown
for an additional 3 generation times and submitted to EDTA treatment and lysis by lactose-penicillin. Absorbance readings (●)
and trichloracetic acid precipitable radioactivity in centrifugal
pellets (O) were measured at close intervals and expressed as
percent of the value obtained at zero time.

steps. Nevertheless the figure shows that the surviving cells have
noticeably less parental DAP than the total population. We can
conclude that diaminopimelate is randomized to about 50%. Figure
11 illustrates phospholipid label with ^{32}P. Melibiose was again
the selective marker. During the penicillin step samples were
treated with penicillinase and were centrifuged. The pellet was
chloroform methanol extracted. Lipid soluble ^{32}P disappeared for
85-90% from the surviving 50% population. Except for a minor
randomization of the order of 10%. Lipid soluble ^{32}P segregates
together with permease.

The same type of distribution was observed when glycerol tritiated on carbon 2 was incorporated into phospholipids. Only gly-

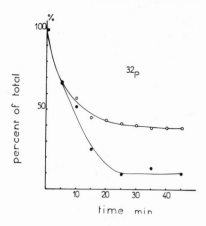

Fig. 11.   Distribution of ^{32}P phospholipids among segregants.   A
culture of <u>E. coli</u> K12 300 P (Lac permease negative) was grown for
6 generation times in medium 63 glycerol plus melibiose at 30°.
^{32}P orthophosphate was added one hour before the end of this induction period.   After washing, the bacteria were grown for an additional three generation times in 63 glycerol, then submitted to the EDTA
treatment and resuspended in 63 melibiose plus penicillin.   Absorbance readings 0-0 and chloroform methanol extractable ^{32}P in centrifugal pellets ●-● were measured at close intervals and expressed
as per cent of the values obtained at the time zero.

cerol which is incorporated as such into phospholipid keeps tritium
on carbon 2 the part going through the normal pathway loses ^{3}H at
the glycerol phosphate dehydrogenase step.   The assay shows that
98 to 99% of the counts are extractable by chloroform methanol.
Therefore we just precipitate the cells with TCA and look for radioactivity.   Again 95% of the radioactivity is lost when 50% of the
bacteria have been lysed.

These results were conflicting with a number of reports about
phospholipid labelling.   The data by Fox which relied on density
label by fatty acids, were particularly disturbing.   Therefore we
also used fatty acid label.   This is shown on figure 12.   The strain

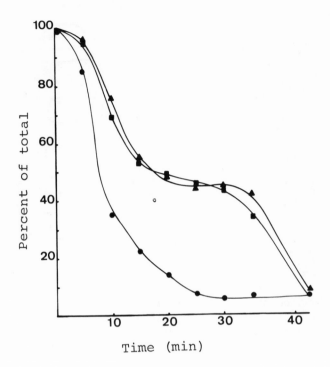

Time (min)

Fig. 12. Segregation of phospholipids labeled with ^{3}H oleic acid and ^{32}P during deinduction of lactose permease in E. coli, strain 1059 0180 dlac (courtesy of P. Overath). ▩ - , absorbance; ●-●, chloroform-methanol-extractable ^{32}P; ▲-▲, chloroform-methanol-extractable ^{3}H.

auxotrophic for unsaturated fatty acid has been double labeled during induction with ^{32}P orthophosphate and with tritiated oleic acid. In this case chloroform extractable ^{32}P was lost from the segregant population while 50% of the tritiated oleic acid remained. It is evident that there is a turnover of fatty acids on the backbone of the phospholipids and this explains at least a part of the contradictory results which have been published. This turnover is, as far as we reckon, complete at the time we are able to score for the distribution of the membrane.

We tried to look for markers which are incorporated for a short time. We induced bacteria for times shorter than one generation time in order to make our model more precise. Figure 13 shows the predictions of the model of membrane growth when induction has been limited to less than one generation time (between the two arrows). When begins the deinduction, contrary to the former figures where the hatched area extended all over to the two poles, here it forms

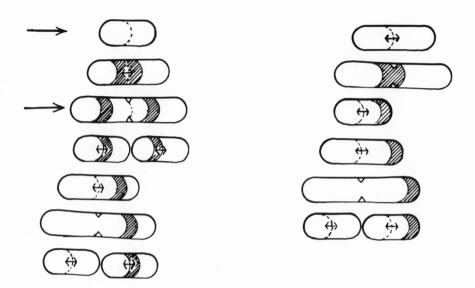

Fig. 13. Distribution of a membrane marker induced for less than one generation time among the progeny of a bacterium with two polar growing zones. Induction pulse between arrows.

two rings. This would result in deinduction one generation earlier than before. Therefore if the permease which is induced is deposited only in the growing zone of the membrane and if this pulse induction has been for less than one generation time segregation should occur earlier. We did this experiment inducing with lactose and melibiose; and invariably we found that segregation appeared at the same time as after full induction. The only possible explanation for these results is that these permeases when induced are deposited everywhere in the membrane and not only in the growing zone. Once they are deposited, they stay in place. The new growing zone removes them from the center and they end up in the final progeny. If so, membrane synthesis is not an all or none phenomenon. A preexisting membrane can receive new components. In order to inquire whether this kind of random deposition of membrane protein is a general phenomenon, we tested all the markers which gave segregation in a series of experiments, and the maltose marker in the first series was found to be the exception. When maltose pulse induction was less than one generation time, segregation occurred one generation time earlier than under conditions of full induction.

The kinetics of maltose permease are shown in figure 14.
After two generations there is 50% segregation. Thus the maltose
system provides a model for a protein which is deposited in growing
membrane exclusively, whereas lactose permease and melibiose permease behave as if they are deposited in preexisting membrane and
subsequently remain where they have been incorporated. Obviously,
when a membrane is synthesized it must contain proteins immediately.
These proteins must be incorporated in the growing membrane faster
than they are incorporated into non-growing membrane. But it was
shown for permeases which, according to our results, are incorporated in preexisting membrane, that their functional temperature
dependence is related to the species of fatty acids incorporated
into the membrane at the time of their synthesis. This results in
an Arrhenius plot with a break at the temperature characteristic
of the fatty acid. This experiment suggests that permease and
fatty acid are simultaneously incorporated, and pushing the rea-
soning, phospholipid is incorporated not only in the growing zone.

At this point one should recall that there are many kinds of
phospholipids. Phosphatidyl ethanolamine is the major component
of E. coli membrane and it is metabolically stable. Since it is

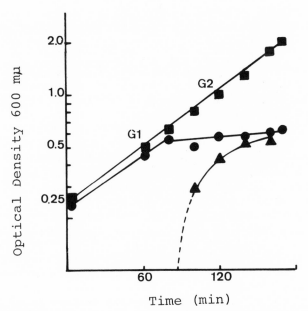

Fig. 14. Segregation kinetics of a membrane marker induced by
maltose for 0.25 generation time.

metabolically stable it remains in the membrane after three generations of induction. Phosphatidyl glycerol has a rather short half-
life and although it composes about 1/3 of the phospholipids, the
label from phospholipid segregation can be attributed to phosphatidyl ethanolamine and phosphatidyl glycerol might obey to a different pattern without modifying substantially the results of phospholipid labeling experiments. For example phosphatidyl glycerol
might be incorporated everywhere as are some of the permeases.

If phospholipid is labeled for a period shorter than one generation time together with the induction of maltose permease and
segregation is ascertained at G2 with penicillin maltose the surviving half population, devoid of maltose permease has also lost its
PE label, while half of the remaining PG label is still present.
The situation is exactly parallel with what is obtained by pulse
induction of two permeases. If pulse induced maltose permease is
the selective marker, pulse induced lactose permease is present in
the half population devoid of maltose permease, whereas a pulse
induced D glucuronate permease segregates together with maltose
permease.

Table II summarizes the results of the dual pattern of deposition of protein markers and phospholipids.

Coming back to the geometry of membrane growth the previous
statements and models have to be specified in the following way.
The growing zones, as represented in the various models are the
sites of about 2/3 to 3/4 of the total elongation of the membrane,
corresponding roughly to the proportion of PE. Membrane moving
away from this zone undergoes a further progressive elongation of
about one 1/4-1/3 of the total, by insertion of PG and of a new
class of membrane proteins. Therefore the chemical composition of
the membrane should be different close to the growing zone from the
bulk composition. It should be possible to distinguish new membrane
from mature membrane.

We also reexamined with our method the segregation of DNA. The
bacteria in our culture conditions behave like the two unit bacteria,
and for many reasons they are probably dinucleated. Therefore, to
obtain segregation, one more cell division must occur to separate
the two original labeled nuclei. Therefore, we labeled with tritiated thymidine one generation before deinducing for three generation times and looked for the DNA label in the two halves of the
population. We find consistently 60 to 65% of the DNA label associated with the permease containing parental membrane and 35 to 40%
associated with the population devoid of the parental membrane.
Therefore, the DNA population is partly randomly segregated. These
are very preliminary results. However, when we are doing a segregation experiment we begin with a sample of 10^9 or 10^{10} cells and
there is no problem of sample statistics. If we find a ratio of

60 to 40 it is not a matter of statistical error. We never found
a ratio higher than 65/35. This situation where neither complete
segregation nor complete randomization is observed might be interpreted by saying that there is a rule and there are exceptions.
The rule is that the DNA strands should stick to the membrane upon
which they have been synthesized but then sometimes they get mixed
and randomize. This is a rather high rate of randomization if a
ratio of 2/3 to 1/3 is reached at the end. There might be a different principle for the segregation of DNA. DNA is double-
stranded, and the two strands differ both by age and by chemical
composition. One is left-handed and the other is right-handed.
They have a different base sequence and are complementary. It is
possible to imagine that the DNA initiation site which we believe
to be in the membrane has a completely asymetric setup such that
when a choice must be made between the two DNA strands during the
next initiation, the basis for such choice is not made on the
grounds of the age of the strands, but according to their chemical
composition. If we hypothesize for a moment that a "left-handed"
strand must remain on the membrane which has been synthesized at
the time of its initiation and that the right-handed strand is
free to go to the other membrane or even must switch over to the
opposite half membrane then we shall have 50% of the DNA label
which remains with the original membrane and the other 50% could
distribute evenly. The result after segregation would be 25/75
distribution. We find 35/65 which is not very different and could
reflect some additional minor randomizing processes. This specu-
lation is interesting because it provides an image of the topography
of E. coli which has complete asymetry as in higher organisms, so
that one can distinguish head and tail, dorsal and ventral part,
right and left side.

This is symbolized in figure 15.

Head and tail would be distinguishable by the fact that the
two poles have a different age, one resulting from the septum of
the last cell division, the other one older. The DNA initiation
point which is attached to the membrane would define a priviliged
direction along the circumference of the cylinder, say the ventral
part. Finally, when DNA duplication starts from the initiation
point, two qualitatively different replication forks appear, which
could, if attached to the membrane, move in opposite directions
along the circumference that contained the initiation point, thus
defining a left hand and a right hand direction. The attachment of
the forks to the membrane is not a prerequisite for the concept of
this asymetry. Adding to this picture the growing zone of the membrane and two opposite gradients of membrane, maturity as suggested
above leaves us with a rather complex but ordered topographical
differentiation along and around the primitive cylinder.

T A B L E   II a

Segregation of pulse induced membrane markers

Inducer and substrate of the membrane marker	Heterogeneity after	
	2 generations	3 generations
Lactose	−	+
Melibiose	−	+
Glucose 6-P	−	+
D-Mannitol	−	+
D-Gluconate	−	+
Maltose	+	
Succinate	+	
Trehalose	+	
D-Glucuronate	+	

T A B L E   II b

Segregation of pulse labeled phospholipids

Label	Cosegregates with pulse induced maltose permease at G2	Cosegregates with pulse induced Lac permease at G3
2 ^{3}H glycerol	+	++
Lipid soluble ^{32}P	+	++
^{14}C palmitic acid	+	++
^{3}H oleic acid	−	−
^{32}P label in phosphatidyl-ethanolamine	++	++
^{32}P label in phosphatidyl-glycerol	−	++

++ more than 90% in bacteria with parental membrane

+ 75–85% in bacteria with parental membrane

− 50% in bacteria with parental membrane

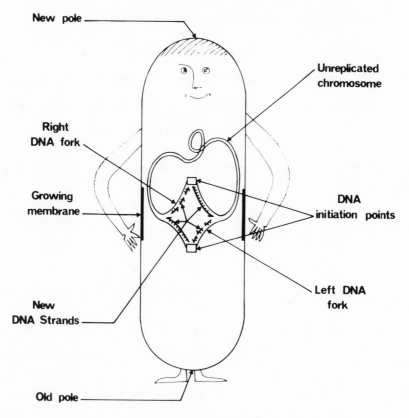

Fig. 15.   Schematic representation of an E. coli cell with discernible "head-tail" "dorsal-ventral" and "right-left" differentiation.

REFERENCES

Autissier, F., A. Jaffe and A. Kepes. 1971. Mol. Gen. Genet. 112: 275.

Autissier, F. and A. Kepes. 1971. Biochim. Biophys. Acta. 249:611.

Autissier, F. and A. Kepes. 1972. Biochimie. 54:93.

Azoulay, E. and F. Pichinoty. 1967. C. R. Acad. Sci. Paris. 264: 107.

Beachey, K. L. and R. M. Cole. 1966. J. Bacteriol. 92:1245.

Chung, K. L., R. Z. Harwirko and P. K. Isaac. 1964. Can. J. Microbiol. 10:43 and 473.

Cohn, M. and K. Horibata. 1959. J. Bacteriol. 78:613.

Cole, R. M. 1964. Science. 43:820.

Donachie, W. D. and K. J. Begg. 1970. Nature. 227:1167 and 1220.

Eberle, H. and K. G. Lark. 1966. J. Mol. Biol. 22:183.

Fitz James, P. 1965. Bacteriol. Rev. 29:293.

Fox, C. F. and E. P. Kennedy. 1966. Proc. Natl. Acad. Sci. U.S. 54:890.

Ganesan, A. J. and J. Lederberg. 1965. Biochem. Biophys. Res. Commun. 18:824.

Jacob, F., S. Brenner and F. Cuzin. 1963. Cold Spring Harbor Symp. Quant. Biol. 28:239.

Jacob, F. and J. Monod. 1961. J. Mol. Biol. 3:318.

Jacob, F., A. Ryter and F. Cuzin. 1966. Proc. R. Soc. London. 164: 267.

Lark, C. and K. G. Lark. 1964. J. Mol. Biol. 10:120.

Lark, K. G. and C. Lark. 1965. J. Mol. Biol. 13:105.

Leal, H. J. and P. Marcovich. 1971. Ann. Inst. Pasteur. 120:1167.

Leive, L. 1965. Biochem. Biophys. Res. Commun. 21:290.

Lin, E. C. C., Y. Hirota and F. Jacob. 1971. J. Bacteriol. 108: 375.

Outtara, A. 1969. Dissertation. University of Paris.

Overath, P., H. U. Schairer and N. Stoppel. 1967. Proc. Natl. Acad. Sci. U.S. 67:606.

Pardee, A. and L. S. Prestidge. 1961. Biochim. Biophys. Acta. 49:77.

Pontefract, R. D., G. Bergeron and F. S. Thatcher. 1969. J. Bacteriol. 97:384.

Puig, J. and E. Azoulay. 1967. C. R. Acad. Sci. Paris. 264:1507 and 1916.

Ryter, A. 1971. Ann. Inst. Pasteur. 121:271.

Ryter, A., Y. Hirota and F. Jacob. 1968. Cold Spring Harbor Symp. Quant. Biol. 33:669.

Ryter, A. and F. Jacob. 1963. C. R. Acad. Sci. Paris. 257:3060.

Ryter, A. and F. Jacob. 1964. Ann. Inst. Pasteur. 107:384.

Schwartz, M. 1967. Ann. Inst. Pasteur. 113:685.

Tsukagoshi, N., P. Fielding and C. F. Fox. 1971. Biochem. Biophys. Res. Commun. 44:497.

Tsukagoshi, N. and C. F. Fox. 1971. Biochemistry. 10:497.

Wilson, G. and C. F. Fox. 1971. J. Mol. Biol. 55:49.

GROWTH OF THE CELL ENVELOPE IN THE E. COLI CELL CYCLE

Y. Hirota, A. Ryter, M. Ricard

Département de Biologie Moléculaire, Institut Pasteur

Paris, France

Uli Schwarz

Friedrich-Miescher Laboratorium der Max-Planck-Gesellschaft

D 74 Tübingen, Germany

I. MORPHOLOGICAL ASPECTS OF MUREIN BIOSYNTHESIS

U. Schwarz

This is a discussion of the biochemistry of cell division, intended also as an introduction to Dr. Hirota's forthcoming contribution.

Role of the Sacculus in Shape Maintenance
It is quite clear that cell division means modification of the bacterial cell at the site of division. The processes which underlie such alterations of cell shape will be the topic of this discussion, concentrated on Escherichia coli as a model system. To understand the mechanisms and processes involved in maintenance and in modification of the cell shape during growth and division, one first has to know which structure is responsible for the retention of cell shape. In the case of bacteria this is known since long; they contain a well defined shape-conserving element, called the sacculus (Weidel and Pelzer, 1964; Osborn, 1969; Higgins and Shockman, 1971; Strominger and Ghuysen, 1967). This sacculus is an integral part of the complex cell envelope which in the case of gram-negatives consists of the inner cytoplasmic membrane, an outer

lipid bilayer, and the sacculus in between (De Petris, 1967).

The sacculus unequivocally is the main shape-conserving structure in the bacterial cell wall. A study of bacterial morphogenesis,
therefore, involves the study of the biosynthesis of the sacculus.
Such analysis became possible after sacculi were isolated in pure
form and after they had been analysed chemically, the pioneer work
for which has been done by Weidel and coworkers in Tübingen (Weidel
and Pelzer, 1964). The isolation of sacculi is facilitated through
their mechanical properties. Sacculi are extremely stable, and one
can shake complex cell envelopes with glass beads in a vibrator and
after further purification obtain pure sacculus molecules. These
are clearly visible in the electron microscope and faithfully reflect
the shape of the cell from which they have been isolated (Weidel
and Pelzer, 1964).

The evidence for the form being functional in the sacculus
comes from the quite early observation that, whenever the sacculus
is destroyed visually, the cell wall loses its defined shape and its
mechanical stability as well. As an example, egg white lysozyme
breaks cell walls and degrades sacculi into small fragments. This
can be demonstrated either chemically or by use of the electron
microscope (Weidel, Frank and Martin, 1960). Upon degradation by
lysozyme, the sacculus is broken down into small fragments. These
fragments can be taken as the basic subunits of the polymer murein
from which the sacculus is tailored. The arrangement of the muro-
peptides in the sacculus was deduced from an analysis of fragments
which were obtained by partial degradation of murein by lysozyme.
According to these analyses the principle of construction of the
sacculus became clear: It is as follows. The bag of the sacculus
is made up from a net-like structure in which long polysaccharide
strands, running parallel to each other, are interlinked by a
short peptide, the whole resulting in a net-like structure which
forms the three dimensional bag of the sacculus (Weidel and Pelzer,
1964).

The shape maintaining function of the sacculus is quite obvious
from the lysozyme experiments. However, with respect to bacterial
morphogenesis we are left with the question of how the form of the
sacculus itself is determined. There are two extreme alternatives.
First, the simplest, that the form of the sacculus could be directed
simply by its chemical composition. In this case there should exist
a strict correlation between morphology and chemistry of the
sacculus. Secondly, the murein could be used as a neutral building
material which is molded by a specific morphogenetic system. We
have tried an experimental approach to test these two alternatives.
The first alternative, the correlation between chemistry and shape,
was tested by an analysis of the composition of murein obtained from
cells growing under different conditions. We found very significant

differences in murein composition but no difference in shape, which
means there is no simple direct correlation between shape and
chemistry of the cells. For the second alternative, we tried to
demonstrate that the existence of a morphogenetic system in the wall
could be detected. We asked whether a modification of the organization of the wall resulted in a morphological change of the sacculus.
Rod-shaped cells of a proper mutant were converted into spherical
cells by degradation of murein under osmotic protection. The spheroplasts then were allowed to make a new sacculus; this now was
spherical as the cells were at the moment when they made it. The
overall chemistry of the sacculus was the same as the chemistry of
the normal cell (Schwarz and Leutgeb, 1971; Henning et al., 1972).

   The most obvious conclusion we drew from this experiment was
that in one and the same cell the sacculus can take different
shapes. It is clear that the sacculus definitely conserves the
shape in the cell; on the other hand the cell clearly lays down
the shape of the sacculus. That is a paradox situation, which we
take as evidence for the existence of a morphogenetic apparatus
within the cell wall which shapes the sacculus. We assume that
this morphogenetic apparatus is part of the cytoplasmic membrane
which normally needs the intact sacculus as mechanical support to
function properly. If the organization of the morphogenetic system
is modified, as it occurs upon degradation of the murein in our
experiment, the shape of the sacculus which it produces is changed
too.

Function of Murein Hydrolases in Morphogenesis

   Since the sacculus is a macromolecule, held together by covalent
bonds throughout, any modification in shape must be initiated by
splitting of some of these bonds. Thus, acceptor sites would be
liberated at which new material could be attached in a sequence of
reactions as extensively studied by Strominger and coworkers (Ghuysen,
Strominger and Tipper, 1968). Consequently, the enlargement and
modification of the sacculus during cell growth and division would
require action of a murein hydrolase system. If this is correct,
the action of this system would determine at which time, at which
place and at which extent the sacculus is modified during cell
growth; murein hydrolases thus were key enzymes in bacterial morphogenesis including cell division. An enzyme system which is active
in morphogenesis should fulfill at least two requirements. First,
the system should be able to determine exactly the place where new
building blocks can be inserted into the growing sacculus. Second,
the system should be ready to act at an exactly fixed time in the
life cycle. As an example, the system should be constructed such
that polar cap formation in a dividing cell is triggered at a given
time in the right place. For operational reasons one would assume
that a system of structurally bound enzymes which can be locally

activated and inactivated at a given time would be superior to a
set of soluble enzymes. The murein hydrolase system from E. coli
has these postulated properties as will be shown.

The demonstration of localized murein hydrolase action was based
on the assumption that a partial block in murein synthesis should
lead to a degradation of the sacculus just in regions of increased
hydrolase activity. In areas of low murein hydrolysis, degradation
would continue to be compensated for some time by the insertion of
new material, still taking place at reduced rate. Sites of high
enzyme activity thus should be revealed as gaps in the sacculus
under the electron microscope.

Partial inhibition of murein synthesis was accomplished by
addition of penicillin at low concentration. From penicillin-
treated cells sacculi were prepared and these were used to determine
the topology of hydrolase action. We found that indeed the sacculus
shows an extremely sharp cut at the place where under normal
conditions the cell divides (Schwarz, Asmus and Frank, 1969).

We assumed that the localized murein hydrolase action is an
essential step in cell division. Thus we expected this step to be
correlated with DNA replication and to be exactly timed in the life
cycle. This is what we find (Hoffmann, Messer and Schwarz, 1972).
DNA replication was inhibited in temperature sensitive mutants and
at the same time penicillin was added to measure the functioning of
the hydrolytic system: Cells become sensitive to osmotic shock and
die if the sacculus is cut. The experiment showed that if DNA
replication is blocked, the cutting mechanism is not triggered. In
addition we did experiments on the timing of the cutting mechanism.
We used a synchronized culture of B/r. To this culture short pulses
of penicillin were given and the number of survivors again was
determined. We found that the cells had a significant maximum of
penicillin sensitivity at about 35 minutes in the life cycle (generation time of about 40 minutes. This is shortly before cell division
takes place.

For functional reasons we expected at least some of the hydrolases to be membrane-bound; we assumed that this would better enable
them to fulfill their morphogenetic function. It is certain that at
least three different hydrolases are bound to the cell envelope.
Even by repeated washings it is not possible to release these
enzymes from the envelope. The only way to liberate them is treatment of the envelopes with a combination of Triton X-100 and salt
(Hartman, Höltje and Schwarz, 1972). Under this condition we are
able to liberate the enzymes from the envelope. Quite recently we
succeeded in working out a suitable modification of the procedure
developed by Mary Jane Osborn which allows the separation of the
outer wall layer from the cytoplasmic membrane. Using this procedure

we succeeded in demonstrating that some murein hydrolases are located
in the cytoplasmic membrane (Hakenbeck, 1973).   Since the substrate
sacculus is not part of this membrane, this indicates a compartmental-
ization of enzymes and substrate which might be an operational basis
for the triggering of localized enzyme action.

According to our working hypothesis, murein hydrolysis and
murein synthesis are coupled with each other.   Therefore, a correlation should be found between the timing of increased hydrolase
activity and the rate of murein synthesis in the life cycle.   We
found, again by using synchronized cultures of E. coli, that the
rate of murein synthesis is not constant over the life cycle.   We
found that the rate of murein synthesis oscillates during the life
cycle; at about the time at which the cells are hypersensitive to
penicillin (that is the time shortly before division) there is an
increased rate of murein synthesis.   Thus there is a coincidence with
respect to the timing of hydrolase action and the rate of murein
synthesis (Hoffman et al., 1972).   The rate of murein synthesis had
been measured in a mutant in which murein can be specifically
labelled.   Pulses of the specific label 2.6 diaminopimelic acid
were given at different times in the life cycle.   The doubling in
the rate of murein synthesis, observed shortly before division, is
a good coincidence with the experiments to be described in more
detail in the second part of this paper (Ryter, Hirota and Schwarz,
1973).   These experiments had been started in order to elucidate the
growth pattern of the sacculus and to compare the topology of
hydrolase action and of murein synthesis by means of high resolution
autoradiography.

Tentative Model for the Regulation of Hydrolase Action
The mechanismus by which the temporal and localized action of
the murein hydrolase system are regulated are not understood.   At
present I can provide only some hypothesis and a patchwork of
experimental information.   Our working hypothesis is as follows:
The murein hydrolases are consistently present at all times of the
life cycle.   They form a uniform mosaic pattern in the cytoplasmic
membrane all over the cell surface.   There is no difference in enzyme
content at the poles compared with the cylindrical parts of the cell
wall.   There exists a barrier between the enzymes and their substrate
preventing uncontrolled enzyme action.   The local breakdown and
reconstitution of this barrier provides the basis for the regulation
of enzyme activity.   We have some experimental results supporting
this hypothesis.

Permanence of murein hydrolases is suggested by the finding
that enzyme activity is found both in exponential and stationary
phase cells.   The half time of enzyme activity after blocking protein
synthesis by chloramphenicol is more than 90 min which under our
experimental conditions equalling 3 generation times.   This finding

matches with our hypothesis.

In order to test whether the pattern of the enzyme is the same all over the surface of the envelope we used a mini-cell producer strain which divides normally but from time to time also makes a so-called mini-cell. Such a cell can be considered as consisting only of two polar caps. This means that such a cell contains the membrane from a topologically defined area of the cell envelope from the cell ends. We made cell walls both from purified mini-cells and rod-shaped cells and compared the specific activity of murein hydrolases. Membranes from long cylindrical cells and from mini-cells showed no significant difference between each other (Bock, Goodell and Schwarz, unpublished).

The existence of a sort of barrier between murein hydrolases and the substrate is suggested by the finding that in intact cells, when incubated in buffer without a carbon source, murein hydrolases do not attach the sacculus. However, in just the same cells the sacculus is exhaustively degraded after previous treatment either with 1 m sodium chloride, or more effective and more astonishing, by 5% trichloroacetic acid (Hartmann, Bock and Schwarz, in press). Whatever the material basis of this hypothetical barrier may be is absolutely unclear.

Finally, I would like to discuss the nature of two murein hydrolases. We have isolated two of these enzymes one of which is strictly membrane-bound. One enzyme is inhibited _in vitro_ by penicillin at low concentration. A similar concentration of penicillin also inhibits cell division quite specifically. The concentration of penicillin required to block cell division also inhibits the enzyme _in vitro_. This enzyme, therefore, seems to be the penicillin target in the living cell. We have a second hydrolase which is inhibited by penicillin only at high concentrations; high concentrations of the drug interestingly block cell elongation (Hartmann _et al_., 1972). Taking these findings one has reason to assume that   one of these enzymes is involved in division and the other one in cell elongation. To be sure of these facts we need mutants. Until now we have not been successful in the search for mutants, but we have screening procedures which might allow us to isolate some mutants which we need to analyze the morphogenetic function of the murein hydrolases into more detail.

REFERENCES

Bock, B., E. W. Goodell and U. Schwarz. Unpublished.

De Petris, S. 1967. J. Ultrastruct. Res. 19:45.

Ghuysen, J. M., J. L. Strominger and D. J. Tipper. 1968. In
   M. Florkin and E. H. Stotz, eds., Comprehensive Biochemistry.
   26A:53.

Hakenbeck, R. 1973. Diplomathesis. Universität Tübingen.

Hartmann, R., J.-V. Höltje and U. Schwarz. 1972. Nature. 235:426.

Hartmann, R., B. Bock and U. Schwarz. Eur. J. Biochem. Submitted
   for publication.

Henning, U., K. Rehn, V. Braun, B. Höhn and U. Schwarz. 1972.
   Eur. J. Biochem. 26:570.

Higgins, M. L. and G. D. Shockman. 1971. Critical Rev. in A. J.
   Laskin and H. Lechevalier, eds., Microbiol. 1:29.

Hoffmann, B., W. Messer and U. Schwarz. 1972. J. Supramolec.
   Struct. 1:29.

Osborn, M. J. 1969. Ann. Rev. Biochem. 38:501.

Ryter, A., Y. Hirota and U. Schwarz. 1973. J. Mol. Biol. 78:183.

Schwarz, U., A. Asmus and H. Frank. 1969. J. Mol. Biol. 41:419.

Schwarz, U. and W. Leutgeb. 1971. J. Bact. 106:588.

Strominger, J. L. and J. M. Ghuysen. 1967. Science. 156:213.

Weidel, W., H. Frank and H. H. Martin. 1960. J. Gen. Microbiol.
   22:158.

Weidel, W. and H. Pelzer. 1964. Adv. Enzymol. 26:193.

II.   THE GROWTH PATTERN OF E. COLI MUREIN

Y. Hirota

This presentation concerns the growth pattern of the murein
sacculus and cell division in E. coli. The introductory remarks,
the ideology and the biochemical aspects of murein synthesis have
been discussed in this volume by Dr. U. Schwarz. We will describe
here recent experiments designed to prove the relationship between
the pattern of murein growth and the process of cell division.

Murein synthesis was analyzed by the incorporation of tritiated
diaminopimelic acid (^{3}H DAP 20 C/mM) into the cell wall. The use

of E. coli strains having nutritional requirements for both lysine
(lys⁻) and diaminopimelic acid (dap⁻) allowed specific labelling of
murein. The distribution of radioactivity within single cells which
had incorporated DAP into murein was measured by the use of two
autoradiographic techniques which will be described in the following
sections. A series of thermosensitive mutants of E. coli, defective
in the process of cell division were available (Hirota, Ricard and
Shapiro). A set of lys⁻ dap⁻ derivatives of these strains was
constructed for this work.

The Methocel Autoradiographic Technique
The method. Studies of segregation of parental molecules in
E. coli, growing in liquid culture, are difficult because the
complete separation of the two daughter cells after division makes
studies in randomly dividing cells impossible to interpret. We
developed a technique, "methocel autoradiography" (Lin, Hirota and
Jacob, 1971) to obviate such a difficulty. Labelled cells were
seeded and grown without radioactive precursors in a highly viscous
medium containing methyl-cellulose (Lederberg, 1956). The high
viscosity of the medium prevents cells from moving apart, thus the
progeny of a single cell is maintained in a linear array with no
apparent disturbance to the normal growth process.

The individual cells labelled with radioactive precursor were
allowed to divide and to form chains of various lengths. The
patterns of distribution of radioactivity among the progeny could
be examined by the use of quantitative autoradiography by counting
the grain number associated with each cell in the chain as it is
schematically shown in Fig. 1. Cells divide synchronously up to a
period of five generation times. Chains containing 2, 4, 8, 16, 32
cells per chain could be obtained, and the cell sizes within a clone
formed (see Plate 1) by synchronized cell division are very homo-
genous.

I would like to point out a possible artifact caused by the
differential rate of growth of cells (of the type shown in Fig. 2)
in a liquid culture. This may become a significant factor,
especially when a bacterial culture is subjected to complex manipula-
tive procedures for the labelling and washing of cells. During the
subsequent growth of the labelled cells in a liquid culture, a
heterogenous cell population in which cells containing different
amounts of radioactivity, will be obtained. In liquid culture,
therefore, it is impossible to ascertain if the variation in the
labelling patterns between individual cells in a culture is due to
the segregation mechanism or to artifacts caused by an irregular
rate of cell division. By the use of the methocel technique, one
can avoid such artifacts.

Results: the dispersed transmission. The results of our

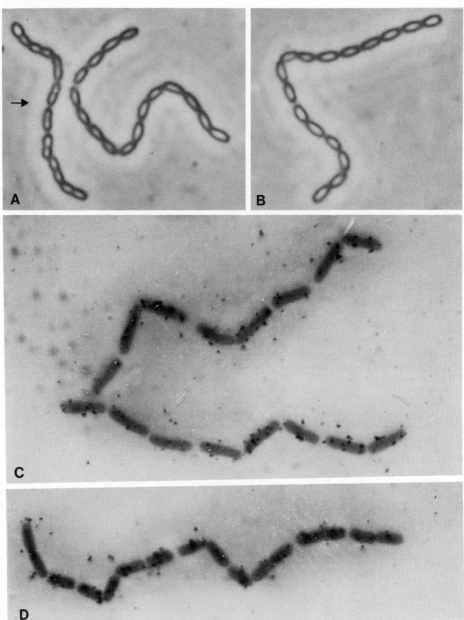

Plate 1.   Bacterial chains formed in the methocel medium.
A and B. Bacterial chains formed in the methocel medium revealed
by phase-contrast microscopy.
A.   Cells of a chain (pointed by an arrow) were at a stage just
before synchronized cleavage.
C and D.   Autoradiogram of chains derived from single bacteria
labeled for several generations with ^{3}H DAP (Lin et al., 1971).

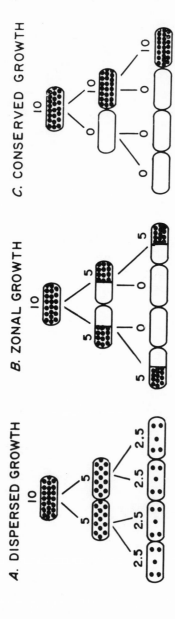

Fig. 1. Diagram showing three possible inheritance patterns of cell membrane material.

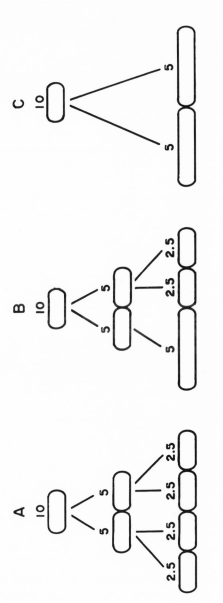

Fig. 2. Schematic representation of cell chain formed in methocel experiment under the normal and abnormal cell division.

The numbers are arbitrary units chosen to represent the radioactivity. Fig. A, B, C, depict the fate of three cells, each of which arose from the same culture and originally contained identical amounts of radioactive material. The cells were subsequently grown in the methocel medium containing no radioactivity. Notice the different modes of cell division which can be discriminated by this system. In a mixed liquid culture it would be impossible to avoid artifacts of the types described in the text.

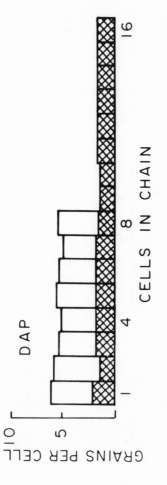

Fig. 3. Distribution of ³H-diaminopimilate in chains of 8 (clear blocks) and 16 (cross-hatched blocks) cells. (See legend to Plate 1).

studies of murein growth obtained by the use of methocel technique
are shown in Plate 1 C, D and Fig. 3.

We conclude that E. coli murein appears to be dispersed and the
"old" structure is diluted out in the course of several bacterial
divisions. Our data, thus obtained, is in full agreement with a
previous work of Van Tubergen and Setlow (1961) but in conflict with
that of others who observed the conserved segregation of parental
murein and lipids into progeny (Kepes, in this issue). Such a
discrepancy could be explained if the results supporting conserved
segregation were generated by the kind of artifact described above.

Our observations from the methocel experiment can be interpreted
as follows: (a) E. coli murein is indeed dispersively segregated
and parental material is diluted by the incorporation of new material
over the entire surface. Alternatively, (b) one or a few growing
zones of the murein-sacculus could exist where new murein is synthesized. However, this "new" murein is rapidly randomized over the
whole cell surface and mixed with pre-existing "old" murein in the
course of cell elongation.

These alternatives have been examined by short-pulse labelling
of E. coli murein with ^{3}H DAP which was followed by the immediate
arrest of cell growth in order to avoid possible randomization.
Murein-sacculi were prepared, and the distribution of radioactivity
within individual sacculi was examined by autoradiography and
electron microscopy.

The Processes of Zone Formation and Mixing of Murein
A series of experiments demonstrated that the radioactivity
of nascent murein appeared on the EM-autoradiographs as a well-
defined growth zone in the central area of the sacculus (Fig. 4-
Plate 2) and this was true regardless the size of the cells.
Essentially, the same growth pattern was obtained from cells growing in synthetic glucose medium and in synthetic succinate medium,
at generation times of 50 min and 80 min, respectively. Our result,
therefore did not support the unit cell model which postulates that
the murein growth zone is displaced towards one pole if cells are
growing with generation time longer than 60 min (Donachie and Begg,
1970).

The presence of a "mixing process" which would account for the
dispersed transmission of murein as seen in the methocel experiment
was examined by a pulse-chase experiment. Such an experiment, shown
in Fig. 5 and Plate 2 D, E and F, demonstrated that radioactive DAP
incorporated into a central zone during a short pulse, as shown
above, was subsequently found to become evenly distributed over
the entire surface of the cell. This mixing process occurred in
less than 1/4 of a generation.

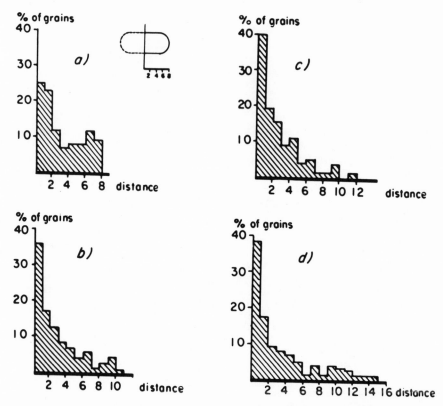

Fig. 4.   The distribution of label over sacculi of strain W7 grown in glucose medium.   An exponentially growing culture of W7 in glucose medium at 37°C with a generation time of 50 min was pulse-labelled with (^{3}H) DAP for 6 min (1/8 of a generation time) and sacculi were immediately prepared.   The distribution of radio-activity over autoradiographs of sacculi was analysed.   The 4 histograms show the location of silver grains on the sacculi (50 to 200 sacculi) of 4 different classes in size.   The distance of every grain from the mid-line of each sacculus was measured.   The frequency of the appearance of grains was plotted on the abscissa and the distance on the ordinate.   (See Ryter, 1968.)

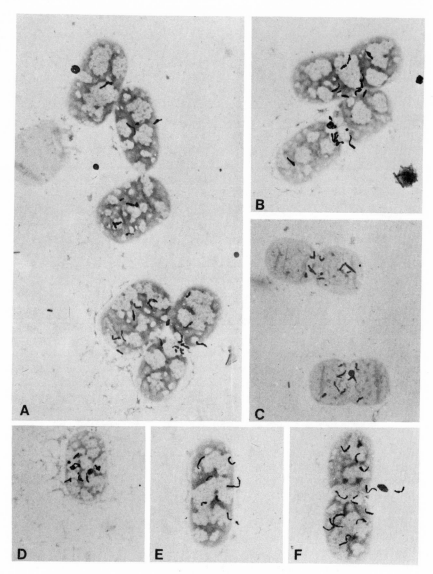

Plate 2.   Autoradiographs of pulse-labelled sacculi of E. coli
W7 grown in glucose medium (see legend to Fig. 4) (Magnification x
22,500).

     Autoradiographs of pulse-labelled sacculi after chase in cold
medium (see text) to 0 min (A, B, C and D) 1/4 of a generation time
(E) 1/2 of a generation time (F).

Thus, the unique features of the growth zone on murein-sacculus
can be documented as follows: (a) the growth zone formation:
practically all bacterial cells carry one growth zone at the central
area of every cell, independent of cell size. The sacculus exactly
reflects the size of the individual cell from which it was isolated
and cell size is correlated with cell age. We explain the results,
as follows. The growth zone is initiated immediately after division
of the mother cell, and murein synthesis continues at the zone until
the cells are cleaved off. The new growth zone is made, de novo, at
the center of nascent cell, regardless, of the rate of cell growth.
A schematic representation of the results is shown in Fig. 6. (b) the
mixing of old and new murein: the new murein inserted at the
growth zone is rapidly randomized over the whole surface of sacculus
and mixed with old murein (this result explains why the dispersed
transmission of DAP-label was observed in the previous studies,
Lin et al., 1971; Van Tubergen and Setlow, 1961).

The Growth Pattern of Murein-Sacculus in E. Coli Mutants Defective in Cell Division

A series of temperature-sensitive conditional mutants of
E. coli which are defective in different steps of cell division
have been isolated and characterized (Hirota et al.) Six
classes of these had thermosensitive mutations in DNA replication,
either in the initiation of DNA replication (2 classes) or elongation of the DNA chain (4 classes). Two classes were defective in
DNA segregation and seven classes in separation. The types of
mutants isolated and characterized are schematically shown in Fig. 7.

The way in which zonal growth of murein is geared to the
processes of cell division can be effectively studied by the use
of mutants defective in the processes of cell division. We shall
present preliminary data about the murein growth in some of these
mutants. A thermosensitive mutant, dna B, defective in DNA elongation was incubated at the non-permissive temperature to arrest DNA
replication. When a ^{3}H DAP pulse was given to cells treated in this
way, the growth zone could not be detected. Instead, the label was
found uniformly distributed over the sacculi (Plate 3, A and B).
The normal growth zone, however, was found in the mutant cells
grown at the permissive temperature. The exact mechanism remains
to be determined. However, whatever the mechanism, DNA replication
seems to be coupled to the formation of the murein growth zone.
It is interesting to cite here the following established fact, as a
common mechanism: the arrest of DNA replication inhibits the division
of cells and prevents the triggering of localized murein hydrolase
activity at the step in the cell division (Schwarz, Asmus and Frank,
1969).

Another thermosensitive mutant defective in the process of
septation, Fts A, forms multiply-nucleated filaments at the non-

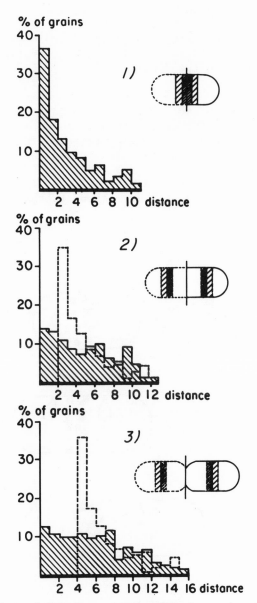

Fig. 5.   Redistribution of label over sacculi after the chase.

The growth conditions, pulse-labelling, chasing of radioactivity, preparation of sacculi, autoradiographic procedure and method of analysis are described in the legend to Fig. 4.

Class b histogram.   Distribution of radioactivity over sacculi of class b immediately after pulse-labelling (no chase control).

Legend to Fig. 5 (cont'd).

Class c and d histograms. Distribution of radioactivity over
sacculi with chase of 1/4 and 1/2 of a generation time, respectively.
The histograms give the distribution of radioactivity over sacculi
of class c and in class d.

As a reference, the profile of the distribution of radioactivity shown in histogram b is also shown in c and d, as a dotted
line. If radioactive molecules integrated into the sacculi remain
fixed at the site of incorporation during the cell growth, the
further growth of sacculi should result in the migration of the
peak over the sacculi towards the cell termini, and the expected
position of the peak should correspond to the dotted line. The data
show no such peak.

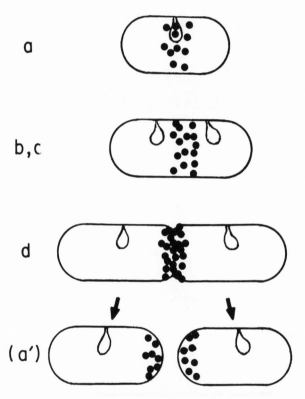

Fig. 6. A schematic representation of the zonal growth of murein
and cell division in E. coli. Arrows in (a) point to site of new
growing zone; (•) nascent murein subunit integrated in the sacculi.

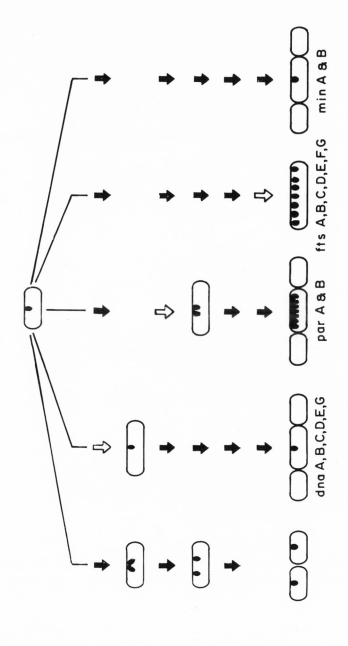

Fig. 7. A schematic representation of the morphological alteration
of thermosensitive mutants altered in the different steps of
cellular division. The open arrow signifies the process(es) altered.
The closed arrow signifies the normal process(es) of cellular
division. The genes are indicated under each figure, respectively.

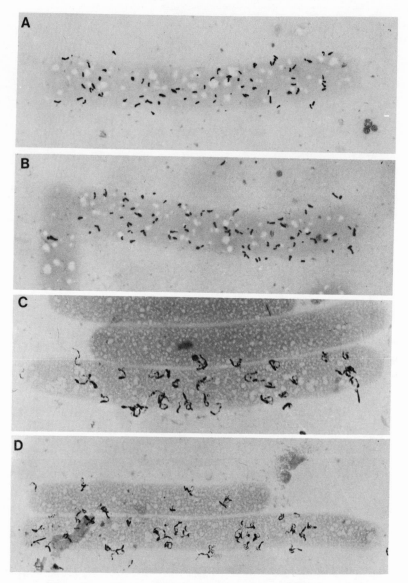

Plate 3. Distribution of labeled DAP during growth of E. coli
mutants thermosensitive for cell division processes.
A and B. dna B mutant grown at non-permissive temperature, pulse-
labeled with ^{3}H DAP (see text). Grain distribution shows no
detectable growth zone.
C and D. Fts A mutant (defective in septa formation) grown at non-
permissive temperature pulse-labeled with ^{3}H DAP (see text). Grain
distribution shows regularly spaced growth zones.

permissive temperature.   When a pulse of ^{3}H DAP was given to the
culture of filamentous cells at the non-permissive temperature,
multiple growth zones were observed.   The hot spots are located
with regular spacing along the filaments (Plate 3, C and D).   Thus,
it seems clear that the area of zone formation can be the potential
division site, and this is a different step to the events of septation.

Discussion

Enzymes for cell division.   The experiments described in the
text support Dr. Schwarz's view of the role of the murein-sacculus
as a "mitotic-apparatus" for E. coli.   Murein growth, both in
timing and in space is exactly correlated with the processes of
cell division.   The scission of co-valent bonds of murein could
serve as a primary trigger reaction for the initiation of cell
division.

In order to examine such an important implication of the in
vivo role of murein-enzymes, both murein-hydrolase(s) and
synthetase(s), in the process of cell division, we are systematically
screening the E. coli mutants defective in cell division for
specific thermosensitive alteration of murein-enzymes.

The operational model.   We summarize the interpretation of our
results as follows:

In the early stage of the cell cycle (at the time when the
events of DNA replication occur), the growth zone of murein is
formed, de novo, at the place where DNA is located in the nascent
cell.   We will call this event "zone-initiation."   It is likely that
DNA replication is a "pace-maker" and serves as a "pointer" for
zone initiation.   Zone initiation is followed by the continuous
addition of new murein materials at both sides of the growth zone.
We name this event, the "murein-elongation."   Thus the two cell-
termini are progressively separated through insertion of new murein
materials at a centrally placed growth zone.   Physical strength of
murein structure, as described by the previous speaker, thus provides
the molecular base in the DNA segregation.   At the time when DNA
replication is terminated, murein-elongation is switched off, and
the next event, "septum formation," follows.   This synthesis is
terminated by the completion of a hemispherical bag of murein.   A
thermosensitive mutant defective in septum formation, fts A, thus
forms multiple growth zones.   When daughter cells are cleaved off,
DNA initiation is triggered and the whole cell cycle recommences.

We propose an operational model for cell division based upon
our observations and some assumptions: (a) DNA is attached to the
murein-sacculus through membrane, and the zone initiation of murein-
sacculus occurs at the attachment site only at the time when DNA

replication begins.

The existence of a functional DNA-membrane-murein complex is still completely hypothetical. However, some evidence for its existence might come from the electron-microscopical demonstration of connections between the cytoplasmic membrane and murein (Bayer, 1968).

(b) The processes of murein-sacculus growth are controlled by the processes of DNA replication through regulatory circuits. The events of DNA replication, DNA initiation and/or DNA-elongation, and DNA-termination, elicit the biosynthesis of specific proteins which regulate operons controlling the sequence of events of murein growth, zone-initiation, murein-elongation, or septum-formation.

Thus, murein-growth may be coupled, in time and in space, with DNA replication. The murein sacculus can serve as a mitotic apparatus to segregate DNA during the growth and cell division of E. coli.

The model can be compared with the replicon model (Jacob, Brenner and Cuzin, 1963) which postulated the attachment of DNA to membrane (Fielding and Fox, 1970; Ganesan and Lederberg, 1965; Ryter, 1968) and the growth of membrane as the mechanisms of DNA replication, DNA segregation, and cell division (Jacob et al., 1963; Ryter, 1968). Our experimental results are in full agreement with the replicon model as to the growth pattern of bacterial cell surface, and we extended the model of replicon towards the role of murein-sacculus as a mitotic apparatus of bacterial cell division.

Future work. A molecular model of cell division based on the growth pattern of murein-sacculus, at the moment, must be fragmentary and pose many questions: i.e. What is the nature of signals for the initiation of murein synthesis? If such a signal given to the operons controlling murein biosynthesis either negatively or positively? How can the DNA-murein-membrane-complex, if it indeed exists, murein-synthesis be triggered precisely at the correct place and only once at the early time of DNA replication? Does DNA-replication function as a pace-maker of cell division?

Whatever these mechanisms would be, the model proposes the regulatory mechanism of murein biosynthesis as an integral part of cell growth and division of E. coli. It is known that the bacterial membrane is the factor where the reactions of murein biosynthesis are operating under precise control, both in time and place. It is clear, therefore, that the critical step of our ultimate understanding of cell division will stem from the progress of the genetics and biochemistry of membrane synthesis and function. Thus the firm basis of genetics and biochemistry of E. coli established during past decades offers substantial advantage as a model system for the study of cell division.

ACKNOWLEDGEMENT

We thank Professor F. Jacob for his continuous interest and encouragement during the course of this study, and Dr. J. Merlie for assistance in preparing this manuscript. We thank Miss S. Cayre for the secretarial work.

This work was supported by grants from the Centre National de la Recherche Scientifique, the Délégation Générale à la Recherche Scientifique et Technique, and Fondation Recherche Médicale to the two of the authors (Y. Hirota and A. Ryter).

REFERENCES

Bayer, M. E. 1968. J. Gen. Microbiol. 53:395-404.

Briles, E. B. and A. Tomasz. 1970. J. Cell. Biol. 47:786-90.

Cole, R. M. 1964. Science. 143:820-22.

Cole, R. M. 1965. Bacteriol. Rev. 29:326-44.

Donachie, W. D. and K. J. Begg. 1970. Nature. 227:1220-24.

Fielding, P. and C. F. Fox. 1970. Biochem. Biophys. Res. Commun. 41:157-62.

Ganesan, A. T. and J. Lederberg. 1965. Biochem. Biophys. Res. Commun. 18:824-35.

Hirota, Y., M. Ricard and B. Shapiro. Biomembranes. 2:13-31.

Jacob, F., S. Brenner and F. Cuzin. 1963. Cold Spring Harbor Symp. Quant. Biol. 28:329-48.

Kepes, A. In this issue.

Lederberg, J. 1956. Genetics. 41:845-71.

Lin, E. C. C., Y. Hirota and F. Jacob. 1971. J. Bacteriol. 108: 375-85.

Ryter, A. 1968. Bacteriol. Rev. 32:39-54.

Ryter, A., Y. Hirota and U. Schwarz. 1973.   J. Mol. Biol. 78:183.

Schwarz, U., A. Asmus and H. Frank. 1969.   J. Mol. Biol.   41:419.

Van Tubergen, R. P. and R. B. Setlow. 1961.   Biophys. J.   1:589–
    625.

CELL DIVISION IN BACTERIA

William Donachie

MRC Molecular Genetics Unit

Edinburgh, Scotland

This discussion will deal with the timing and localization of division in bacterial cells. Cell division involves a number of biochemically different processes which take place co-ordinately both in space and in time. Thus there are interactions between DNA replication and the various lipid, protein and mucopeptide syntheses involved in septum formation. This interaction is so important that it is not possible to discuss cell division without considering DNA replication. Therefore I wish to start by summarizing the way in which DNA replication is controlled in the cell cycle of E. coli (a similar system of control probably operates also in Bacillus subtilis). (Most of the following discussion refers to E. coli B/r/1 because it is only in this strain that the timing of DNA replication has been elucidated for a large number of growth condition. However, there is sufficient information available from other strains to make it virtually certain that the same general rules apply to them also.)

Let us consider a hypothetical experiment which demonstrates the general behaviour of a cell which has a single chromosome. The control of DNA replication in bacteria is essentially concerned with the control of initiation of rounds of chromosome replication, and the times at which these initiation events take place are correlated with the mass of the cell. We start the experiment with a single cell with one chromosome and a mass (M_i, Donachie, 1968) such that it is ready to initiate DNA synthesis immediately. In our experiment, this cell is inoculated into a medium (minimal salts + glycerol) in which cell mass can double every 60 minutes. The pattern of chromosome replication that occurs is as follows. At zero time, DNA replication begins at the chromosome origin and a pair of replication forks move away from each other in both directions around the chromosome until they meet at a terminus on the opposite side (Masters &

431

Broda, 1971; Bird, Louarn, Martuscelli & Caro, 1972; Hohlfeld &
Vielmetter, 1973; Prescott & Kuempel, 1972). This process takes
approximately 40 minutes and produces two chromosomes (Clark &
Maaløe, 1967; Helmstetter, 1967; Helmstetter & Cooper, 1968; Cooper
& Helmstetter, 1968). Nothing further happens until the cell attains twice its initial unit mass ($2.M_i$) after 60 minutes. At this
time initiation of new rounds of replication takes place simultaneously at the origins of both chromosomes, or, in other words,
whenever the ratio of number of unit mass equivalents to number of
chromosome origins reaches 1. This process is identical in faster
growing cells, with peculiar, but by now well known consequences.
Let us illustrate this by taking the example of the same cell inoculated into a richer medium (minimal salts + casamino acids) in
which it can grow twice as fast, doubling in mass every 30 minutes.
The cell is exactly the same to begin with; it has a mass = M_i and
a single chromosome with a single origin. The first round of chromosome replication is initiated at zero time and the first replication forks meet at the terminus 40 minutes later. However, by
30 minutes the cell will already have doubled its mass ($2.M_i$) and
will therefore start new rounds of replication at the two copies of
the chromosome origin (formed as soon as the first round of replication began). The result is that between 30 and 40 minutes there are
three pairs of replicating forks on the chromosome. Thus, in media
where the mass doubling time is less than 40 minutes, successive
rounds of chromosome replication will overlap, giving a so-called
"dichotomously replicating" chromosome (Yoshikawa, O'Sullivan &
Sueoka, 1964; Oishi, Yoshikawa & Sueoka, 1964). A cell inoculated
into the richest medium (broth) will double in mass every 20 minutes
and initiate DNA replication every 20 minutes, so that the chromosome
is dichotomous throughout the cell cycle. Let us now consider the
relationship between this process and cell division.

In Figure 1 we show the results of a similar "thought" experiment but in this case we have measured the mass/cell ratio, which
will be halved at every division. It is time now also to introduce the concept of the "unit cell" (Donachie & Begg, 1970). A
unit cell is simply the cell of mass = M_i with its single unreplicated chromosome. (As we shall see, larger cells behave in many
ways as if they consisted of a group of independent unit cells.)
If, as before, we inoculate a unit cell into a medium in which its
mass will double every 60 minutes, then the mass of the cell will
double in 60 minutes, after which the cell will divide, so that
the mass of the cell will fluctuate, in this medium, between 1 and
2 unit cell equivalents. At 60 minutes, there will be two copies
of the chromosome, one per daughter cell. If we consider the case
in which the unit cell is inoculated into a richer medium (mass
doubling time 30 minutes) then we might expect the first division
to come also after the first doubling in cell mass. However, by
30 minutes the first round of chromosome replication has not been
completed so that, if division were to take place at that time,

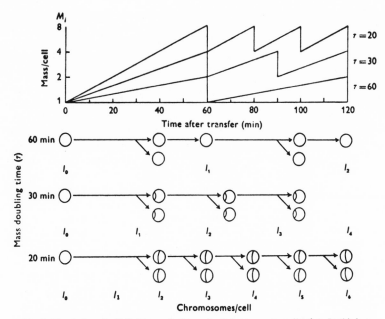

Fig. 1. The timing of cell division and of rounds of chromosome replication. In this hypothetical experiment three single cells, each of mass $1 \cdot M_i$, are inoculated into three different media, in which the mass doubling times are 60, 30 and 20 min respectively. In each case the first division takes place 60 min after inoculation into the new medium. In consequence, cell size at this first division is a function of the growth rate in the particular medium. (In fact, log (mass/cell) $= k/\tau$.) Each successive division takes place 60 min after each successive doubling of the initial unit cell mass (M_i). After the first division, therefore, the interval between successive divisions is equal to the mass doubling time.

At 0 min, each cell has a single unreplicated chromosome and starts replicating it immediately. Successive rounds of chromosome replication are initiated at each successive doubling of the unit cell mass (M_i). $I_0, \ldots I_n$, represent the times of successive initiations in the three cell lines.

FIGURE 1. (Caption as in Figure 3 in W.D. Donachie, N.C. Jones and R.M. Teather. Symp. Soc. Gen. Microbiol. <u>23</u>: 14, 1973.) Copy attached.

only one sister cell could receive the partly replicated chromosome while the other cell would get no DNA. In fact, as one might now expect, the cell does not divide until there are at least two chromosome copies present. The first division does not actually take place until 60 minutes, i.e. 20 minutes after the completion of the first round of chromosome replication (Helmstetter & Cooper, 1968; Cooper & Helmstetter, 1968). Because the cell will have completed two mass doublings by this time, the mass of the cell will therefore fluctuate between 2 and 4 unit cell equivalents in successive cell cycles. Similarly, if the unit cell is inoculated into broth (mass doubling time 20 minutes) the cell will first divide 60 minutes later when its mass will have reached 8 unit cell equivalents. Thus the mass of a cell growing with a doubling time of 20 minutes will

fluctuate between 4 and 8 unit cell equivalents. This relationship between cell size and growth rate was first described quantitatively by Schaechter, Kjeldgaard *Maaløe (1958). This explanation however did not come until ten years later (Donachie, 1968)
after the description of the chromosome replication cycle by Helmstetter and his colleagues. In our experiment the first division
therefore always comes after 60 minutes, regardless of the nature
of the medium or of the growth rate of the cells. Examination of
Figure 2 will show that each successive division (in every medium)

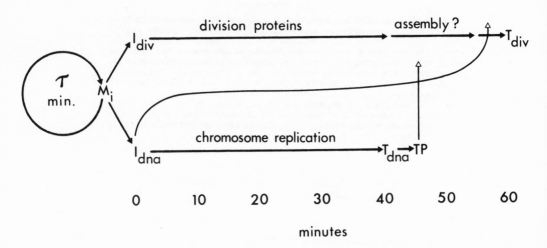

FIGURE 2. Model for the control of the timing of cell division in
E. coli. The initiation mass (M_i) is attained every mass doubling
time (T min.), and at this time two parallel sequences of events are
initiated. "I_{div}" indicates the initiation of a sequence of events,
required for division, which takes 60 minutes to complete and consists of at least three sequential processes. Initiation of chromosome replication (I_{dna}) sets up conditions which block the completion of a late step in the division sequence. However, termination
of chromosome replication (T_{dna}) allows the synthesis of termination
protein (TP) which relieves this inhibition, so that division will
normally occur at the termination of the division sequence at 60 min.

also follows 60 minutes after each doubling in number of unit cell
equivalents (i.e. 60 minutes after cell mass reaches $2^n.M_i$, where
n is any integer, 0, 1, 2,...n). The cardinal rule governing cell
division is therefore that it takes place about 60 minutes after
each doubling in number of unit cell equivalents. Note also that
one division event (one septum) will take place per number of unit
cell equivalents present 60 minutes earlier (this can also be worked
out from Figure 1).

I now want to concentrate on the events taking place in the
interval between the attainment of unit cell mass and subsequent
division. The most interesting point which we have made so far is
that the length of this interval is largely independent of the
growth rate of the cell. For this reason, I think it is fair to
say that the events leading to cell division resemble a clock, in-
asmuch as the rate at which they take place is independent of other
events in the cell, such as the rates of macromolecular syntheses.
As we have seen, chromosome replication is another clock-like process, since the time (C) required for replication forks to travel
from origin to terminus is about 40 minutes in most media. Both
the division clock and DNA replication are initiated at the same
time and it therefore once seemed very likely that the division
"clock" was simply a reflection of the process of DNA replication.
In support of this idea is the well known dependence of cell division on continuing DNA replication. Thus, many bacteria, including
E. coli, form long aseptate filaments if the cells are grown under
conditions where DNA replication is specifically inhibited (for
example after UV irradiation). Clark (1968) and Helmstetter &
Pierucci (1968) using synchronous cultures, showed further that each
division required the prior completion of a round of chromosome
replication. However, I will attempt to show that despite this and
other evidence, it is unlikely that chromosome replication is normally responsible for the timing of cell division.

I will start by considering which other events are required for
cell division to take place. Pierucci & Helmstetter (1969) have
shown that cell division also requires a prior period of protein
synthesis. Thus, if chromosome synthesis is initiated, but protein
synthesis is then inhibited, the round of chromosome replication is
able to be completed (see Maaløe and Kjeldgaard, 1966) but cell division will not take place. In a series of experiments of this kind
using synchronous cultures, Pierucci & Helmstetter found that about
40 minutes of protein synthesis was required before each successive
division. It is extremely interesting to note that it is not a
particular amount of protein synthesis which is required for division but rather a particular duration of synthesis. Approximately
40 minutes of prior protein synthesis is needed, no matter how fast
the cell is growing or what the rate of protein synthesis is. In
fast growing cells these periods of protein synthesis must be overlapping in time which in turn suggests that they must be spatially
separated from each other within the cell. Here then is another
clock-like process concerned with division.

If protein or DNA synthesis is inhibited during a period of about 20 minutes before the expected time of cell division, then that division will take place normally. The duration of this interval is also constant at different growth rates (Helmstetter, 1967; Helmstetter & Cooper, 1968; Cooper & Helmstetter, 1968) and therefore indicates the existence of a third clock-like process, which in this case involves neither RNA, DNA nor protein synthesis. This in turn indicates that both the required 40 minutes of DNA replication and the 40 minutes of protein synthesis are completed 20 minutes before division (and therefore must both have begun at about the same time, namely at a cell mass equal to $2^n.M_i$).

How are these three clocks interrelated and how is the time of division determined?

To try to answer these questions, I will first return to a consideration of the nature of the interaction between DNA replication and cell division. Why does failure to terminate a chromosome round interfere with division? Before discussing the experimental evidence on this point, it must be pointed out that failure to terminate rounds does not always prevent cell division. An example of this is provided by B. subtilis, in which DNA synthesis may be specifically inhibited without preventing the continuation of cell growth and division (Gross, Karamata & Hempstead, 1968; Donachie, Martin & Begg, 1971). In populations of wild-type cells in which DNA synthesis has been prevented, cell division continues at almost the normal rate so long as growth itself continues. As a result, up to half of the resultant population will come to consist of cells which contain no DNA (although otherwise normal in size and appearance). In wild-type E. coli, as we have said, inhibition of DNA rounds prevents division but certain mutants exist in which, like B. subtilis, inhibition of DNA synthesis does not prevent division (Hirota, Jacob, Ryter, Buttin & Nakai, 1968; Inouye, 1969, 1971). A similar mutant has been described in Salmonella typhimurium (Spratt & Rowbury, 1971). Some of these mutants are temperature sensitive for DNA synthesis; e.g. CRT46 (Kohiyama, Cousin, Ryter & Jacob, 1966) is a mutant which is able to complete rounds of replication at the restrictive temperature but which is unable to initiate new rounds. On shifting these cells from the permissive to the restrictive temperature, DNA synthesis continues until all rounds of chromosome replication are completed and then stops; cell growth and division continue and eventually large numbers of DNA-less cells are produced. In this mutant, termination of preexisting rounds of replication is required for subsequent division but the single termination which takes place at the restrictive temperature is sufficient for several successive divisions. Inouye (1971) has shown that the rec A$^-$ mutation confers the ability to divide in the absence of DNA synthesis in much the same way as does B. subtilis. In both these coli mutants, the DNA-less cells which are formed appear to be normal in size and this therefore is a strong argument for

the existence of a timing mechanism for division which is independent of DNA replication.

To return now to the mechanism by which DNA synthesis is coupled to cell division in wild-type E. coli, I would like to discuss
some recent experiments carried out in Edinburgh (N.C. Jones & W.D.
Donachie, Nature, in press). These experiments were designed to
study the nature of the events which take place between termination
of chromosome rounds and cell division. To do this, a round of
chromosome replication was synchronised by a modification of the
standard "lining up" procedures. The strain used, E. coli Kl2 Km7
T⁻ (a thymine requiring derivative of the Km7 strain of Matzura,
Molin & Maaløe, 1971) is partially resistant to rifampicin and, more
importantly for us, resumes RNA synthesis almost immediately after
removal of an inhibitory amount of rifampicin. This allows one to
add and remove rifampicin to stop and start RNA and protein synthesis
at will. To produce a synchronous round of chromosome replication
exponentially growing asynchronous cells were treated with rifampicin
for a period of time sufficient to allow all chromosome rounds to be
completed. Rifampicin was then removed and the cells allowed to
resume RNA and protein synthesis. At the same time thymine was
removed from the medium so that DNA synthesis could not take place.
When the mass of the cells had doubled, so that every cell had
reached a mass of M_i or more, thymine was added back. Every cell
then initiated a round of chromosome replication at the same time.
This synchronous round was shown to be completed about 40 minutes
later. In the absence of further treatments, a synchronous wave of
cell division took place at about 45-50 minutes (Donachie, Hobbs &
Masters, 1968; Donachie, 1969; Leighton & Donachie, 1970). Under
these conditions therefore, the delay between completion of chromosome rounds and division is reduced to 5-10 minutes, instead of
the 20 minutes found in the undisturbed cell cycle. We have described above how neither RNA, DNA nor protein synthesis seems to
be required during an interval (D. Helmstetter & Cooper, 1968) of
about 20 minutes before division. We would therefore guess that
neither RNA nor protein synthesis would be required during the 5-10
minute interval between termination and division in the present experiment. However, it was found that the addition of chloramphenicol,
tetracycline or rifampicin during this interval did prevent cell
division. Inhibition of RNA and protein synthesis for various periods of time between initiation of chromosome replication and division
showed that the required period of RNA and protein synthesis must
take place between 40 and 50 minutes, i.e. at about the time of
termination of the synchronous chromosome rounds. Moreover, we have
shown that this period of synthesis must follow the completion of
chromosome replication (or, at least of the replication of a late
segment of the chromosome) by allowing the synchronous round to
proceed for only 20 minutes and then interrupting it (by removing
thymine) for a further 45 minutes. As expected, the cells did not
divide so long as termination was prevented (Donachie, 1969) but did

so about 20 minutes after the interrupted rounds were allowed to
complete. However, once again the addition of rifampicin before the
time of completion of the delayed rounds prevented cell division and
therefore the required RNA and protein cannot be made until at least
the major part of the chromosome has been replicated. Our working
hypothesis is therefore that the replication of a late segment of
the chromosome, perhaps the terminus itself, initiates the transcription of a gene or genes with the eventual synthesis of a protein
or proteins specifically required for one of the final stages in
septum formation. In addition, we know that this protein, which we
have called "termination protein" (Donachie, Jones and Teather, 1973),
although normally synthesized immediately after the completion of
chromosome replication, can also be made at a later time, if its
synthesis is prevented at the normal time. Replication of the
relevant chromosome segment therefore allows the production of the
termination protein immediately but these conditions persist until
the synthesis has taken place. Thereafter, this synthesis must
once again be repressed, until the next termination event again
induces it. This represents a very unusual control system and we
hope that further experiments will help to make the mechanism
clearer. The nature of the termination protein is also unknown but
differential labeling experiments (Jones, unpublished) have shown
that one, or perhaps two, proteins of the cell membrane are synthesized specifically at about the time that the termination protein is being formed. Messer (this symposium) has presented evidence that the replication of a very short terminal segment of the
chromosome requires protein synthesis and the question therefore
immediately arises as to whether our termination protein is required for such replication. However, experiments in which the
synthesis of termination protein was allowed to take place in the
presence of nalidixic acid (to inhibit DNA synthesis) demonstrated
that cell division took place whether or not further DNA synthesis
was allowed. We therefore think it unlikely that the termination
protein plays any part in DNA replication.

     I will now summarize this section by presenting our current model
of the way in which the timing of cell division is determined. This
model, we believe, is consistent with the experimental observations
but it includes many assumptions which are based on rather sketchy
evidence. Nevertheless, I believe it is probably correct in its
main outline. (It should also be emphasized that this model does
not attempt to include all known events concerned with division but
only those which are probably important for its timing in the normal
cell cycle.) The model proposes that whenever a cell attains a mass
(or volume) equal to $2^n.M_i$, then DNA replication is initiated at 2^n
chromosome origins. At the same time, 2^n separate processes are
initiated which will lead to the formation of 2^n septa 60 minutes
later. The nature of this "division clock" is not known but it
requires protein synthesis for the first 40 minutes. The completion
of this period of protein synthesis in turn initiates a final process

which takes about 20 minutes to complete. Since neither RNA, DNA
nor protein synthesis is required during this period, we assume,
for the moment, that it involves some sort of assembly of preformed
components into a septum or septum precursor. The final completion
of the septum, however, requires the participation of the termination protein. In the normal cell cycle this is synthesized immediately after the completion of the chromosome rounds, i.e. at
40-45 minutes, and is therefore available 15-20 minutes before the
final stage of division takes place. This model, therefore, states
that the rate limiting process for division in the undisturbed cell
cycle is the completion of the events in the division clock and not
those dependent on chromosome replication.

In B. subtilis we assume that either termination protein is not
required for division, or that the equivalent of the coli termination protein is not dependent on chromosome replication for its
synthesis. In the various coli mutants we think it most likely
that division is no longer dependent on termination protein. The
argument for this conclusion is based on the observation that cell
division is blocked in cells which contain an irradiated plasmid,
such as F, Col I or Pl (Monk, 1969). In the case of UV-irradiated
Pl, replication of the phage genome is blocked until the thymine
dimers are excised and the damage repaired. During this time, the
host cell is able to grow but unable to divide. Uvr⁻ cells, which
are unable to repair the damage in the incoming plasmid, are perma-
nently prevented from dividing (R. Teather, personal communication).
In contrast, recA⁻ cells are probably not prevented from dividing
by the introduction of an irradiated plasmid (Rosner, Kass &
Yarmolinsky, 1968). Thus, it seems possible that division is inhibited by initiation of DNA replication, perhaps in association
with a specific site on the cell membrane, and that this inhibition
is relieved only after that round of replication has been completed.
The UV-irradiated plasmids should be able to initiate replication
but the progress of this DNA synthesis would be halted by the first
thymine dimer. Thus, division would also be blocked until the DNA
was repaired and replication completed. In the case of the chromosome itself, we must make the additional postulate that termination
protein is required to release the division process from the inhibition set up at the time of initiation of chromosome replication.
The recA⁻ mutant cell would then be insensitive to the inhibition of
division in this strain would not be dependent on the synthesis of
termination protein. In the CRT46 strain the situation is somewhat different and cell division at the restrictive temperature is
inhibited until those rounds of chromosome replication which had
been initiated at the permissive temperature have been completed
(Hirota et al., 1968). No further initiations take place at the
restrictive temperature and division is therefore presumably not
inhibited for that reason. One would predict that division in this
strain at the restrictive temperature would still be inhibited by
the introduction of a UV-irradiated plasmid (unlike the division of
recA⁻ cells.

Our view of the interaction between DNA replication and division is therefore that initiation of chromosome rounds (or the replication of any other membrane associated plasmid) sets up an inhibition of division which is relieved only by the completion of these rounds (perhaps by the dissociation of the replicated DNA from a specific membrane site; Donachie, 1969). Termination protein would then be required specifically to bring about the release of this inhibition. At the moment we know only of termination protein required for the release of cell division from the inhibition set up by chromosome replication but there may also be specific termination proteins required to release the inhibition set up by the replication of ether replicons. This model for the control of the timing of cell division is summarized in the diagram in Figure 2.

In the final part of this talk, I would like to discuss the way in which the localization of the division site is determined. In untreated cells, the site of division is always exactly in the centre of the long axis of the cylinder. In principle, the spatial information necessary for the location of this site could be generated in one of two ways. One way would be if the cell could in some way measure the distance between its two poles and make a division site equidistant between them as soon as the two poles were a certain minimum distance apart. This might occur if, for example, each cell pole inhibited the formation of a new pole within a certain distance. Such a model would make no prediction about the mode of growth of the cell surface but it would suggest, at least in its simplest form, that division should always begin at a particular constant cell length. In fact, cell length at division increases with increasing growth rate. This need not, however, be a fatal observation for this model; one could suppose that the division process was initiated at a constant length but that it took a significant fraction of the generation time to complete. The cell length at the end of this period would then depend on the growth rate. (If for example the division process was initiated at a constant cell length 20 minutes before final cell separation, then the expected ratio of cell lengths at separation would be 1.6: 1 for cells growing with generation times of 20 versus 60 minutes. This is quite similar to the observed ratio of cell lengths.) Another way in which the location of the division site could be determined is by the generation of discontinuities between sections of the cell surface which are actively growing and older sections which have stopped growing. The clearest example of this kind of growth is provided by Streptococcus species (for reviews see Cole, 1965; Higgins & Shockman, 1971). The basic unit in this organism is in the form of a diplococcus, i.e. two partially-completed spheres with a central furrow where they are joined together. Growth of the cell wall takes place by addition of new material at this furrow. The two spheres are completed by the formation of a cross wall at this position and wall growth then resumes at the equator of each new spherical cell. As it does so, a visible raised equatorial ring formed by a thicker portion of the

wall splits into two. Growth at the central furrow then results
in the progressive insertion of new wall material between the two
halves of the ring, so that they move apart as the cell grows.
When the two new hemispheres are completed, at the end of the next
cell cycle, these rings are each located at the equators of the
newly formed sister spheres. Growth begins again at these sites
and the process repeats. Thus the position of the site of wall
growth and the position of the site of cell division are identical
and new growth and division sites arise at the junction between
wall which has been made in a previous cell cycle. The means by
which these successive sites of growth are initiated at particular times are unknown, but it is clear how they come to be located at the centres of the cells.

Unfortunately, the type organism for molecular biology is E.
coli, rather than Streptococcus, and it has proved very much harder
to determine the way in which the surface of the coli cell grows.
One complicating factor is that E. coli, is a gram-negative organism and therefore has a much more complicated cell envelope than do
gram-positive organisms such as Streptococcus. The E. coli cell
envelope consists of an outer lipo-protein membrane, which also
contains the lipo-polysaccherides, a mucopeptide layer, in the form
of a single molecule which is responsible for the shape of the cell,
and an inner lipo-protein membrane. It is by no means certain that
all three layers grow in the same way. Attempts to uncover the way
in which this envelope grows have been numerous and various but I
will not attempt to review them all here. Two of the most important
of these pieces of work have been presented at this symposium. Dr.
Kepes described experiments which elegantly demonstrate that various permeases and other enzymes which are components of the inner
membrane are segregated during cell growth and division as if new
membrane was being inserted at only one or two sites per cell. Dr.
Hirota has described the work on the growth of the mucopeptide layer
which he has done in collaboration with Doctors Ryter and Schwarz.
This work demonstrates that, in all cells, there is a narrow central
band in which diaminopimelic acid is incorporated into mucopeptide.
After initial incorporation at this site, the labeled DAP is then
translocated by some unknown mechanism to other positions in the
mucopeptide layer. These final positions seem to be randomly located over the cell surface. Inhibition of DNA synthesis, which
prevents cell division as we have discussed above, also prevents
the incorporation of DAP into a localized central band. It therefore seems likely that this band represents the location of the
future site of cell division. Since mucopeptide synthesis and
envelope growth continue at a more or less normal rate in the absence of DNA replication, it also seems clear that net growth in
the mucopeptide layer does not require incorporation of DAP to be
localized initially. These results, unfortunately, appear to be
at variance with earlier results (Donachie & Begg, 1970). In these
earlier experiments the location of the potential division site was

taken to be the point at which the mucopeptide layer was autolyzed
when net synthesis was inhibited by penicillin (Schwarz, Asmus &
Frank, 1969). It was indeed found that this site was located in
the centre of cells which were growing under conditions where cell
size was relatively large (as in rich media). However, in slow
growing cells in poorer media (i.e. in smaller cells: see above)
it was found that this site varied in position according to cell
length (and hence according to stage in the cell cycle). It was
clear that, in these smallest cells, this site appeared first next
to one pole and that this pole was the one formed in the cell division immediately preceding. As the cell grew in length, the site
maintained a constant distance from the older of the two poles and,
therefore, moved progressively further away from the newer pole.
When the cell had doubled in length and was ready to divide, this
site was therefore in the exact centre of the cell. Hirota, Ryter
& Schwarz did not find this to be the case for the site of initial
DAP incorporation in cells which were growing slowly in poor medium.
The resolution of this paradox will obviously require further experiments; it should be noted in this regard that they used a K12 strain
while we studied B/r and 15 strains.

Another approach to the elucidation of the way in which the cell
surface grows was suggested by a Paper by Leal & Marcovitch (1971).
This paper described the kinetics of the appearance of phenotypic
resistance to phage T6 after transfer by mating of a gene conferring
resistance (tsx-) to cells which had previously been sensitive to
the phage (tsx+). Phenotypically resistant cells began to appear
only after a few divisions of the recipient cells, suggesting that
the receptor sites for T6 segregated at division as one or two large
blocks per cell. Thus, the receptors for T6, which are located in
the outer membrane of the cell, would be behaving in much the same
way as the permeases in the inner membrane, as described by Autissier,
Jaffe & Kepes, 1971. The advantage of studying the T6 receptors is
that their spatial localization can be revealed by determining the
positions of attached T6 phages. The mating system used by Leal &
Marcovitch could not be used in such studies because the recombinant
cells represent only a small fraction of the total population of
male and female cells. Accordingly, a system was devised (Begg &
Donachie, in preparation) in which the synthesis of receptor sites
was made temperature sensitive. Thus, in a particular strain, T6
receptors are synthesized about 10 times faster at one temperature
than at another. The spatial distribution of receptors can then be
determined in cells which have grown for part of their cycle at one
temperature and for the other part at another. This is done by
adding T6 phages at high multiplicity, fixing immediately and examining the cells with the electron microscope. Studies of the pattern of receptors has now been made under a number of conditions
and some conclusions can be made. Firstly, newly synthesized receptors are incorporated with equal probability into all parts of
the outer membrane (in this behaviour they again resemble most of

the permeases of the inner membrane: Autissier, 1971; Kepes, 1972).
Secondly, once inserted into the membrane, the sites do not move.
Thirdly, performed membrane areas are conserved; new membrane being
made at only a few localized sites. The results also indicate that
new membrane is synthesized at the cell poles (the number of poles
at which growth takes place appears to be a function of cell size).
This surprising result is consistent with the observations previously
reported by us (Donachie & Begg, 1970) but make it necessary to modify the growth model which we advanced to explain our observations
at that time. This modified "unit cell" model is set out in Figure 3.

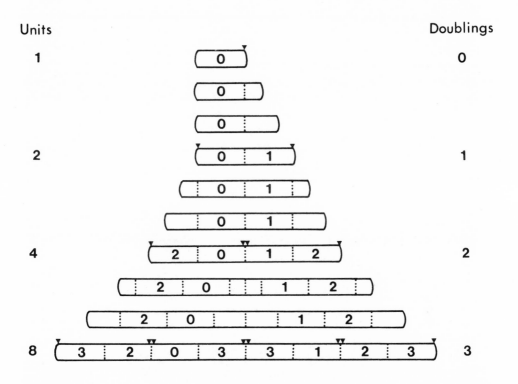

FIGURE 3. Model for the growth of the cell envelope in E. coli. The
model is based on the observed segregation pattern of receptor sites
for phage T6 (Begg & Donachie, unpublished) and on the number and positions of potential division sites (Donachie & Begg, 1970). The sequence starts with a single unit cell (top) which grows (at a rate pro
portional to the number of growth sites) without dividing (this being
achieved experimentally by the addition of low concentrations of penicillin). Unit cells are indicated as sections separated by dotted
lines and numbered according to the generation (mass doubling) at which
they were completed. Arrowheads indicate sites of new growth at the
time of unit cell doublings.

In making this model we have assumed that the growth of the outer membrane (as deduced from the distributions of T6 receptor sites) is correlated with the positions of the potential division sites (as determined by the positions of penicillin induced autolysis of the mucopeptide layer). This is a gratuitous assumption at the moment but one which we make for the sake of producing a provocative model which will suggest further experiments. This model (like the original unit cell model proposed in 1970) proposes that the smallest cell, the "unit cell", grows at one pole but that a cell of size two unit cell equivalents grows at both poles. The location of potential division sites is determined by the locations of the junctions between sections of the cell envelope of different ages. Unfortunately, this new model does not explain the localization of the DAP incorporation site as determined by Hirota, Ryter & Schwarz. The way in which the cell envelope of E. coli grows remains unknown. It is becoming increasingly clear, however, that at least some of the components of some of the envelope layers are synthesized at a small number of sites per cell, and that some of the structure of these layers is conserved intact for many generations. Thus, in E. coli as in Streptococcus, the localization of division sites could be determined by the localization in the envelope of junctions between sections of different ages.

REFERENCES

1. Autissier, F., Jaffe, A. and Kepes, A. (1971) Molec. Gen. Genet. 112: 275.
2. Bird, R.E., Louarn, J., Martuscelli, J. and Caro, L.G. (1972) J. Mol. Biol. 70: 549.
3. Clark, D.J. (1968) Cold Spring Harbor Symp. Quant. Biol 33: 823.
4. Clark, D.J. and Maale, O. (1967) J. Molec. Biol. 23: 99.
5. Cole, R.M. (1965) Bacteriol. Rev. 29: 326.
6. Cooper, S. and Helmstetter, C.E. (1968) J. Mol. Biol. 31: 519.
7. Donachie, W.D. (1968) Nature 219: 1077.
8. Donachie, W.D. (1969) J. Bacteriol. 100: 260.
9. Donachie, W.D. and Begg, K.J. (1970) Nature 227: 1220.
10. Donachie, W.D., Hobbs, D.G. and Masters, M. (1968) Nature 219: 1079.
11. Donachie, W.D., Jones, N.C. and Teather, R. (1973) Symp. Soc. Gen. Microbiol. 23: 9.
12. Donachie, W.D., Martin, D.T.M. and Begg, K.J. (1971) Nature New Biol. 231: 274.
13. Gross, J.D., Karamata, D. and Hempstead, P.G. (1968) Cold Spring Harbor Symp. Quant. Biol. 33: 307.
14. Helmstetter, C.E. (1967) J. Mol. Biol. 24: 417.
15. Helmstetter, C.E. and Cooper, S. (1968) J. Mol. Biol. 31: 507.
16. Helmstetter, C.E. and Pierucci, O. (1968) J. Bacteriol. 95: 1627.
17. Higgins, M.L. and Shockman, G.D. (1971) CRC Critical Reviews in Microbiology 1: 29.

18. Hirota, Y., Jacob, F., Ryter, A., Buttin, G. and Nakai, T. (1968) J. Mol. Biol. 35: 175.
19. Hohlfeld, R. and Vielmetter, W. (1973) Nature, in press.
20. Inouye, M. (1969) J. Bacteriol. 99: 843.
21. Inouye, M. (1971) J. Bacteriol. 106: 539.
22. Jones, N.C. and Donachie, W.D. (1973) Nature, in press.
23. Kohiyama, M.D., Cousin, D., Ryter, A. and Jacob, F. (1966) Ann. Inst. Pasteur 110: 465.
24. Leal, J. and Marcovich, H. (1971) Ann. Inst. Pasteur 120: 467.
25. Leighton, P.M.L. and Donachie, W.D. (1970) J. Bacteriol. 102: 810.
26. Maale, O. and Kjeldgaard, N.O. (1966) The Control of Macro-molecular Synthesis, W.A. Benjamin, New York and Amsterdam.
27. Masters, M. and Broda, P.M.A. (1971) Nature New Biol. 232: 137.
28. Matzura, H., Molin, S. and Maale, O. (1971) J. Molec. Biol. 59: 17.
29. Monk, M. (1969) Molec. Gen. Genet. 106: 14.
30. Oishi, M., Yoshikawa, H. and Sueoka, N. (1964) Nature 204: 1069.
31. Pierucci, O. and Helmstetter, C.E. (1969) Fed. Proc. 28: 1755.
32. Prescott, D.M. and Kuempel, P.L. (1972) Proc. Nat. Acad. Sci. U.S. 69: 2842.
33. Rosner, J.L., Kass, LR. and Yarmolinsky, M.B. (1968) Cold Spring Harbor Symp. Quant. Biol. 33: 785.
34. Schaechter, M., Maale, O. and Kjeldgaard, N.O. (1958) J. Gen. Microbiol. 19: 592.
35. Spratt, B.G. and Rowbury, R.J. (1971) Molec. Gen. Genet. 114: 35.
36. Yoshikawa, H., O'Sullivan, A. and Sueoka, N. (1964) Proc. Nat. Acad. Sci. U.S. 52: 973.

PARTICIPANTS IN THE

NATO ADVANCED STUDY INSTITUTE

CORTINA D'AMPEZZO, ITALY

Alberts, Bruce.   Biological Sciences, Princeton University, Prince-
    ton, New Jersey, U.S.A.

Allin, P.   Institut du Radium Biologie, Faculté des Sciences, Orsay
    91, France.

Autissier, Françoise.   Biomembranes, Institut de Biologie Molécu-
    laire, Paris 5ème, France.

Baril, Earl.   Worcester Foundation for Experimental Biology,
    Shrewsbury, Mass. 01545, U.S.A.

Bautz, Ekhart.   Zoologisches Institut, Universität Heidelberg,
    Heidelberg, West Germany.

Berg, D.   Institut de Biologie Moléculaire, 30 Q. Ecole de Médecine,
    1211 Genève 4, Switzerland.

Bernardi, G.   Institut de Biologie Moléculaire, Paris, 5ème, France.

Bertazzoni, V.   Institut de Biologie Moléculaire, Paris, 5ème,
    France.

Blumenthal, Alan.   Department of Biochemistry, Stanford Medical
    School, Stanford, California 94305, U.S.A.

Bollum, Fred.   Department of Biochemistry, University of Kentucky,
    School of Medicine, Lexington, Kentucky, U.S.A.

Bonhoeffer, Friedrich.   Friedrich-Miescher Laboratorium der Max Planck
    Gesellschraft, Tübingen, Germany.

Brown, Neal C.   Department of Cell Biology and Pharmacology, Uni-
    versity of Maryland, School of Medicine, 600 W. Redwood St.,
    Baltimore, Maryland 21201, U.S.A.

Caro, Lucien.   Institut de Biologie Moléculaire, Quai Ecole de
    Médecine, 1121 Genève 4, Switzerland.

Comings, David E.   Department of Medical Genetics, City of Hope
          National Medical Center, Duarte, California 91010, U.S.A.

Devoret, R.   Lab. d'Enzymologie CNRS, 91, Gif-sur-Yvette.

Donachie, W. D.   MRC Molecular Genetics Department, University of
          Edinburgh, Edinburgh, Scotland.

Dürwald, H.   Max Planck Institut für Medizinische Forschung,
          Abteilung für Molekulare Forschung, Heidelberg, West
          Germany.

Emmerson, P.   Department of Biochemistry the University, Newcastle/
          Tyne, VE1 7RU, England.

Ferdinand, F.   Friedrich-Miescher Lab., Max Planck Institut, 74
          Tübingen, West Germany.

Forterre, P.   Institut de Biologie Moléculaire, Paris 5ème, France.

Frenkel, Gerald.   Department of Genetics, Weizmann Institute,
          Revhovot, Israel.

Gallwitz, D.   Philipps University, Medizine Forschung, 355 Marburg,
          Lahnberge

Glaser, Donald A.   Department of Molecular Biology, University of
          California, Berkeley, California 94720, U.S.A.

Green, Elizabeth.   Natick Laboratories, Natick, Mass., U.S.Ą.

Gurgo, Corrado.   Institute for Molecular Virology, St. Louis
          University, School of Medicine, 3681 Park Avenue,
          St. Louis, Missouri 63110, U.S.A.

Henry, C.   Max Planck Institut, 74 - Tübingen, West Germany.

Henry, N.   Faculté des Sciences, ULB, 1640 Rhode-St.-Genèse,
          Belgique.

Henry, T.   Max Planck Institut, 74 - Tübingen, West Germany.

Hirota, Y.   Service Genetique, Institut Pasteur, Paris, 15ème,
          France.

Holland, I. B.   School of Biological Sciences, University of
          Leicester, Department of Genetics, Leicester, LE1 7RH,
          England.

Huberman, Joel.   Department of Biology, Massachusetts Institute of
        Technology, Cambridge, Massachusetts, U.S.A.

Jaenisch, Rudolf.   Department of Biology, Princeton University,
        Princeton, New Jersey 08540, U.S.A.

Johnson, Lee.   Department of Biology, Massachusetts Institute of
        Technology, Cambridge, Mass. 02139, U.S.A.

Joseleau-Petit, Danielle.   Biomembranes, Institut de Biologie
        Moléculaire, Paris 5ème, France.

Juricek, Diane.   Department of Biology, Emory University, Atlanta,
        Georgia 30322, U.S.A.

Kaerner, Hans C.   Max Planck Institut für Medizine, Heidelberg,
        West Germany.

Kasamatsu, Harumi.   Biology Division, California Institute of
        Technology, Pasadena, California.

Kaufman, Elliot.   B-11 Moffet Labs, Department of Biochemical
        Sciences, Princeton University, Princeton, New Jersey
        08540, U.S.A.

Kepes, Adam.   Biomembranes, Institut de Biologie Moléculaire, Faculté
        des Sciences de Paris, Paris 5ème, France.

Klenow, H.   Univ. Biokemiske Inst. B., 2100 København Ø, Denmark.

Knippers, Rolf.   Friedrich-Miescher Laboratorium der Max Planck
        Gesellschaft, Tübingen, Germany.

Kohiyama, Masamichi.   Biomembranes, Institut de Biologie Moléculaire,
        Faculté des Sciences de Paris, Paris, 5ème, France.

Kolber, Alan.   Institut fur Virusforschung, Deutsches Krebs Forschungs
        Zentrum, 69 - Heidelberg, West Germany.

Kornberg, Arthur.   Biochemistry Department, Stanford University,
        Stanford, California, U.S.A.

Kornberg, Thomas.   Department of Biology, Massachusetts Institute
        of Technology, Cambridge, Massachusetts, U.S.A.

Kraus, S.   Zentralinst. für Mikrobiologie, 69, Jena, East Germany.

Leibowitz, Paul.   Department of Molecular Biology and Microbiology,
        Tufts University, School of Medicine, Boston, Mass., U.S.A.

Levine, Arnold.   Department of Biology, Princeton University,
          Princeton, New Jersey, U.S.A.

Liebat, J. C.   Institut de Microbiologie, Faculté des Sciences
          Orsay, France.

Lindahl, G.   Institut Pasteur, 25, rue du Docteur Roux, Paris 5ème,
          France.

Ljskengquist, E.   Karolinska Inst. für Microbiologie, 10401
          Stockholm 60, Sweden.

Lubochinsky, B.   Faculté des Sciences, Poitiers, France.

McGavin, S.   University of Dundee, Isotope Lab., Dundee, Scotland.

Masters, Millicent.   MRC Molecular Genetics Unit, Edinburgh,
          Scotland.

Mather, Jennie.   c/o Dr. H. Stern, Biology Department, University
          of California, San Diego, La Jolla, California 92037, U.S.A.

Messer, W.   Max Planck Institut, 1 Berlin 33 (Dahlem), Germany.

Milewski, Elisabeth.   Institut de Biologie Moléculaire, Paris,
          5ème, France.

Mizuno, Nobuko.   Veterans Administration Hospital, Experimental
          Surgery, Minneapolis, Minnesota 55417, U.S.A.

Moses, Robb.   Department of Biochemistry, Baylor College of Medicine,
          1200 Moursand Ave., Houston, Texas 77025.

Newton, A. A.   Department of Biochemistry, Univ. Cambridge,
          Tennis Court Road, Cambridge, England.

Op den Kamp, J.   Biochemisch Lab. der Riyks Univ., Utrecht, Holland.

O'Sullivan, M. A.   Biochemical Sciences Div., Moffett Laboratories,
          Princeton University, Princeton, New Jersey, U.S.A.

Otto, B.   Friedrich-Miescher Lab., Max Planck Institut, 74 -
          Tübingen, West Germany.

Perez-Bercoff, R.   Lab. Pasteur, Inst. du Radium, 26, rue d'Ulm,
          Paris 5ème, France.

Razin, A.   Dept. of Cellular Biology, Hebrew University, Hadassah
          Medical School, Jerusalem, Israel.

Recondo, A. M. de.   Institut du Cancer, Villejuif 91, France.

Reichard, P.   Karolinska Inst., Biochemical Dept., S10401 Stockholm
        60, Sweden.

Riva, S.   Department of Microbiology, Lepetit SA - via Durando 38,
        20100 Milano, Italy.

Rougeon, F.   Institut de Biologie Moléculaire, Paris, 5ème, France.

Sauer, Gerhart.   Institut für Virusforschung, Deutsches Krebs-
        forschungszentrum, Heidelberg, West Germany.

Schaechter, Moselio.   Department of Microbiology, Tufts University,
        Boston, Massachusetts, U.S.A.

Schultz, Stephen R.   Department of Basic Research, Eastern Pennsyl-
        vania Psychiatric Institute, Henry Ave. and Abbotsford Rd.,
        Philadelphia, Pennsylvania 19129, U.S.A.

Schwarz, Uli.   Friedrich-Miescher Laboratorium der Max Planck
        Gesellschaft, Tübingen, Germany.

Seamon, Ken.   Biomembranes, Institut de Biologie Moléculaire, Paris
        5ème, France.

Sgaramella, Vittorio.   Department of Genetics, Stanford University
        School of Medicine, Stanford, California 94305, U.S.A.

Sherratt, D.   Univ. of Sussex, Falmer Brighton, Sussex BN1 9QG,
        England.

Siccardi, A.   Instituto di Genetica Pavia, Via S. Epitano 14, Pavia,
        Italy.

Skalka, Anna Marie.   Department of Cell Biology, Roche Institute of
        Molecular Biology, Nutley, New Jersey 07110, U.S.A.

Spaereu, U.   Univ. of Tromsø, Instit. of Medical Biology, 9000 Tromsø,
        Norway.

Spardi, S.   Lab. Genetica Biochemica, Via S. Epitano 14, 27100, Pavia,
        Italy.

Staudenbauer, Walter L.   Max Planck Institut für Biochemie,
        Goethestrasse, München, Germany.

Sueoka, Noboru.   Department of Biology, University of Colorado,
        Boulder, Colorado.

Tata, J. R.   National Institute of Health, Mill Hill, London, England.

Truesdell, Suzan.   Human Genetics Department, University of Michi-
          gan, Ann Arbor, Michigan 48104, U.S.A.

Tye, Bik Twoon.   Department of Biology, Massachusetts Institute of
          Technology, Cambridge, Massachusetts, U.S.A.

Tyrsted, G.   Biokemisk Inst. B., DK-2100, Copenhagen Ø, Denmark.

Upholt, W. B.   Univ. van Amsterdam, Afdeling Medioche Enzymologie,
          E. C. Huygenstr. 20 Amsterdam, The Netherlands.

Van Gool, D.   Janssenslab. Voor Genetica, Universiteit te Leuven,
          B-3030 Heverlee, Belgique.

Vinograd, Jerome.   Biology Division, California Institute of Tech-
          nology, Pasadena, California, U.S.A.

Vosberg, H. P.   Max Planck Institut für Medizinische Forschung,
          Abteilung für Molekulare Biologie, Heidelberg, Germany.

Waldeck, Waldimer.   Institut for Virus Research, German Cancer
          Research Institut, Heidelberg, West Germany.

Wanka, F.   Lab. Chemical Cytology, Univ. Nijmegen, Nijmegen,
          Netherlands.

Weintraub, Harold.   MRC Laboratory of Molecular Biology, Hills
          Road, Cambridge, England.

Wilkins, B. M.   Dept. of Genetics, The University, Leicester,
          LE1 7RH, England.

Winnacker, E. L.   Biokemiska - Karolinska Institute, Solnavägen 1,
          S-10401 Stockholm 60, Sweden.

Womack, John E.   Dept. Plant Sciences, Texas A and M University,
          College of Agriculture, College Station, Texas 77843, U.S.A.

Worcel, Abe.   Dept. of Biochemical Sciences, Moffet Laboratories,
          Princeton University, Princeton, New Jersey 08540, U.S.A.

Zaritsky, A.   Univ. Institute of Microbiology, Øster Farimagsgade
          2A, DK-1353 Copenhagen K, Denmark.

Zechel, K.   Max Planck Institut für Biochemie, Göethestrasse 31,
          München, West Germany.

INDEX

Actinomycin D, 16, 32, 68
"Activated" DNA, 261, 277
Alkaline phosphatase, 3
Alpha amylase, 294, 330
Amethopterin, 323
Amino acid starvation, 41, 205, 433
Ammediol buffer, 261
ATP analogues, 29
ATP effect, 23, 162
Autoradiography, 301, 321, 411
  of DNA, 305
  methocel technique, 414

Bacillus megaterium, 65
Bacillus subtilis, 63
  and bidirectional DNA
      replication, 242
  cell division, 436
  DNA of, 215
  temperature-sensitive mutant
      of, 241
Bacteriophage,
  M13, 14, 117, 312
    and asymmetric DNA
        replication, 126
    and gene 5 protein, 126
  T7, 158, 164, 172
    DNA concatemers, 160
    and DNA membrane complex, 166
    DNA polymerase, 165
    gene products, 158
    in vitro DNA replication, 167

Bacteriophage (cont'd)
  T4
    capsid initiator protein, 149
    DNA, 141, 208, 377
    and gene 5 product, 18, 134
    gene products, 141
    head morphogenesis, 149
    and host cell membrane, 149
    models for DNA replication,
        141
    polynucleotide ligase of (see
        Polynucleotide ligase)
    temperature-sensitive mutant
        in gene 32, 141
  φX174, 18, 30, 71, 87, 312, 339
    and discontinuous DNA repli-
        cation, 71
    and polynucleotide ligase, 77
    and replication complex, 81
    and replicative intermediates,
        94
    and tails, 106
    DNA dimers, 106, 360
  QB, 307
  MU-1, 225
  P1, 225, 439
  P2, 39
  T6, 442
  P22, 57
    oligomers of, 57
  lambda (λ)
    denaturation mapping, 186
    DNA, 181, 211, 253
    DNA replication, 194
    mutants in recombination genes,
        191